For Br

Birder, friend,
and fellow traveller!

bo

28-1-2015

Sharks

Sharks

An Eponym Dictionary

Michael Watkins and Bo Beolens

Published by Pelagic Publishing
www.pelagicpublishing.com
PO Box 725, Exeter EX1 9QU, UK

Sharks: An Eponym Dictionary

ISBN 978-1-907807-93-0 (Hbk)
ISBN 978-1-907807-94-7 (ePub)
ISBN 978-1-907807-18-3 (Mobi)
ISBN 978-1-907807-037-7 (PDF)

British Library Cataloguing in Publication Data
A catalogue record for this book is available from the
British Library.

Cover image: At certain times of the year scalloped
hammerhead sharks (*Sphyrna lewini*) gather in large
numbers at Cocos Island off Costa Rica. This island is
known for its sharks and great fish populations.
© Ethan Daniels/shutterstock.com

Typeset by XL Publishing Services, Exmouth

CONTENTS

INTRODUCTION

The naming of sharks (and other cartilaginous fishes) is a difficult matter … (who was responsible? – surely *not* McCavity?)

Firstly, apologies for adopting and adapting T.S. Eliot's *The Naming of Cats* but it seems a convenient way of describing the content of this book. It covers all the people who are associated with the vernacular and scientific names of extant Sharks, Rays, Skates and Chimeras. We have ignored species only found in fossil remains. It includes not only those after whom species have been named but also the authors of the original descriptions of them.

Therefore some entries relate to people after whom taxa have been named and others refer to people who have described taxa. Taxa named after the person are always shown before those that he or she has described. In each case taxa are set out starting with family, then a gap to genera, then a gap to species, then a gap to sub-species. Where there are, for example, more than one species they will be set out in order of description date starting with the earliest; if a date appears against more than one taxon, then they are arranged in the alphabetical order of the scientific name.

If a heading is shown in bold italics, e.g. *Zahn*, then it is a mini-biography that is not about a person but something else, such as a trawler after which a taxon has been named.

Who is it for?

Vernacular names of animals often contain a person's name (such names are called 'eponyms'). Furthermore, many scientific names contain the Latinised name of its discoverer, or some other person thought worthy of the honour, whether it be a binomial for a species or a trinomial for a subspecies. Indeed some genera names are also eponyms. So this book is for the amateur ichthyologist, the student of zoology or anyone else interested in taxonomy, nomenclature of fish in general and the Elasmobranch fishes in particular.

Describers and namers

New species are first brought to the notice of the scientific community in a formal, published description of a type specimen, essentially a dead example of the species, which will eventually be lodged in a scientific collection. The person who describes the species will give it its scientific name, usually in Latin but sometimes in Latinised ancient Greek! Sometimes the 'new' animal is later reclassified and then the scientific name may be changed. This frequently applies to generic names (the first part of a binomial name), but specific scientific names (the second part of a binomial), once proposed, usually cannot be amended or replaced – there are precise and complicated rules governing any such changes.

We may have missed a few recently published taxonomic changes (although we have tried to be comprehensive up until the final proof of the book), but we have put the name of the original describer after every entry. Because alterations to taxonomy have been so radical, and so swiftly changing, we decided not to put brackets around changed entries.

Since fish live in an environment into which humans can only venture if they are suitably equipped to survive, it is axiomatic that there may be a huge number of new species to be discovered. This is evidenced by the number of entirely unknown fish that were stranded on the beaches near Fukushima in Japan after the earthquake and tsunami of 11 March 2011.

Fish, therefore, are a class of animals where the rate of discovery of new species seems most unlikely to abate, unlike the discovery rates of mammals and birds. Therefore the reader may find that more recently named species and sub-species are not in this book.

Although we have used current scientific names as far as possible, these are not always as universal as the casual observer might suppose. There is no 'world authority' on such matters.

There are no agreed conventions for English names and indeed the choice of vernacular names is often controversial. Often the person who coined the scientific name will also have given it a vernacular name, which may not be an English name if the describer was not an English-speaker. On the other hand, vernacular names have often been added afterwards, frequently by people other than the describers. In this book, therefore, when we refer to an animal having been NAMED by someone, we mean that that person gave it the ENGLISH name in question. We refer to someone as a DESCRIBER when they were responsible for the original description of the species and hence for its scientific name. As stated above, it is the describer's name which is given after the scientific name in the biographies.

How to find an entry

At the back of this book there is an appendix that gives the basic identification details for all the fishes named in this book and their authors. This appendix also acts as an index and shows the scientific name, vernacular name and the name of either the only or, if more than one person was involved, the senior author of the original description.

Some of the Appendix entries are in **bold type** and they show that there is an entry under the name of the eponym which may be in regard to either the vernacular name or the scientific name or both the vernacular and scientific names. As examples, some clues for you to follow:

> *Narcine westraliensis* **McCain's** Skate – you should look up under M for an entry for McCain
> *Bathyraja eatonii* **Eaton's** Skate – you should look up under E for an entry for Eaton
> *Bathyraja fedorovi* Cinnamon Skate – you should look up under F for an entry on Fedorov
> *Irolita* Softnose Skate genus – you should look up under I for an entry on Irolita
> *Irolita waitii* Southern Round Skate – you will find entries for both Irolita and for Wait
> *Carcharhinus tilstoni* **Whitley's** Blacktip Shark – you will find entries for Whitley and for Tilston

The last column is for the name of an author and you will find an entry for him or her. Very often taxa are described by more than one person, for instance:

> Brazilian Soft Skate *Malacoraja obscura* Carvalho, Gomes & Gadig, 2005

To avoid cluttering up the appendix/index it appears there as:

> Brazilian Soft Skate *Malacoraja obscura* Carvalho, but under the entry for Carvalho you will find the names of his co-authors and so know where to look for entries on them.

Spelling of scientific names

Some Latin rendering of names may confuse because of slight alphabetical differences. For example, any name beginning 'Mc' may, in a binomial, be rendered 'mac'. It is wise, therefore, to search for both spellings, especially as names now tend to be used in full with a simple gender tag. Confusion may also arise as other alphabets (such as Cyrillic) have fewer or larger numbers of letters, some of which are often interchangeable when written in the modern English version of the Roman alphabet. Examples are V and W, J and Y, and letter combinations such

as Cz and Ts for Czar and Tsar.

By convention, diacritical marks, such as accents in French and the tilda used in Spanish and Portuguese, have to be ignored in scientific names and the phonetic sense of them expressed in other ways. The Scandinavian letters å, ä, ö and ø are normally expressed as aa, ae, oe and oe and the German ö and ü as oe and ue. In the English name, either spelling is acceptable. We have tried to ensure correct accents in the names of people, their book titles etc.

Sources and acknowledgements

As many of the species of fish mentioned in this book were described in the recent past, many (well over 130) of the individuals included were able to help us; often they vetted their own entry, corrected our mistakes and virtually wrote their own entries. We are very grateful to them.

Where we had to delve for the biographical details behind a name, especially if it was an eponym and not connected to a describer, then the best thing to do was to read the etymology (if any) in the original description. A certain number of these are freely available to be read on the internet but the majority are not. The cost of buying sources becomes extortionate and so the answer is libraries. The principal library to which we had access was the Natural History Museum Library in South Kensington, where we could read many of the journals and books that we had identified as original sources. We would like to thank the staff at that library for their forbearance and help. Obituaries published in journals were very useful as a source of biographical detail. As the number and variety of written sources is so large, we are not including a formal bibliography (which would in any case be an oddity in a dictionary!), but at the end of this introduction you will find a selection of the variety of publications we have consulted.

When any source material could not be found, we were able to call on help and advice from friends and contacts in many countries. We wish to acknowledge in particular the generosity and help of those who went out of their way to help by supplying details of other people (including some who are deceased):

Dr Arcady V. Balushkin, Curator of Fishes and Head of the Ichthyology Laboratory, Zoological Institute of the Russian Academy of Science, St Petersburg
Dr Gregor Caillet, Emeritus Professor, Moss Landing Marine Laboratories, California
Dr Hugo Patricio Castello, Buenos Aires, Argentina
Dr De La Cruz-Agüero, National Polytechnic Institute La Paz, Baja California, Mexico

Dr Sherine Sonia Cubelio, Centre for Marine Living Resources & Ecology (CMLRE), Cochin, India

Dr Dave Ebert, Moss Landing Marine Laboratories, California

The Editor of the website www.neglectedscience.com

John Harrison, WAFIC, Perth, Australia

Dr Jim Harvey, Moss Landing Marine Laboratories, California

Hajime Ishihara, based in Amman, Jordan

Dr Anne-Flore Laloë, Marine Biological Association, Plymouth

Dr John Edward McCosker, California Academy of Sciences, San Francisco

Dr John D. McEachran, Texas A & M University

Jonathan Mee, Issham Aquatics, Jeddah, Saudi Arabia

Dr Yefim Izrailevich Kukuev, Atlantic Scientific Research Institute AtlantNIRO, Russia

Dr Peter Rask Møller, Zoological Museum, University of Copenhagen

Dr Giuseppe Notarbartolo di Sciara, the Tethys Research Institute, Milan, Italy

Heather L. Prestridge, Texas A & M University

C. Raghunathan, Zoological Survey of India, Port Blair, Andaman & Nicobar Islands

Dr Florian M. Steiner, University of Innsbruck, Austria

Daniel R. Yagolkowski, Buenos Aires, Argentina

Short list of written sources

American Society of Ichthyologists and Herpetologists

Annals of the South African Museum

Australian Fisheries

Cladistics

Copeia

Cybium

Environmental Biology of Fishes

Ichthyological Society of Japan

Japanese Journal of Ichthyology

Journal of Applied Ichthyology

Journal of Fish Biology

Journal of Ichthyology

Journal of Marine Biological Association

Marine Biology

Marine Biodiversity

Northwestern Naturalist

Proceedings of the Biological Society Washington

Proceedings of the Royal Society London

Proceedings of the Society of Natural Sciences in Hamburg

Revista de Museo Argentino Ciencas Naturales

South African Society of Maritime Science

Zootaxa

A

Abbott

Least Brook Lamprey *Lampetra aepyptera* Abbott,
1860

Dr Charles Conrad Abbott (1843–1919) was an
American physician, naturalist and, primarily, an
archaeologist. He served as an army surgeon in
the Union forces during the American Civil War
before receiving his degree from the University of
Pennsylvania (1865). He was Assistant Curator at
the Peabody Museum, Cambridge, Massachusetts
(1876–1889).

Abe

Bareskin Dogfish *Centroscyllium kamoharai* Abe,
1966

Tokiharu Abe (1911–1996) was a Japanese ichthyologist working at the University of Tokyo. His
reputation was established with his taxonomic study
of the puffer fish, particularly the genus *Takifugu*
which he described (1949). Consequently a number
of fish taxa, such as *Tetraodon abei*, were named in his
honour. He was an honorary member of the American Society of Ichthyologists and Herpetologists.
He wrote a great many articles, including *Taxonomic
Studies of the Puffers (Tetraodontidae, Teleostei) from
Japan and Adjacent Regions* (1952). He died of a cere-
bral haemorrhage.

Abernethy

Blackbelly Lanternshark *Etmopterus abernethyi*
Garrick, 1957 NCR
[Junior Syn *Etmopterus lucifer*]

Fred Abernethy (d.1995) was chief engineer on
board the 'Holmwood' when it was captured and
sunk by German raiders operating around New
Zealand and in the Pacific (late 1940). He was held
captive on board one of the raiders until released on
to Emirau Island (east of New Guinea). He wrote
of his experiences in *A Captive's Diary* (1985). He
later worked on the research vessel M.T. 'Thomas
Currell' and contributed greatly to the collection of
New Zealand elasmobranchs. He was on the New
Zealand Chatham Islands Expedition (1954). Garrick
named the fish when commercial fisherman Richard
Baxter (q.v.) (1956) caught it.

Acero

Gorgona Guitarfish *Rhinobatos prahli* Acero & Franke,
1995
Hagfish sp. *Eptatretus wayuu* Mok, Saavedra-Diaz &
Acero, 2001
Hagfish sp. *Eptatretus ancon* Mok, Saavedra-Diaz &
Acero, 2001
[Syn. *Quadratus ancon*]

Dr Arturo Acero Pizarro (b.1954) is a marine biologist and ichthyologist from Santa Marta, Colombia.
The Universidad Jorge Tadeo Lozano, Cartagena
awarded his bachelor's degree (1977), the University
of Miami, Coral Gables his master's (1983) and the
University of Arizona, Tucson his doctorate (2004).
He worked as a full-time instructor at the Universidad Jorge Tadeo Lozano (1977–1978) and was
employed as a marine biologist by the Instituto de
Investigaciones de Punta de Betín (1981–1986). Since
1986 he has been an associate professor, Universidad Nacional de Colombia, Caribbean Campus.
He wrote *On an Interesting Specimen of Cownose Ray*
(Pisces: Rhinopteridae) *from the Colombian Caribbean*
(1982) and *Primer Registro de Torpedo peruana Chirichigno* (Elasmobranchii: Torpedinidae) *para el Pacífico
Colombiano* (1999). He has described more than 20
species of bony fish from Colombian and Mexican
waters.

Achenbach

Freshwater Stingray sp. *Potamotrygon labradori*
Castex, Maciel & Achenbach, 1963 [Junior Syn.
Potamotrygon motoro]
Freshwater Stingray sp. *Potamotrygon menchcai*
Achenbach, 1967
[Junior Syn. *Potamotrygon falkneri*]

Guillermo Martinez Achenbach (1911–1990) was
an Argentine ichthyologist and Director of the Provincial Museum of Natural Sciences 'Florentino
Ameghino' (1944–1978).

Ackley

Ackley's Ray *Raja ackleyi* Garman, 1881
[Alt. Ocellate Ray]

Lieutenant (later Rear Admiral) Seth Mitchell
Ackley (1845–1908) was the officer commanding the
Coast Survey steamer 'Blake' (1877). He joined the
navy (1862), entering the US Naval Academy. After
graduating (1866) he was commissioned as a second
lieutenant. His first command (1876–1877) was the
'RS Wyoming'. He was stationed at Olongapo and

Cavite and also served as a member of the Naval War College at Newport and the General Board of the US Navy.

Addison

Ornate Sleeper Ray *Electrolux addisoni* Compagno & Heemstra, 2007

Mark Ramsay Addison (b.1967) is the South African owner-operator of the underwater filming and expedition company 'Blue Wilderness' founded in 1997. He and his company filmed some of the most amazing sequences, such as the sardine run, in the BBC's 'Blue Planet', for which they won an Emmy. He originally intended to become a lawyer and studied at Rand Afrikaans University (1988–1990). His father, Brent Addison, was a marine scientist so he was exposed to the sea life off southern Africa from an early age. He collected the holotype. He has been described as part-man, part-shark, part-comedian. After graduating, Mark founded what became South Africa's largest dive charter boating company (1990–1996). He said of the naming, 'They just ran out of ideas and named it after me'. Interestingly, the first part of the binomial of the Ornate Sleeper Ray is named after the Electrolux™ company because of the ray's well-developed electrogenic properties and its vigorous sucking action when feeding.

Afuer

Peruvian Butterfly Ray *Gymnura afuerae* Hildebrand, 1946

Not an eponym but derived from the Spanish word for 'exterior'.

Agassiz

Sand Shark genus *Odontaspis* Agassiz, 1838
River Shark genus *Glyphis* Agassiz, 1843
Weasel Shark genus *Hemipristis* Agassiz, 1843

Rio Skate *Rioraja agassizii* Müller & Henle, 1841
Megatooth Shark *Carcharodon megalodon* Agassiz, 1843

Jean Louis Rudolphe Agassiz (1807–1873) was a Swiss-American geologist, glaciologist and zoologist whose speciality was ichthyology. He studied at Zurich, Heidelberg and Munich, where he qualified as a physician (1830), and in Paris under Cuvier (1831). While still a student he was tasked with working on the Spix and Martius Brazilian freshwater fish collection. He became Professor of Natural History at the Lyceum de Neuchâtel (1832).

He was the first person to propose scientifically that the Earth had been subject to an ice age and to study ice as a subject, having lived in a special hut built on a glacier in the Alps (1837). He went to the USA (1846) to study American natural history and geology and to deliver a course of zoology lectures. He visited again (1848) and remained there for the rest of his life, becoming Professor of Zoology and Geology at Harvard, where he founded and directed the Museum of Comparative Zoology (1859–1873). Latterly he took up studies of Brazilian fishes again and led the Thayer expedition to Brazil (1865). He established the Marine Biological Laboratory (1873). Three reptiles are named after him.

Ahmad

Circle-blotch Pygmy Swellshark *Cephaloscyllium circulopullum* Yano, Ahmad & Gambang, 2005
Sarawak Pygmy Swellshark *Cephaloscyllium sarawakensis* Yano, Ahmad & Gambang, 2005

Dr Amiruddin Ahmad is an ichthyologist who teaches biology at the University of Malaysia. He received his bachelor's degree in botany (1995), an MSc in environmental science (2002) and his doctorate in ecology (2012). The primary goal of his research is to investigate the pattern of species diversity and analyse the community composition and assemblage structure of the fauna of Malaysia.

Akaje

Whip Stingray *Dasyatis akajei* Müller & Henle, 1841
[Alt. Japanese Red Stingray]

The original description has no etymology. It is probable that it is an adaptation of a Japanese name for this species. Müller and Henle have adopted such names in other cases. (See **Zuge**)

Alastair

Australian Hagfish sp. *Eptatretus alastairi* Mincarone & Fernholm, 2010

Alastair Graham (see **Graham, A**) was honoured for help and hospitality offered to the second author.

Alava

Ridgeback Skate *Dipturus amphispinus* Last & Alava, 2013

Moonyeen Nida R. Alava (b.1964) is a Filipino marine biologist and independent marine consultant. Her bachelor's (1984) and master's (1993) degrees, in biology and marine biology respectively, were awarded by Silliman University. She has a post-graduate certificate in integrated ecosystem management (2002) from the Watson Institute for International Studies, Brown University, and PhD units on marine sciences (2010) from the Marine Science Institute, University of the Philippines. She has done work for the Deutsche Gesellschaft für Internationale Zusammenarbeit, Conservation International, Global Marine Species Assessment – Coral Triangle, Coastal Conservation and Education Foundation, and Worldwide Fund for Nature, focusing on biodiversity research, conservation and management, in particular of marine threatened species. She was a co-author of the *Checklist of Philippine Chondrichthyes* (2005) and is lead author of *A Field Guide to Sharks, Batoids and Chimaeras of the Philippines* (in preparation).

Albert

Ratfish sp. *Hydrolagus alberti* Bigelow & Schroeder, 1951

Dr Albert Eide Parr (1890–1991) was an oceanographer and marine biologist born and raised in Norway where he took his first degree at the University of Oslo. He was in the Norwegian Merchant Marines and undertook postgraduate research at Bergen Museum. Yale awarded his doctorate. He worked at the New York Aquarium (1926), from where he was recruited to curate Harry Payne Bingham's fish collection, which he donated to Yale; Parr went with it. At Yale he became Director of the oceanographic laboratory (1930s) and rose through the academic ranks, becoming Professor of Oceanography (1938) and Director of the Peabody Museum (1938–1942). He became Director of the AMNH (1942–1959) and reorganised many departments. He was editor of the authors' *Fishes of the Western North Atlantic* monographs, and was honoured for his many contributions to ichthyology.

Alcock

Pale-spot Whipray *Himantura alcockii* Annandale, 1909
Arabian Catshark *Bythaelurus alcockii* Garman, 1913

Blind Ray genus *Benthobatis* Alcock, 1898

Ornate Dogfish *Centroscyllium ornatum* Alcock, 1889
Prickly Skate *Fenestraja mamillidens* Alcock, 1889
Bristly Catshark *Bythaelurus hispidus* Alcock, 1891
Dark Blind Ray *Benthobatis moresbyi* Alcock, 1898
Indian Ringed Skate *Okamejei powelli* Alcock, 1898
Quagga Catshark *Halaelurus quagga* Alcock, 1899
Travancore Skate *Dipturus johannisdavisi* Alcock, 1899

Major Dr Alfred William Alcock (1859–1933) joined the Indian Medical Service in 1885, having qualified as a physician at the University of Aberdeen. He was appointed Surgeon-Naturalist to the Indian Marine Service. As a naturalist he was mainly interested in fishes. He was based in Calcutta, dividing his time between the Indian Museum and the survey ship 'Investigator'. He published a number of papers on the ichthyology of the Bay of Bengal, after which he returned to England. He wrote *A Naturalist in the Indian Seas* (1902). The catshark was named after him because he was the first (1896) to note that it was a distinct species. Two reptiles are also named after him.

Alexandrine

Alexandrine Torpedo *Torpedo alexandrinsis* Mazhar, 1987

Named after the city of Alexandria in Egypt, not after a person.

Alfred

Reef Manta Ray *Manta alfredi* JLG Krefft, 1868
[Alt. Reef Manta Ray]

Prince Alfred, Duke of Edinburgh (1844–1900) was a son of Queen Victoria who ruled as Duke of Saxe-Coburg and Gotha (1893) in the Prussian Empire. He previously served in the Royal Navy and commanded 'HMS Galatea' in a circumnavigation (1867–1868). During this voyage he visited Australia twice, surviving an assassination attempt (1868) when he was shot in the back (the would-be assassin was promptly arrested, tried and hanged). The 'Sydney Illustrated News' in which Krefft published his article, stated that Prince Alfred had been presented with photographs of the Manta Ray.

Al-Hassan

Pita Skate *Okamejei pita* Fricke & Al-Hassan, 1995

Dr Laith Abdul Jalil Jawad Al-Hassan (b.1948) is an Iraqi-born New Zealand biologist and ichthyologist. He is a freelance fish biodiversity expert and consultant who has worked in Libya, Yemen and more recently Oman, as well as being a researcher in New Zealand. The University of Basrah, Iraq awarded his bachelor's degree (1971), the University of Bristol his master's (1980) and the University of Auckland his doctorate. He was an assistant professor, University of Basrah and an assistant professor, Zoology Department, University of Garyounis, Libya.

Among over 220 scientific published articles and papers is the co-written *A Bibliography of the Fishes of the Tigris-Euphrates Basin* (1988).

Alia

Smalleye Pygmy Shark *Squaliolus aliae* Teng, 1959

Huang A-li is the name of Teng's wife. He named the shark, which was first caught in Taiwanese waters, after her '...for her continuous encouragement and assistance over the past 20-some years'.

Alis

Skate sp. *Notoraja alisae* Séret & Last, 2012

Alis is the name of a ship used by the Institut de Recherche pour la Développement to undertake research off the coast of New Caledonia. The ship is in turn named after a local wind.

Allen

Allen's Skate *Pavoraja alleni* McEachran & Fechhelm, 1982

Milne Epaulette Carpetshark *Hemiscyllium michaeli* Allen & Dudgeon, 2010
Walking Shark *Hemiscyllium galei* Allen & Erdmann, 2008
Bamboo Shark sp. *Hemiscyllium henryi* Allen & Erdmann, 2008

Dr Gerald R. Allen (b.1942) is an American-born Australian ichthyologist. The University of Hawaii awarded his PhD (1971), after which he began work (1972) as an ichthyologist at the Australian Museum, Sydney. He then moved to be Curator of Fishes at the Department of Ichthyology, Western Australian Museum, Perth (1974–1997). He worked (1997–2003) for Conservation International preparing distribution maps for all known reef fishes. He continues as a Research Associate of the Western Australian Museum. He has a particular interest in freshwater fish in New Guinea and Northern Australia. He has taken part in many collecting trips such as the Western Australian Museum expedition to the Kimberley Coast (1991) aboard 'North Star IV'. During these trips he has logged over 7,500 dives. Among 32 books and over 300 papers he co-wrote *The Marine Fishes of North-Western Australia: A Field Guide for Anglers and Divers* (1988). In 2013 he won the Bleeker Award and published *Reef Fishes of the East Indies* (3 volumes). His other interests include underwater photography and pursuits outside the marine environment include bird watching, rock climbing, mountaineering and bicycle racing, for which he has been eight-time state veteran champion. He was President of the Australian Society for Fish Biology (1979–1981). To date he has described 13 new genera and 456 species and was honoured in the skate name having often sent fish species to the authors.

Almeida

Ghost Shark sp. *Hydrolagus lusitanicus* Moura, Figueiredo, Bordalo-Machado, Almeida & Gordo, 2005

Ana Cristina Almeida is a microbiologist who works at the Instituto Nacional de Investigação Agrária e das Pescas, Lisbon.

Álvarez

Mexican Brook Lamprey *Tetrapleurodon geminis* Álvarez, 1964

Dr José Álvarez del Villar (1908–1988) collected the holotype (1962) when at the Hydrobiology Laboratory, National School of Biological Sciences, Mexico City, which awarded his doctorate. He wrote *Claves para la Determinación de Especies en los Peces de las Aguas Continentales Mexicanas* (1950).

Amanda

Freshwater Stingray sp. *Potamotrygon amandae* Loboda & Carvalho, 2013

Amanda Lucas Gimeno (1984–2006) was a Brazilian biologist who graduated (2005) from the University of São Paulo, where the senior author had been one of her undergraduate colleagues. She was killed when there was a collapse of an external awning of an amphitheatre at the State University of Londrina during the 26th Brazilian Congress of Zoology.

Ancon

Hagfish sp. *Eptatretus ancon* Mok, Saavedra-Diaz & Acero, 2001
[Syn. *Quadratus ancon*]

This refers to the research vessel 'Ancon' from which the holotype was collected.

Anderson

Florida Torpedo *Torpedo andersoni* Bullis, 1962

William Wyatt Anderson (1909–1993) was an American ichthyologist who was a fisheries research biologist at the State Game and Fish Commission, Coastal Fisheries Division, Brunswick, Georgia, USA, and a close friend of the describer. He was honoured as a '…colleague and mentor, whose labors have contributed immeasurably to our knowledge of the marine fauna of the southeastern United States.' They co-wrote a number of papers such as *Searching the Sea Bed by Sub* (1970).

Andriashev

Little-eyed Skate *Bathyraja andriashevi* Dolganov, 1985

Professor Dr Anatoly Petrovich Andriyashev (1910–2009) was a Russian marine biologist. Leningrad State University awarded his bachelor's degree (1933), his doctorate (1937) and his further doctorate of science (1951). He became a researcher at the Zoological Institute of the Russian Academy of Sciences (1943) and a professor there (1970). He wrote *Fishes of the Northern Seas of the USSR* (1964).

Anikin

Siberian Brook Lamprey *Lethenteron kessleri* Anikin, 1905

V.P. Anikin was Curator of the Zoological Museum of Tomsk University (1898). He undertook a collecting expedition (1902) in the Narym Territory. Among other papers he wrote *Description of New Fish Species from Asia*, which includes the description of the lamprey (1905).

Annandale

Annandale's Skate *Rajella annandalei* Weber, 1913
Annandale's Guitarfish *Rhinobatos annandalei* Norman, 1926
Annandale's Skate *Rajella annandalei* Stehmann, 1970

Smalleye Stingray *Dasyatis microps* Annandale, 1908
Brown Numbfish *Narcine brunnea* Annandale, 1909
Pale-spot Whipray *Himantura alcockii* Annandale, 1909
Honeycomb Whipray *Himantura fava* Annandale, 1909
Jenkins' Whipray *Himantura jenkinsii* Annandale, 1909

Dr 'Thomas' Nelson Annandale (1876–1924) was a zoologist (primarily entomologist and herpetologist) and anthropologist who became Superintendent of the Indian Museum, Calcutta, which still houses his insect and spider collection. He was instrumental in establishing a purely zoological survey, not combined with anthropology, undertaking several expeditions (from 1899), most notably the Annandale-Robinson expedition that collected in Malaya (1901–1902). He went to India as Deputy Superintendent at the Museum (1904), and Director (1907). He became the first director of the Zoological Survey of India (1916–1924). He was also noted for his work on the biology and anthropology of the Faroe Islands and Iceland. He wrote or co-wrote a number of scientific papers (1903–1921), including *The Aquatic and Amphibious Molluscs of Manipu* (1921). He was honoured in the name of the guitarfish for his contributions to Indian ichthyology, which included an account of this species (1909). Six amphibians, four reptiles and a mammal are named after him.

Antuna-Mendiola

White-margin Fin Houndshark *Mustelus albipinnis* Castro-Aguirre, Antuna-Mendiola, González-Acosta & De La Cruz-Agüero, 2005

Alberto Antuna-Mendiola is a Mexican biologist and ichthyologist who graduated from the Universidad Autónoma de Baja California, Mexico (2004) and is now at the Ichthyology Department, Interdisciplinary Centre of Marine Sciences, National Polytechnic Institute.

Applegate

Carpet Sharks *Orectolobiformes* Applegate, 1972
Blind Sharks *Brachaeluridae* Applegate, 1974

Snouted Eagle Ray *Myliobatis longirostris* Applegate & Fitch, 1964

Dr Shelton Pleasants Applegate (1928–2005) was an expert on both living and fossil sharks. After service in the US Navy he took a bachelor's degree at the University of Richmond and a master's at the University of Virginia; the University of Chicago awarded his doctorate. At the time of his death he was an associate professor at Harvard and professor at the University of Mexico, Mexico City, having previously worked at the Smithsonian, the Los Angeles County Museum of Natural History, Duke University, the University of Chicago and Arkansas State University.

Ara

Roughtail Catshark *Galeus arae* Nichols, 1927

This species is named after the yacht 'Ara', owned by William K. Vanderbilt, which collected the first two specimens.

Archey

Sherwood Dogfish *Scymnodalatias sherwoodi* Archey, 1921

Sir Gilbert Edward Archey (1890–1974) was a zoologist, ethnologist and museum director who was born in England and taken to New Zealand as a child. He took both his bachelor's degree and doctorate at Canterbury University College, Christchurch. He was Assistant Curator, Canterbury Museum (1914–1923), though First World War service in the New Zealand Army interrupted his career, as did similar Second World War service. He was Director, Auckland Institute (1924–1964) and noted for his work on moas and on Maori art. He wrote *The Moa, a Study of the Dinornithiformes* (1941). An amphibian is named after him.

Arlyza

Fine-spotted Leopard Whipray *Himantura tutul* Borsa, Durand, K-N Shen, Arlyza, Solihin & Berrebi, 2013

Dr Irma Shita Arlyza (b.1974) is an Indonesian geneticist. The Pertanian Bogor Institute awarded both her master's degree (2003) and her doctorate (2013). Since 2005 she has worked at the Oceanography Research Centre, Indonesian Institute of Sciences, Jakarta. She co-wrote *Resurrection of New Caledonia Maskray* Neotrygon trigonoides (Myliobatoidei: Dasyatidae) *from Synonymy with* N. kuhlii *Based on Cytochrome-oxydase I Gene Sequences and Spotting Patterns* (2013).

Aschliman

Roughnose Legskate *Cruriraja hulleyi* Aschliman, Ebert & Compagno, 2010

Dr Neil C. Aschliman is an assistant professor in the Biology Department, St Ambrose University, Iowa. His bachelor's degree (2005) was awarded by Texas A & M University and his doctorate by Florida State University, Tallahassee (2011). The American Society of Ichthyologists and Herpetologists gave him the 2010 Frederick H. Stoye Award in General Ichthyology. He wrote *A New Framework for Interpreting the Evolution of Skates and Rays (Chondrichthyes: Batoidea)* (2011) and is the lead author of *Body Plan Convergence in the Evolution of Skates and Rays (Chondrichthyes: Batoidea)* (2012) and of *Phylogeny of Batoidea*, a chapter in *Biology of Sharks and Their Relatives* (2012, second edition, edited by Carrier *et al.*).

Atlantis

Atlantic Legskate *Cruriraja atlantis* Bigelow & Schroeder, 1948

This is named after the Woods Hole Oceanographic Institution research vessel 'Atlantis', which collected three new species of skate along the coasts of Cuba, including this one.

Atz

Atz's Numbfish *Narcine atzi* Carvalho & Randall, 2003

Dr James Wade Atz (1915–2013) was Curator Emeritus Vertebrate Zoology and Dean Bibliographer, Department of Ichthyology at AMNH, which he joined (1964), having been Associate Curator of the New York City Aquarium (c.1959–1964). He was a curator at the AMNH (1970–1981). He was also Adjunct Professor of Biology, Graduate School of Arts and Science, New York University, where he had earned his master's degree (1952) and his doctorate. Among his many publications are several books on aquarium and tropical fish and his *Dean Bibliography of Fishes 1968* (1971). For the benefit of the American Jewish community he produced a definitive list of kosher and non-kosher fishes. He was reported to have a wicked sense of humour and enjoyed winding up the well-known chat-show host Johnny Carson on how fish have sex and how some change sex to adjust the male/female sex ratio. He was honoured '...for his many contributions to different aspects of ichthyology, and for his unparalleled enthusiasm for the study of fishes.'

Ayres

Lamprey sp. *Lampetra ayresii* Günther, 1870

Shovelnose Guitarfish *Rhinobatos productus* Ayres, 1854
Seven Gill Shark sp. *Notorynchus maculatus* Ayres, 1855
Pacific Electric Ray *Torpedo californica* Ayres, 1855
Pacific Angelshark *Squatina californica* Ayres, 1859

Dr William Orville Ayres (1817–1887) was an American ichthyologist who qualified as a physician at Yale and was one of Audubon's friends. He was the first Curator of Ichthyology at the California Academy of Sciences (1854–1864) but gave up ichthyology after suffering ruthless criticism from Theodore Nicholas Gill at the Smithsonian. Ayres moved to Chicago (1871), leaving (1878) to live in New Haven and to teach medicine at Yale.

Ayson

Blind Electric Ray *Typhlonarke aysoni* A Hamilton, 1902

Lake Falconer Ayson (1855–1927) gave the holotype to the describer after the vessel 'Doto' trawled it up on a research cruise of which Ayson was the leader. Ayson had been a farm labourer, a rabbit inspector, an acclimatisation officer and finally Chief Inspector of Fisheries.

B

Baldwin, CC

Jaguar Catshark *Bythaelurus giddingsi* McCosker,
Long & CC Baldwin, 2012

Dr Carole C. Baldwin is a research zoologist and
marine biologist at the Smithsonian, which she
joined (1992) and became Curator of Fishes (2001).
James Madison University, Virginia awarded her
bachelor's degree (1981), the College of Charleston,
South Caroline her master's (1986) and the College
of William and Mary, Virginia her doctorate (1992).
She co-wrote *A New Species of Soapfish (Teleostei:
Serranidae: Rypticus), with Redescription of R. subbi-
frenatus and Comments on the Use of DNA Barcoding
in Systematic Studies* (2012) as well as the Smithsonian's sustainable seafood cookbook *One Fish, Two
Fish, Crawfish, Bluefish* (2003).

Baldwin, ZH

Oman Bullhead Shark *Heterodontus omanensis* ZH
Baldwin, 2005

Zachary H. Baldwin is an ichthyologist who is a
PhD student at the Richard Gilder Graduate School
of the American Museum of Natural History. He has
described several new species in papers such as *A
New Species of Parapercis (Teleostei: Pinguipedidae) from
Madagascar* (2012).

Bancroft

Bancroft's Numbfish *Narcine bancroftii* Griffith & CH
Smith, 1834
[Alt. Brazilian/Lesser Electric Ray, Syn. *Narcine
umbrosa*]

Lesser Devil Ray *Mobula hypostoma* Bancroft, 1831

Dr Edward Nathaniel Bancroft (1772–1842) was a
British surgeon at St George's, London (1805–1811).
He was educated at St John's College, Cambridge
where he graduated as bachelor of medicine
(1794). He became an army physician (1795) and
saw service in Portugal, Egypt and the Windward
Islands. He returned to study at Cambridge, where
he obtained his doctorate (1804). He became a physician in Kingston, Jamaica (1811) 'for health reasons',
where he later became the Deputy Inspector General
of Army Hospitals in Jamaica. He was the author

of several scientific papers, such as *On Several Fishes
of Jamaica* (1831). A fine watercolourist, he not only
studied electric fish but also painted them; the scientific description of the numbfish was based on his
illustration.

Baranes

Elat Electric Ray *Heteronarce bentuviai* Baranes &
Randall, 1989
Seychelles Gulper Shark *Centrophorus seychellorum*
Baranes, 2003
Seychelles Spurdog *Squalus lalannei* Baranes, 2003

Dr Albert 'Avi' Baranes (b.1949) was born in Alexandria, Egypt. He lived in Paris (1956–1970) and the
Faculté des Sciences, University of Paris awarded
his bachelor's degree. He is both Honorary Consul
of France and Director of the Marine Biology Laboratory at Eilat, part of the Hebrew University of
Jerusalem, where he was awarded both his master's
and doctorate in oceanography (1986). He has published 50 papers on sharks and deep-sea fishes of
the Red Sea, such as, with Ben-Tuvia (q.v.), *Two Rare
Carcharhinids,* Hemipristis elongatus *and* Iago omanensis, *from the Northern Red Sea* (1979); *Ichthyofauna
of the Rocky Coastal Littoral of the Israeli Mediterranean*
(2007); and two books on sharks. His main research
interests are sharks of the Red Sea and the taxonomy
and biology of deep Red Sea fish. He is also Grand
Master of the Grand Lodge of Free Masons in Israel
(2013–2015).

Barbour

Ratfish sp. *Hydrolagus barbouri* Garman, 1908
Cuban Ribbontail Catshark *Eridacnis barbouri* Bigelow
& Schroeder, 1944

Dr Thomas Barbour (1884–1946) was an American zoologist. He graduated from Harvard (1906)
and obtained his PhD there (1910). He became an
associate curator of reptiles and amphibians at the
Harvard Museum of Comparative Zoology, and was
its Director (1927–1946). He became Custodian of the
Harvard Biological Station and Botanical Garden,
Soledad, Cuba (1927). He was executive officer
in charge of Barro Colorado Island Laboratory,
Panama (1923–1945). During his time at the museum
he explored in the East Indies, the West Indies,
India, Burma, China, Japan, and South and Central
America. He was famously jovial good company and
would invite all and sundry to eat and converse next
door to his office in the 'Eateria', where his secretary,
Helen Robinson, prepared food for his thousands of

guests. Something of an all-rounder, he wrote many articles and books, including *The Birds of Cuba* (1923) and *Naturalist at Large* (1943). He also co-wrote *Checklist of North American Amphibians and Reptiles*. He was honoured for the 'constant assistance' he gave the authors in their studies of western North Atlantic sharks. His special area of interest was the herpetology of Central America. Twenty-four reptiles are named after him, as are three birds, two mammals and four amphibians.

Barnard

Bigthorn Skate *Rajella barnardi* Norman, 1935

Eightgill Hagfish *Eptatretus octatrema* Barnard, 1923
Fivegill Hagfish *Eptatretus profundus* Barnard, 1923
Soft Skate *Malacoraja spinacidermis* Barnard, 1923
Saldanha Catshark *Apristurus saldanha* Barnard, 1925

Dr Keppel Harcourt Barnard (1887–1964) was a UK-born South African invertebrate zoologist particularly interested in marine crustaceans. He graduated from Cambridge then studied law, but being more interested in science was for a short time honorary naturalist at the Plymouth Marine Biology Laboratory. He worked at the South African Museum, Cape Town (1911–1964), initially as a marine biologist, becoming Assistant Director (1924) and then Director (1946–1956). He undertook many collecting expeditions in South Africa and Mozambique, and in three donkey-trek expeditions in Namibia (1924–1926). A keen mountaineer, he was Secretary of the Mountain Club of South Africa (1918–1945). He wrote on molluscs and entomology. Two reptiles are named after him. He was probably honoured in the skate name for his monograph on South African marine fishes, which Norman repeatedly cited, and for the loan of rays from the South African Museum. He also showed 'kindly interest' in Norman's work. He wrote *A Monograph of the Marine Fishes of South Africa* (1925).

Barnett

Galápagos Ghostshark *Hydrolagus mccoskeri* Barnett, Didier, Long & Ebert, 2006

Lewis Abraham Kamuela Barnett (b.1981) is an American population ecologist who has worked in a variety of areas in fish ecology and fisheries science while employed by NOAA Fisheries at the Alaska Fisheries Science Center, Southeast Fisheries Science Center, and the Southwest Fisheries Science Center. Oregon State University awarded his bachelor's degree (2003), Moss Landing Marine Laboratories his master's (2008); he is working towards his doctorate at the University of California Davis. Among his publications is the co-written *Comparative Demography of Skates: Life-history Correlates of Productivity and Implications for Management* (2013) and, as a member of the International Union for Conservation of Nature (IUCN)/Species Survival Commission (SSC) Shark Specialist Group, has written and co-written threat assessments of sharks and chimaeras for the Red List of Threatened Species.

Baro

Iberian Pygmy Skate *Neoraja iberica* Stehmann, Séret, Costa & Baro, 2008

Dr Jorge Baro Domínguez is a Spanish marine biologist. The University of Malaga awarded his bachelor's degree (1982), his master's (1984) and his doctorate (1996). He taught in a secondary school (1984–1986) and then joined the Spanish Oceanographic Institute (1986), was co-ordinator of its Mediterranean Fisheries Program (2003–2008) and became Director of its Oceanographic Centre, Malaga (2008).

Basilewsky

Chinese Ray *Raja chinensis* Basilewsky, 1855

Dr Stepan Basilewsky was a Russian physician and ichthyologist. He wrote a paper on Chinese fish, *Ichthyographia Chinae Borealis. Nouveaux Memoirs de la Société Impériale des Naturalistes de Moscou* (1885).

Bass

Lined Catshark *Halaelurus lineatus* Bass, D'Aubrey & Kistnasamy, 1975
Thorny Lantern Shark *Etmopterus sentosus* Bass, D'Aubrey & Kistnasamy, 1976

Dr A. John Bass was a zoologist and ichthyologist at the National Museum of New Zealand Te Papa Tongarewa, Wellington. He was attached to the Oceanic Institute, Durban, South Africa (1970). He wrote *Problems in Studies of Sharks in the Southwest Indian Ocean* (1978).

Bauchot

Rosette Torpedo *Torpedo bauchotae* Cadenat, Capapé & Desoutter, 1978

Dr Marie-Louise Bauchot née Boutin (b.1928) is a French ichthyologist who was Curator of Ichthyology and Assistant Manager at the Muséum National d'Histoire Naturelle, Paris (1952–1991). Several other Indo-Pacific fishes are named after her. Amongst other works she wrote a catalogue of the fish type specimens at the MNHN and *La Vie des Poissons* (1967). She collaborated on a number of books and papers, such as *Les Poissons* (1954) with her husband Roland Bauchot, who was also a biologist and primarily a herpetologist and ichthyologist.

Baxter

New Zealand Lantern Shark *Etmopterus baxteri*
    Garrick, 1957

Richard Baxter was a retired New Zealand fisherman who took part in 'experimental line fishing' to collect shark specimens for the Zoology Department of Victoria University College, Wellington (1955). He 'collected' one from 500 fathoms (about 915 metres) close inshore in the Kaikoura region. He also caught another 'new' shark *Etmopterus abernethyi* that was actually of an already known species, *Etmopterus lucifer*.

Beamish

Vancouver Lamprey *Entosphenus macrostomus*
    Beamish, 1982

Dr Frederick William Henry Beamish (b.1935) was Commissioner and Chairman of the Great Lakes Fishery Commission (1989–2004). He took his degrees at the University of Toronto. He was (2000–2012) Director of the Canadian Wildlife Federation. He joined the Zoology Department, University of Guelph (1965) as Assistant Professor and was Chairman of the department (1974–1979). He was Professor at the Institute of Ichthyology, University of Guelph (1991–2000) and is now Professor Emeritus. In 'retirement' he became a contract professor in the biology department of Burapha University, Thailand. Most of his papers are on lampreys, such as *Biology of the Southern Brook Lamprey* (1982). He has donated his library and papers to the University of Guelph.

Bean, BA

Skate genus *Dactylobatus* BA Bean & Weed, 1909

Blind Torpedo *Benthobatis marcida* BA Bean & Weed,
    1909
Skilletskate *Dactylobatus armatus* BA Bean & Weed, 1909

Barton Appler Bean (1860–1947) was the younger brother of T.H. Bean (below). He worked under his

brother at the US National Museum (1881), becoming Assistant Curator of Fishes there (1890–1932) until his retirement. During this time he occasionally worked as an investigator for the United States Fish Commission. He also made trips to New York, Florida and the Bahamas. He wrote more than 60 papers, including *Fishes of the Bahama Islands* (1905) and, with others, *The Fishes of Maryland* (1929). He died falling from a bridge. At least four fishes are named after him.

Bean, TH

Chimaera Genus *Harriotta* Goode & TH Bean, 1895

Alaska Skate *Bathyraja parmifera* TH Bean, 1881
Mexican Lamprey *Tetrapleurodon spadiceus* TH Bean,
    1887
Pacific Longnose Chimaera *Harriotta raleighana*
    Goode & TH Bean, 1895
Deep-water Catshark *Apristurus profundorum* Goode
    & TH Bean, 1896
Boa Catshark *Scyliorhinus boa* Goode & TH Bean, 1896

Tarleton Hoffman Bean (1846–1916) was an American ichthyologist and elder brother of B.A. Bean (above). Columbian (George Washington) University awarded his MD (1876) and Indian University his MS (1883) unexamined, on the basis of his accomplishments. He was a forester (his first interest was botany), fish culturist, conservationist, editor, administrator and exhibitor. He worked as a volunteer at the Fish Commission laboratory in Connecticut (1874) where he met Baird and George Brown Goode (q.v.). He spent two decades (1875–1895) working in Washington at the National Museum (1878–1888) and the Fish Commission (1888–1895). After resigning from the Fish Commission, he was Director of the New York Aquarium (1895–1898). He then worked on forestry or fishery exhibits at the World's Fairs in Paris (1900) and St Louis (1905). He was a fish culturist for New York State (1906) until his death in a motor accident. He wrote nearly 40 papers with Goode, culminating in *Oceanic Ichthyology* (1896). Nine fish species and one genus are named after him.

Beebe

Spiny Skate *Bathyraja spinosissima* Beebe & Tee-Van,
    1941
Ecuador Skate *Dipturus ecuadoriensis* Beebe & Tee-
    Van, 1941
Pacific Chupare *Himantura pacifica* Beebe & Tee-Van,
    1941

Dr Charles William Beebe (1877–1962) was an American naturalist, ornithologist, marine biologist, entomologist, explorer and author. He began his working life looking after the birds at the Bronx Zoo (New York) and became Curator of Ornithology, New York Zoological Society (1899–1952), and Director, Department of Tropical Research (1919). He conducted a series of expeditions for the New York Zoological Society including deep dives in the Bathysphere including a record descent of 923 metres (3,028 feet) off Nonsuch Island, Bermuda (1934). He set up a camp (1942) at Caripito in Venezuela for jungle studies and (1950) bought 92 hectares (228 acres) of land in Trinidad and Tobago, which became the New York Zoological Society's Tropical Research Station (Asa Wright Nature Centre). He married Helen Elswyth Thane Ricker (1900–1981), who wrote romantic novels (pen name Elswyth Thane). He was known for his scientific writing for the public as well as academia. Many of his writings were popular books on his expeditions, such as *Two Bird-lovers in Mexico* (1905) and *Beneath Tropic Seas* (1928). He made enough money from them to finance his later expeditions. Several fishes, two amphibians and a bird are named after him.

Bemis

Pale Ghost Shark *Hydrolagus bemisi* Didier, 2002

Dr William E. Bemis is a vertebrate anatomist, currently Professor of Ecology and Evolutionary Biology at Cornell University, which awarded his first degree (1976). The University of Michigan awarded his master's (1978) and the University of California, Berkeley his doctorate (1983). He is also Emeritus Professor, Biology Department, University of Massachusetts Amherst. His chief interest is the evolution of bony and cartilaginous fishes. Among his many papers and longer works is the co-written *Functional Vertebrate Anatomy* (2001). Didier honoured him as her '…longtime mentor and friend, and a leader in ichthyological research'.

Bennett, ET

Bennett's Stingray *Dasyatis bennetti* Müller & Henle, 1841
[Syn. *Trygon bennetti*]

Bennett's Shovelnose Guitarfish *Glaucostegus typus* ET Bennett, 1830
[Alt. Giant/Green/Common/Brown Shovelnose Ray, Australian Guitarfish Syn. *Rhinobatos typus*]

Spotted Ratfish *Hydrolagus colliei* Lay & ET Bennett, 1839

Edward Turner Bennett (1797–1836) was a British zoologist and elder brother of the botanist John Joseph Bennett. He practised as a surgeon in London but was chiefly interested in zoology. He tried to establish an entomological society (1822) which became incorporated in the Linnaean Society and eventually the Zoological Society of London, of which he was Secretary (1831–1836). He wrote a number of works including *The Tower Menageries* (1829), but his most significant work was his contribution to the section on fish in *Zoology of Beechey's Voyage* (1839). His anonymous contributions to batoid literature, which appear in a (1830) memoir on the life of Thomas Stamford Raffles, are cited several times by the authors of the stingray description. Five mammals and a bird are named after him.

Bennett, MB

Eastern Banded Catshark *Atelomycterus marnkalha* Jacobsen & MB Bennett, 2007

Professor Michael 'Mike' Brian Bennett (b.1959) is a British-born zoologist. The University of Leeds awarded his BSc in zoology (1980) and the University of Wales, Bangor his PhD (1984). He undertook five years of postdoctoral research at Leeds before lecturing there and then at the University of Queensland (1990), where he still teaches human anatomy, functional anatomy and fish biology. A particular interest is the biology of sharks and rays. He says he probably spends too much time birdwatching and too little fishing. He has published many papers on different aspects of zoology such as *Unifying Principles in Terrestrial Locomotion: Do Hopping Australian Marsupials Fit In?* (2000); the co-written *Redescription of the Genus* Manta *with Resurrection of* Manta alfredi *(Krefft, 1868)* (Chondrichthyes; Myliobatoidei; Mobulidae) (2008); and *Community Composition of Elasmobranch Fishes Utilizing Intertidal Sand Flats in Moreton Bay, Queensland, Australia* (2011).

Ben-Tuvia

Elat Electric Ray *Heteronarce bentuviai* Baranes & Randall, 1989

Dr Adam Ben-Tuvia (1919–1999) was born in Krakow, Poland and died in Jerusalem. He initially studied agriculture at the Jagiellonian University of Krakow (1937–1939), interrupted by the Second World War. After the war he settled in Palestine, changing his studies to biology and zoology at the Hebrew University of Jerusalem, where he was awarded his MSc (1947). He was employed (1946–1964) at the Sea Fisheries Research Station, Haifa (Ministry of Agriculture). There he specialised in fisheries biology, exploratory fishing and fish systematics. During this time he finished his PhD at the Hebrew University (1955). He then worked (1964–1969) at the FAO Department of Fisheries, Fishery Resources and Exploitation Division before returning to the Sea Fisheries Research Station in Haifa (1969–1971). When his mentor Professor Heinz Steinitz died (1972), he moved to the Zoological Department of the Hebrew University of Jerusalem as Associate Professor, becoming full Professor (1985) and Emeritus (1988). He taught ichthyology and took care of the extensive fish collection. He wrote numerous scientific papers in the field of systematics, biology and zoogeography. He described nine fish species as new to science. Altogether seven species of marine fish are dedicated to him.

Berg, FGC

Blotched Sand Skate *Psammobatis bergi* Marini, 1932

Dr Frederico Guillermo Carlos (Friedrich Wilhelm Carl) Berg (1843–1902) was a Latvian entomologist and naturalist. After a number of years working in commerce, he became a conservator of entomological specimens at the Riga Museum. He joined the Museo Nacional, Buenos Aires (1873) and went on expeditions to Patagonia (1874) and Chile (1879). He worked at the Museo Nacional de Historia Natural, Montevideo (1890–1892) and was Director, Museo Nacional, Buenos Aires (1892–1901). A mammal, an amphibian and three birds are named after him.

Berg, LS

Bottom Skate *Bathyraja bergi* Dolganov, 1983

Ukranian Brook Lamprey *Eudontomyzon mariae* LS Berg, 1931
Korean Lamprey *Eudontomyzon morii* LS Berg, 1931
Bullhead Sharks *Heterodontiformes* LS Berg, 1940
Skates & Relatives *Rajiformes* LS Berg, 1940
Mackerel Sharks *Lamniformes* LS Berg, 1958
Saw Sharks *Pristiophoriformes* LS Berg, 1958

Lev (Leo) Semionovitch (Semenovich) Berg, (1876–1950) was a Russian geographer and zoologist born in Bender, Moldavia. He established the foundations of limnology in Russia with his systematic studies on the physical, chemical and biological conditions of fresh waters, particularly of lakes. His work in ichthyology was also noted in the palaeontology, anatomy and embryology of Russian fish. He wrote *Natural Regions of the USSR* (1950). He was honoured in the skate name having been the first to give a description of it. At least seven marine fishes and some freshwater species are named after him.

Berrebi

Fine-spotted Leopard Whipray *Himantura tutul* Borsa, Durand, K-N Shen, Arlyza, Solihin & Berrebi, 2013

Dr Patrick Berrebi is an ichthyologist at Laboratoire Génome et Populations, Université Montpellier 2, where he is Director of Research in the team concerned with the evolution of fishes. He co-wrote *Nuclear Markers of Danube Sturgeons Hybridization* (2011).

Besednov

Shortlip Electric Ray *Narcine brevilabiata* Besednov, 1966
Tonkin Numbfish *Narcine prodorsalis* Besednov, 1966

L.N. Besednov (b.1932) is a Russian ichthyologist who graduated from Moscow State University (1956). He has written a number of papers including *A New Shark Species from the Tonkin Gulf* – Negogaleus longicaudatus (1966).

Besnard

Polkadot Catshark *Scyliorhinus besnardi* S Springer & Sadowsky, 1970

Professor Wladimir Besnard (1890–1960) was the founder and first director of the Instituto Oceanográfico da Universidade de São Paulo. Born in France of Russian origin, he studied zoology and comparative anatomy in Kiev. The Governor of São Paulo State invited him to Brazil where he organised and headed what became the Instituto. There he researched marine biology and physical oceanography. A Brazilian oceanographic research vessel is named 'Professor Wladimir Besnard' after him and he is honoured in the names of at least eight marine organisms. The type of the catshark was trawled by the institute's research vessel of the day.

Bickell

Sixgill Stingray *Hexatrygon bickelli* Heemstra & MM Smith, 1980

David 'Dave' Bickell is a South African journalist, formerly the angling correspondent of the 'Eastern

Province Herald'. The holotype of the sixgill stingray was a nearly perfect, 105 cm-long female specimen washed up on a beach at Port Elizabeth in South Africa (1980), where Bickell discovered it. He became a volunteer worker for the protection of turtles.

Bigelow

Bigelow's Ray *Rajella bigelowi* Stehmann, 1978

Blurred Lantern Shark *Etmopterus bigelowi* Shirai & Tachikawa, 1993

Catshark genus *Cephalurus* Bigelow & Schroeder, 1941

Ray genus *Breviraja* Bigelow & Schroeder, 1948

Skate genus *Cruriraja* Bigelow & Schroeder, 1948

Electric Ray genus *Diplobatis* Bigelow & Schroeder, 1948

Chimaera Genus *Neoharriotta* Bigelow & Schroeder, 1950

Skate genus *Springeria* Bigelow & Schroeder, 1951

Skate genus *Pseudoraja* Bigelow & Schroeder, 1954

Cuban Ribbontail Catshark *Eridacnis barbouri* Bigelow & Schroeder, 1944

Broadgill Catshark *Apristurus riveri* Bigelow & Schroeder, 1944

Pacific Sleeper Shark *Somniosus pacificus* Bigelow & Schroeder, 1944

Lightnose Skate *Breviraja colesi* Bigelow & Schroeder, 1948

Atlantic Legskate *Cruriraja atlantis* Bigelow & Schroeder, 1948

Cuban Legskate *Cruriraja poeyi* Bigelow & Schroeder, 1948

Shorttail Skate *Amblyraja jenseni* Bigelow & Schroeder, 1950

Spinose Skate *Breviraja spinosa* Bigelow & Schroeder, 1950

Blackfin Pygmy Skate *Fenestraja atripinna* Bigelow & Schroeder, 1950

Cuban Pygmy Skate *Fenestraja cubensis* Bigelow & Schroeder, 1950

Gulf of Mexico Pygmy Skate *Fenestraja sinusmexicanus* Bigelow & Schroeder, 1950

Yucatán Skate *Leucoraja yucatanensis* Bigelow & Schroeder, 1950

Leaf-nose Legskate *Anacanthobatis folirostris* Bigelow & Schroeder, 1951

Ratfish sp. *Hydrolagus alberti* Bigelow & Schroeder, 1951

Spreadfin Skate *Dipturus olseni* Bigelow & Schroeder, 1951

Prickly Brown Ray *Dipturus teevani* Bigelow & Schroeder, 1951

Speckled Skate *Leucoraja lentiginosa* Bigelow & Schroeder, 1951

Gulf Hagfish *Eptatretus springeri* Bigelow & Schroeder, 1952

Fanfin skate *Pseudoraja fischeri* Bigelow & Schroeder, 1954

Sooty Ray *Rajella fuliginea* Bigelow & Schroeder, 1954

Lined Lantern Shark *Etmopterus bullisi* Bigelow & Schroeder, 1957

Rough Legskate *Cruriraja rugosa* Bigelow & Schroeder, 1958

Hookskate *Dactylobatus clarkia* Bigelow & Schroeder, 1958

San Blas Skate *Dipturus garricki* Bigelow & Schroeder, 1958

Hooktail Skate *Dipturus oregoni* Bigelow & Schroeder, 1958

American Legskate *Anacanthobatis americanus* Bigelow & Schroeder, 1962

Longnose Legskate *Anacanthobatis longirostris* Bigelow & Schroeder, 1962

Broadfoot Legskate *Cruriraja cadenati* Bigelow & Schroeder, 1962

Bullis Skate *Dipturus bullisi* Bigelow & Schroeder, 1962

Plain Pygmy Skate *Fenestraja ishiyamai* Bigelow & Schroeder, 1962

Atlantic Pygmy Skate *Gurgesiella atlantica* Bigelow & Schroeder, 1962

Purplebelly Skate *Rajella purpuriventralis* Bigelow & Schroeder, 1962

Finspot Ray *Raja cervigoni* Bigelow & Schroeder, 1964

Bahama Skate *Raja bahamensis* Bigelow & Schroeder, 1965

Green Lantern Shark *Etmopterus virens* Bigelow, Schroeder & S Springer, 1953

African Lantern Shark *Etmopterus polli* Bigelow, Schroeder & S Springer, 1953

Fringefin Lantern Shark *Etmopterus schultzi* Bigelow, Schroeder & S Springer, 1953

Dr Henry Bryant Bigelow (1879–1967) was an American zoologist. After graduating at Harvard (1901) he worked with the ichthyologist Alexander Agassiz (q.v.) and accompanied him on several major marine expeditions including one on the 'Albatross' (1907). He was Curator of the Museum of Comparative Zoology (1905), then joined the Department of Invertebrate Zoology, Harvard (1906–1966). He founded the Woods Hole Oceanographic Institution (1930) and became its first director (1931–1940). He published several books and over 100 scientific papers, including the co-written *Fishes of the Gulf Main* (1953). At least 26 species, including more than 15 marine organisms, are named after him, as is the

NOAA research vessel 'Henry B. Bigelow'. He was honoured in the name of the lantern shark as he had first described it but misidentified it as *E. pusillus*.

Bischoff

Hagfish sp. *Eptatretus bischoffii* AF Schneider, 1880

Dr Theodor Ludwig Wilhelm Bischoff (1807–1882) was a German physician and biologist who taught anatomy at Heidelberg (1835–1843). He was Professor of Anatomy and Physiology at Giessen (1843–1855) and Professor of Anatomy at Munich University (1854). His particular study was embryology and he wrote a number of detailed papers on different mammal ova (1842–1854). He was a colleague of Schneider's.

Blainville

Longnose Spurdog *Squalus blainville* Risso, 1827

Requiem Shark genus *Carcharhinus* Blainville, 1816
Catshark genus *Scyliorhinus* Blainville, 1816
Tope Shark genus *Galeorhinus* Blainville, 1816
Bull Shark genus *Heterodontus* Blainville, 1816
Basking Shark genus *Cetorhinus* Blainville, 1816
Squaliform Shark genus *Echinorhinus* Blainville, 1816
Dogfish Shark genus *Squalidae* Blainville, 1816
Shark genus *Aetobatus* Blainville, 1816

Professor Henri Marie Ducrotay de Blainville (1777–1850) was a French zoologist and anatomist. He was one of Cuvier's bitterest rivals and his successor to the chair of comparative anatomy at the Muséum National d'Histoire Naturelle, and in the Collège de France. He applied his principles of anatomy to bird classification and had much influence in establishing skeletal evolution as one determinant of the classification of species. He wrote *Manuel de Malacologie et de Conchyliologie* (1825–1827). His most valuable contribution was to herpetology where he elevated reptiles and amphibians to separate classes but he did name several shark taxa. Five mammals, a bird and a reptile are named after him.

Blanc

Smooth Freshwater Stingray *Dasyatis garouaensis* Stauch & Blanc, 1962
[Alt. Niger Stingray, Syn. *Potamotrygon garouaensis*]

Maurice Blanc (b.1923) has written a number of papers and longer works, including *Secrets et Merveilles de la Vie des Poissons* (1974), several under the auspices of the Food and Agriculture Organi-

zation of the United Nations. The name *garouaensis* refers to the type locality Garoua, Cameroon.

Bleeker

Bleeker's Whipray *Himantura bleekeri* Blyth, 1860

Saw Shark genus *Pristiophoridae* Bleeker, 1837
Sicklefin Weasel Shark genus *Hemigaleus* Bleeker, 1852
Gulper Shark genus *Centrophoridae* Bleeker, 1859

Longcomb Sawfish *Pristis zijsron* Bleeker, 1851
White-cheeked Shark *Carcharhinus tjutjot* Bleeker, 1852
Hooktooth Shark *Chaenogaleus macrostoma* Bleeker, 1852
Sicklefin Weasel Shark *Hemigaleus microstoma* Bleeker, 1852
Hasselt's Bamboo Shark *Chiloscyllium hasseltii* Bleeker, 1852
Stingray sp. *Himantura pareh* Bleeker, 1852
Round Whipray *Himantura pastinacoides* Bleeker, 1852
Whitenose Whipray *Himantura uarnacoides* Bleeker, 1852
Leopard Whipray *Himantura undulate* Bleeker, 1852
Zonetail Butterfly Ray *Gymnura zonura* Bleeker, 1852
Ornate Eagle Ray *Aetomylaeus vespertilio* Bleeker, 1852
Japanese Eagle Ray *Myliobatis tobijei* Bleeker, 1854
Starspotted Smooth-Hound *Mustelus manazo* Bleeker, 1854
Grey Reef Shark *Carcharhinus amblyrhynchos* Bleeker, 1856
Borneo Shark *Carcharhinus borneensis* Bleeker, 1858
Japanese Angelshark *Squatina japonica* Bleeker, 1858
Pygmy Devil Ray *Mobula eregoodootenkee* Bleeker, 1859
Smallfin Gulper Shark *Centrophorus moluccensis* Bleeker, 1860
Bigeye Skate *Okamejei meerdervoortii* Bleeker, 1860
Lusitanian Cownose Ray *Rhinoptera peli* Bleeker, 1863
[Syn. *Rhinoptera marginata*]
Tasselled Wobbegong *Eucrossorhinus dasypogon* Bleeker, 1867

Dr Pieter Bleeker (1819–1878) was an ichthyologist and army surgeon commissioned (1841) by the Dutch East India Company. He was apprenticed to an apothecary (1831–1834) and became interested in anatomy and zoology. He qualified as a surgeon at Haarlem (1840) and went to Paris, working in hospitals while attending Blainville's (q.v.) lectures. He was stationed in the Dutch East Indies (Indonesia) (1842–1860) and acquired 12,000 fish specimens, most of which are today in the Rijksmuseum van Natuurlijke Historie, Leiden. He returned to Holland (1860), taking with him his collection, which was sold after his death. He wrote over 500 scientific papers describing 511 new genera and 1,925 new species! He wrote a monumental 36-volume work, *Atlas Ichthyologique des Indies Orientales Neerlandaises* (1862–1877). Blyth frequently cites Bleeker's study of the cartilaginous fishes of Bengal in his description of the whipray. Three reptiles are named after him.

Bloch

Winghead Shark *Eusphyra blochii* Cuvier, 1817
Bluntnose Guitarfish *Rhinobatos blochii* Müller & Henle, 1841

Shark Ray genus *Rhina* Bloch & JGT Schneider, 1801

European Brook Lamprey *Lampetra planeri* Bloch, 1784
Greenland Shark *Somniosus microcephalus* Bloch & JGT Schneider, 1801
Smoothnose Wedgefish *Rhynchobatus laevis* Bloch & JGT Schneider, 1801
Porcupine Ray *Urogymnus asperrimus* Bloch & JGT Schneider, 1801
Blind Shark *Brachaelurus waddi* Bloch & JGT Schneider, 1801
Bowmouth Guitarfish *Rhina ancylostoma* Bloch & JGT Schneider, 1801
Fanray *Platyrhina sinensis* Bloch & JGT Schneider, 1801
Longnose *Dasyatis guttata* Bloch & JGT Schneider, 1801
Scaly Whipray *Himantura imbricate* Bloch & JGT Schneider, 1801
Smooth Butterfly Ray *Gymnura micrura* Bloch & JGT Schneider, 1801
Longheaded Eagle Ray *Aetobatus flagellum* Bloch & JGT Schneider, 1801
Nieuhof's Eagle Ray *Aetomylaeus nichofii* Bloch & JGT Schneider, 1801
[Alt. Barbless Eagle Ray, Banded Eagle Ray, Syn. *Raja niehofii, Myliobatus nieuhofii*]
Gulper Shark *Centrophorus granulosus* Bloch & JGT Schneider, 1801
Blackspotted Numbfish *Narcine timlei* Bloch & JGT Schneider, 1801
Numbray *Narke dipterygia* Bloch & JGT Schneider, 1801

Marcus Élieser Bloch (1723–1799) was a German physician and naturalist specialising in ichthyology. At first unable to read or write German, he became tutor to a Jewish surgeon's family in Hamburg, where he learned Latin, German and the basics of anatomy. He was subsequently able to move to Berlin and study for his doctorate in natural history and medicine and later practise medicine there (1747). He wrote several books and the 12-volume magnum opus *Allgemeine Naturgeschichte der Fische* (The Natural History of Fish) (1782–1795), which included over 400 beautifully hand-painted illustrated plates. He then set out to catalogue every fish species. J.G.T. Schneider completed this work after Bloch died as *Blochii System Ichthyologia ME* (1801), which described 1,519 species. Bloch's collection included around 1,500 specimens including a preserved specimen of the guitarfish that the authors studied. At least nine fishes are named after him.

Blyth

Bleeker's Whipray *Himantura bleekeri* Blyth, 1860
Blackedge Whipray *Himantura marginata* Blyth, 1860

Edward Blyth (1810–1873) was an English zoologist and author. He was Curator of the Museum of the Asiatic Society of Bengal (1842–1864) and wrote its catalogue. He wrote a series of monographs on cranes, gathered together and published as *The Natural History of Cranes* (1881). Hume said of him, 'Neither neglect nor harshness could drive, nor wealth nor worldly advantages tempt him, from what he deemed the nobler path. [He was] ill paid and subjected ... to ceaseless humiliations.' In a similar tribute Arthur Grote wrote, 'Had he been a less imaginative and more practical man, he must have been a prosperous one... All that he knew was at the service of everybody. No one asking him for information asked in vain.' Twenty-six birds, five mammals, two reptiles and an amphibian are named after him.

Boardman

Australian Sawtail Catshark *Figaro boardmani* Whitley, 1928
[Alt. Banded Shark]

William Boardman was a naturalist at the Australian Museum who was a friend and colleague of the describer and who collected the holotype from a trawler. He made regular collecting trips such as one of ten days on North West Island, Queensland with Melbourne Ward (1929). He wrote a number of scientific papers such as *Two New Species of the Genus Notoscolex (Oligochaeta) from Ulladulla, New South Wales* (1931) in the records of the Australian Museum and another with Whitley, *Marine Animals from Low Isles, Queensland* (1929).

Bocage

Squaliform Sleeper Shark genus *Centroscymnus* Barbosa du Bocage & Brito Capello, 1864

Squaliform Shark genus *Scymnodon* Barbosa du
  Bocage & Brito Capello, 1864

Lowfin Gulper Shark *Centrophorus lusitanicus* Barbosa
  du Bocage & Brito Capello, 1864
Portuguese Dogfish *Centroscymnus coelolepis*
  Barbosa du Bocage & Brito Capello, 1864
Longnose Velvet Dogfish *Centroselachus crepidater*
  Barbosa du Bocage & Brito Capello, 1864
Knifetooth Dogfish *Scymnodon ringens* Barbosa du
  Bocage & Brito Capello, 1864

José Vicente Barboza du Bocage (1823–1907) was a
Portuguese zoologist and politician. He studied at
the University of Coimbra (1839–1846). He lectured
in the zoology department of what is now the University of Lisbon (1851–1880) and became (1858)
Scientific Director and Curator of Zoology of the
National Zoological Museum of Lisbon, which was
(1905) named in his honour. He greatly expanded the
collection with items brought back from the colonies
and wrote a book about how to collect and prepare
such specimens. He retired from most scientific work
(1880) except the directorship of the museum and
went into politics becoming the Minister of Navy
and Ultramarine Possessions and later Minister of
Foreign Affairs (1883–1886). He published numerous works, including over 200 papers, on mammals,
herpetology and birds as well as fishes, such as *Diag-
nose de Algumas Espécies Inéditas da Família Squalidae
que Frequentam os Nossos Mares* (1864). Twelve birds,
six mammals, ten reptiles and four amphibians are
named after him.

Boeseman

Speckled Catshark *Halaelurus boesemani* S Springer &
  D'Aubrey, 1972
Boeseman's Skate *Okamejei boesemani* Ishiyama, 1987
Freshwater Ray sp. *Potamotrygon boesemani* Rosa,
  Carvalho & Wanderley, 2008

Sharpsnout Stingray *Dasyatis geijskesi* Boeseman, 1948

Dr Marinus Boeseman (1916–2006) was an ichthyologist at the Department of Zoology, Rijksmuseum
van Natuurlijke Historie, Leiden, becoming Curator
of Fishes (1947–1981). Universiteit Leiden awarded
his master's degree (1941). He was in the Dutch
resistance during the Second World War and was
arrested (1943) but survived imprisonment at
Dachau, though for years his health was so badly
affected that he could not work. He collected fishes
in El Salvador (1953) and travelled in New Guinea

(1954–1955). He caught polio (1957) and suffered
from a permanent disability in his right arm, but
still took part in collecting expeditions to *inter alia*
Surinam and Trinidad. In his etymology of the skate
Ishihara thanks him for making him aware of problems in the systematics of Japanese *Raja*. The citation
for the ray honours him as the person who 'contributed substantially to our knowledge of both South
American ichthyology (including chondrichthyans)
and zoological history.' A reptile and an amphibian
are named after him.

Bollman

Equatorial Ray *Raja equatorialis* Jordan & Bollman, 1890

Charles Harvey Bollman (1868–1889) was an
American zoologist and a museum assistant at the
University of Indiana. Among the papers he wrote
are *A Review of the Centrachidæ, or Fresh-water Sunfishes, of North America* (1888) and *The Myriapoda of
North America* (1893). He was in charge of explorations for the Fish Commission in Georgia, USA. He
was described at his death as a 'young naturalist of
great promise'.

Bonaparte

Smallnose Fanskate *Sympterygia bonapartii* Müller &
  Henle, 1841

Chimaeras *Chimaeridae* Bonaparte, 1831
Rays *Rajidae* Bonaparte, 1831
Thresher Sharks family *Alopiidae* Bonaparte, 1838
Angel Sharks family *Squatinidae* Bonaparte, 1838
Torpedo Electric Rays family *Torpedinidae* Bonaparte,
  1838
Sawfishes family *Pristidae* Bonaparte, 1838
Tropical Stingray family *Myliobatidae* Bonaparte, 1838

Carpet Shark genus *Orectolobus* Bonaparte, 1834
Ray genus *Glaucostegus* Bonaparte, 1846

Pelagic Stingray *Pteroplatytrygon violacea* Bonaparte,
  1832

Atlantic Torpedo *Torpedo nobiliana* Bonaparte, 1835
[Alt. Dark Electric Ray]
Smoothback Angelshark *Squatina oculata* Bonaparte,
  1840

Prince Charles Lucien Bonaparte (originally Jules
Laurent Lucien) (1803–1857) was a nephew of
Napoleon Bonaparte and a renowned biologist and
zoologist. He was much travelled and spent many
years in the United States cataloguing birds. He has

been described as the 'father of systematic ornithology'. He eventually settled in Paris and commenced his *Conspectus Generum Avium*, a catalogue of all bird species, which was unfinished on his death. Its publication was heralded as a major step forward in achieving a complete list of the world's birds. He also wrote *American Ornithology* (1825) and *Iconografia della Fauna Italica – Uccelli* (1832). Swainson described Bonaparte as 'destined by nature to confer unperishable benefits on this noble science'. His treatise on the fauna of Italy introduced several new sharks and rays to science. Twenty birds and a mammal are named after him.

Bond

> Miller Lake Lamprey *Entosphenus minimus* Bond & Kan, 1973

Professor Dr Carl E. Bond (1920–2007) was an American ichthyologist. Oregon State College awarded his bachelor's (1947) and master's (1948) and the University of Michigan awarded his PhD (1963). He began his teaching and research career in Oregon State College's Fisheries and Wildlife Department (1949) as Assistant Aquatic Biologist. Specialising in the study of freshwater fishes, he became one of the world's leading authorities on sculpin cottidae. Bond was also involved in research internationally, working in Latin America and Africa, as well as India and Iran on Peace Corps projects (1967–1971). He was also Assistant Dean of the Graduate School (1969–1974). He became Professor Emeritus of Fisheries (1984). He wrote *Biology of Fishes* (1979).

Bonnaterre

> Common Thresher *Alopias vulpinus* Bonnaterre, 1788
> Draughtsboard Swellshark *Cephaloscyllium isabellum* Bonnaterre, 1788
> Leafscale Gulper Shark *Centrophorus squamosus* Bonnaterre, 1788
> Kitefin Shark *Dalatias licha* Bonnaterre, 1788
> Bramble Shark *Echinorhinus brucus* Bonnaterre, 1788
> Nurse Shark *Ginglymostoma cirratum* Bonnaterre, 1788
> Epaulette Carpetshark *Hemiscyllium ocellatum* Bonnaterre, 1788
> Sharpnose Sevengill Shark *Heptranchias perlo* Bonnaterre, 1788
> Bluntnose Sixgill Shark *Hexanchus griseus* Bonnaterre, 1788
> Porbeagle *Lamna nasus* Bonnaterre, 1788
> Devil Fish *Mobula mobular* Bonnaterre, 1788

> Spotted Wobbegong *Orectolobus maculatus* Bonnaterre, 1788

Pierre Joseph Bonnaterre (1752–1804) was a French naturalist. He was a contributor to the *Tableau Encyclopédique et Méthodique*, principally the sections on cetaceans, mammals, birds, reptiles, amphibians, insects and (1788) fishes. He is also credited with identifying about 25 new species of fish, and assembled illustrations of about 400 in his encyclopedia work. One claim to fame is that he was the first scientist to study the feral child Victor of Aveyron.

Bonnet

> West African Catshark *Scyliorhinus cervigoni* Maurin & Bonnet, 1970

Dr Marc Bonnet was a French ichthyologist who led an expedition on the research vessel 'Thalassa' to the waters off Spanish Sahara and Mauretania (1971). He co-wrote *Poissons des Côtes Nord-ouest Africaines (Campagnes de la «Thalassa» 1962 et 1968)* (1970).

Bordalo-Machado

> Ghost Shark sp. *Hydrolagus lusitanicus* Moura, Figueiredo, Bordalo-Machado, Almeida & Gordo, 2005

Pedro Alexandre Costa Pinto de Bordalo-Machado is a Portuguese marine biologist at the Instituto de Investigação das Pescas e do Mar, Lisbon. The University of Lisbon awarded his bachelor's degree (1996) and the Universidade Técnica de Lisboa his master's (2002). Among his published papers is the co-written *Deep-water Shark Fisheries off the Portuguese Continental Coast* (2005).

Borsa

> Fine-spotted Leopard Whipray *Himantura tutul* Borsa, Durand, K-N Shen, Arlyza, Solihin & Berrebi, 2013

Dr Philippe Borsa (b.1960) is a population geneticist. The Université Pierre-et-Marie Curie, Paris awarded his PhD (1990). He joined the Institut de recherche pour le développement in Montpellier as a research scientist (1992). In the last ten years, he has been based successively in New Caledonia and in Indonesia, where he has been conducting phylogeographic research on Indo-Pacific shorefishes, including stingrays (family Dasyatidae). He co-wrote *Mitochondrial Haplotypes Indicate Parapatric-like Phylogeographic Structure in Blue-Spotted Maskray* (Neotrygon kuhlii) *from the Coral Triangle Region* (2013).

Bory de Saint-Vincent

Australian Ghost Shark *Callorhinchus milii* Bory de
Saint-Vincent, 1823

Jean Baptiste Bory de Saint-Vincent (1778–1846)
was a French naturalist who was sent as naturalist
aboard Baudin's expedition to Australia (1798), but
left the ship at Mauritius to spend two years exploring Réunion and other Indian Ocean islands. On the
voyage he befriended Milius. He wrote a number of
accounts of his journeys, including *Voyage dans les
Îles d'Afrique* (1803).

Bosc

Clearnose Skate *Raja eglanteria* Bosc, 1800

Louis Augustin Guillaume Bosc (1759–1828) was
a naturalist and botanist who became President of
the French Natural History Society (1790). After his
friend Mme Roland was guillotined (1793), he had to
hide in the Forest of Montmorency, returning to Paris
after Robespierre's fall. After the coup d'état (1799)
he could only support himself by mass-producing
articles for scientific periodicals. He became inspector of the gardens of Versailles and publicly owned
nurseries. Two reptiles are named after him.

Bougainville

Short-snouted Shovelnose Ray *Aptychotrema
bougainvillii* Müller & Henle, 1841

Admiral Baron Hyacinthe Yves Philippe Potentien
de Bouganville (1781–1846) was a French naval
officer in command of the corvette 'Espérance'.
He was a midshipman on 'Le Géographie' under
Baudin (1800–1803) and later took part in a circumnavigation (1824–1826). Two mammals and a bird
are named after him.

Bougis

Rondelet's Ray *Raja rondeleti* Bougis, 1959

Dr Paul Bougis (1920–2009) was a French ichthyologist. The Université de Paris awarded his doctorate.
He was Deputy Director of the Zoological Station at
Villefranche-sur-Mer then became Director (1956).
He bought an oceanographic vessel 'N.O. Korotneff'
and (1974) combined all the separate research laboratories into the 'Station Marine Villefranche'. He
wrote *Marine Plankton Ecology* (1977).

Boulenger

Houndshark genus *Scylliogaleus* Boulenger, 1902

Oman Cownose Ray *Rhinoptera jayakari* Boulenger, 1895
Flapnose Houndshark *Scylliogaleus quecketti*
Boulenger, 1902

George Albert Boulenger (1858–1937) was a Belgian-
British zoologist at the British Museum, London.
He graduated from the Free University in Brussels
(1876) and worked at the Muséum des Sciences
Naturelles, Brussels as Assistant Naturalist studying amphibians, reptiles and fishes, until moving
to London (1880) and taking British nationality
(1882). He began work at the BMNH cataloguing
the amphibian collection. He became First-class
Assistant there (1882), a position he held until retiring (1920). His output was prodigious: nearly 2,600
species described including 1,096 fishes, and 877
scientific papers aided by his almost photographic
memory. He was also a violinist and polyglot.
He retired to grow and study roses, producing 34
papers and 2 books on botanical subjects. He was a
member of the American Society of Ichthyologists
and Herpetologists and was elected its first honorary member (1935). Belgium conferred (1937) on him
the Order of Léopold, the highest honour awarded
to a civilian. Forty amphibians and 75 reptiles are
named after him.

Brauer

Smallbelly Catshark *Apristurus indicus* Brauer, 1906

Dr August Bernhard Brauer (1863–1917) was a
German zoologist, herpetologist, and ichthyologist.
He graduated from Humboldt-Universität, Berlin in
natural sciences (1885) and took his doctorate there
(1895). He received his habilitation at the University
of Marburg, where he subsequently worked as a
lecturer (1892). He collected in the Seychelles (1894–
1895) and was on the deep-sea Valdivia expedition
(1898), describing (1908) the fishes they collected.
He became a professor at Berlin University (1905)
and Director of the university's Zoological Museum
(1906). He was appointed Professor of the Zoological University, Berlin (1914). He wrote the 19-part
work on freshwater fishes, *Die Süßwasserfauna
Deutschlands* (1909–1912). A mammal, a reptile and
an amphibian are named after him.

Braun

Braun's Whale Shark *Rhincodon typus* A Smith, 1829
[Alt. Whale Shark]

Camrin D. Braun is an American ichthyologist. The
College of Idaho awarded his bachelor's degree

(2011) and King Abdullah University of Science and Technology, Saudi Arabia, his master's (2013). He is currently working for his doctorate at the Woods Hole Oceanographic Institution, USA. He wrote *Movements of Juvenile Whale Sharks* (Rhincodon typus) *in the Red Sea* (2010); it is only subsequent to the publication of that thesis that his name has been occasionally attached to this fish. It is interesting to come across an eponym 'emerging' and being able to pinpoint, almost exactly, when it was first coined!

Breder

> Vermiculate Electric Ray *Narcine vermiculatus* Breder, 1928

Dr Charles Marcus Breder Jr (1897–1983) was an experimental and behavioural ichthyologist whose doctorate was honorary, awarded by the University of Newark (1938). After leaving high school he worked for the US Bureau of Fisheries (1919–1921) and for the New York Aquarium (1921–1944) and became Acting Director (1937) and Director (1940). He was Curator and Chairman of the Department of Fishes at the American Museum of Natural History (1944–1965) and was an interim director of the Mote Marine Laboratory (1967). He was a leader in the study of cave fishes and was a dominant figure in this area of research in the 1940s and 1950s. He led the expedition known as 'The Aquarium Cave Expedition to Mexico' (1940). He died from Alzheimer's disease.

Brito Capello

> Portuguese Dogfish genus *Centroscymnus* Barbosa du Bocage & Brito Capello, 1864
> Squaliform Shark genus *Scymnodon* Barbosa du Bocage & Brito Capello, 1864
> False Catshark genus *Pseudotriakis* Brito Capello, 1868
>
> Lowfin Gulper Shark *Centrophorus lusitanicus* Barbosa du Bocage & Brito Capello, 1864
> Portuguese Dogfish *Centroscymnus coelolepis* Barbosa du Bocage & Brito Capello, 1864
> Longnose Velvet Dogfish *Centroselachus crepidater* Barbosa du Bocage & Brito Capello, 1864
> Knifetooth Dogfish *Scymnodon ringens* Barbosa du Bocage & Brito Capello, 1864
> False Catshark *Pseudotriakis microdon* Brito Capello, 1868

Vice-Admiral Hernandogildo Carlos de Brito Capello (1839–1917) served at sea and was based in Angola (1860–1863 and 1866–1869). He explored Africa (1877–1886) and was Governor of Angola (1886–1891). He led a number of expeditions in Angola, including one that traversed Africa to the sea in Mozambique (1878–1886). He described it in *De Angola à Contracosta* (1886). A bird is named after him. (See **Capello**)

Brum

> Freshwater Stingray sp. *Potamotrygon brumi* Devincenzi & Teague, 1942
> [Junior Syn. *Potamotrygon brachyura*]

Baltasar Brum Rodríguez (1883–1933) was an Uruguayan politician and President of Uruguay (1919–1923). When he was Minister of Public Instruction, he was described by the senior author as the only politician with responsibility for education who showed any interest in the Museum of Natural Sciences in Montevideo, where Devincenzi was Director. Brum committed suicide by publicly shooting himself as a protest at the dictatorial rule of one of his successors.

Budker

> Atlantic Weasel Shark genus *Paragaleus* Budker, 1935

Paul Budker (1900–1992) was a French ichthyologist specialising in sharks but also interested in cetaceans. He founded and directed a Whaling Research Centre, which is run in connection with the National Museum of Natural History and the Laboratory of Colonial Fisheries (Paris), where he was Deputy Director. He published more than 70 works on tropical fishing, sharks and whales, including *Whales and Whaling* (1958) and *The Life of Sharks* (1972).

Buen

> Angel Shark genus *Squatiniformes* Buen, 1926

(See **de Buen**)

Bullis

> Lined Lantern Shark *Etmopterus bullisi* Bigelow & Schroeder, 1957
> Bullis Skate *Dipturus bullisi* Bigelow & Schroeder, 1962
>
> Bahamas Sawshark *Pristiophorus schroederi* S Springer & Bullis, 1960
> Florida Torpedo *Torpedo andersoni* Bullis, 1962
> Dwarf Sicklefin Chimaera *Neoharriotta carri* Bullis & JS Carpenter, 1966

Harvey Raymond Bullis Jr (1924–1992) was a US malacologist who was head of the Bureau of Commercial Fisheries Exploratory Base at Pascagoula, Mississippi and Chief, Gulf Fisheries Exploration & Gear Research, US Fish and Wildlife Service. He wrote a number of pamphlets and scientific papers, including *Gulf of Mexico Shrimp Trawl Designs* (1951). At least eight fishes and marine organisms are named after him. He was honoured in the name of the lantern shark for furnishing the authors with it and other fishes collected off the coasts of Florida, and for the skate for 'providing the batoid fishes collected during the exploratory cruises of US Fish and Wildlife Service vessels in the Gulf of Mexico, Caribbean and along the South American coast'.

Bürger

Blackspotted Catshark *Halaelurus buergeri* Müller and
Henle, 1838
Inshore Hagfish *Eptatretus burgeri* Girard, 1855

Heinrich Bürger (1804/06–1858) (his definitive birth date is unknown: the oldest birth date is the one given by Bürger himself, but it appears likely that he used it to appear two years older than he really was) was a German-Jewish physicist and biologist employed by the Dutch government. He studied mathematics and astronomy at Göttingen (1821–1822) and called himself 'doctor' but there is no evidence of his obtaining a PhD. He first travelled to Batavia (Jakarta), Java (1824) and gained a third class degree of apothecary (1825). The Dutch Government appointed him (1825) Assistant to the German physicist P.F. von Siebold. He and Siebold collected natural history specimens for the Leiden Museum (1820s), where Temminck and then Schlegel were Director; they used much of the material Bürger collected for their *Fauna Japonica*. Bürger and Siebold collected together throughout the Dutch East Indies (Indonesia), then in Japan (1837–1838). Bürger was honourably discharged from government service (1843) and set up a rice and sugar business as well as mining and insurance. He collected the holotype of the hagfish. Two amphibians and at least one other fish are named after him.

Burgess

Broad-snouted Lantern Shark *Etmopterus burgessi*
Schaaf-Da Silva & Ebert, 2006

Cylindrical Lantern Shark *Etmopterus carteri*
S Springer & Burgess, 1985

Dwarf Lantern Shark *Etmopterus perryi* S Springer &
Burgess, 1985

West Indian Lantern Shark *Etmopterus robinsi*
Schofield & Burgess, 1997
Tailspot Lantern Shark *Etmopterus caudistigmus* Last,
Burgess & Séret, 2002
Pink Lantern Shark *Etmopterus dianthus* Last, Burgess
& Séret, 2002
Lined Lantern Shark *Etmopterus dislineatus* Last,
Burgess & Séret, 2002
Blackmouth Lantern Shark *Etmopterus evansi* Last,
Burgess & Séret, 2002
Pygmy Lantern Shark *Etmopterus fusus* Last, Burgess
& Séret, 2002
False Lantern Shark *Etmopterus pseudosqualiolus* Last,
Burgess & Séret, 2002
Shortfin Smooth Lanternshark *Etmopterus joungi*
Knuckey, Ebert & Burgess, 2011

George H. Burgess (b.1949) is an American ichthyologist. The University of Rhode Island awarded his bachelor's degree and the University of Florida his master's (1978). He also teaches ichthyology and marine biology. He is Director of the Florida Program for Shark Research at the Florida Museum of Natural History. His research interests are the life history, ecology, systematics, fishery management and conservation of elasmobranchs; shark attack; systematics and biogeography of fishes; management and conservation of aquatic ecosystems. His fieldwork includes the survey of the marine ichthyofauna of southwestern Florida, particularly the Florida Keys and Straits of Florida and the movements of reef-dwelling elasmobranchs through reef passages in Belize. He is Vice-Chair of the IUCN/ SSC Shark Specialist Group. Among his many publications are *Description of Six New Species of Lantern-sharks of the Genus Etmopterus (Squaloidea: Etmopteridae) from the Australasian Region* (2002) and *Management of Sharks and their Relatives* (2000). He was honoured in the name of the lantern shark for his contributions to the systematics of *Etmopterus*.

Bussing

Denticled Roundray *Urotrygon cimar* López &
Bussing, 1998

William 'Bill' A. Bussing (1933–2014) was an American ichthyologist at the School of Biology, Universidad de Costa Rica where he taught for over four decades, retiring as Professor Emeritus (1991) but he still remained Curator of Fishes at the Museo de Zoología. Military service in Korea interrupted his education but the University of Southern California awarded his master's in biology (1965).

He described more than 50 new species. He wrote widely alone or with others, including 7 books, such as *Peces Costeros del Caribe de Centro América Meridional* (2010), and around 70 scientific papers.

Byrne

Deep-water Ray *Rajella bathyphila* Holt & Byrne, 1908
Straightnose Rabbitfish *Rhinochimaera atlantica* Holt & Byrne, 1909

Lucius Widdrington Byrne (1875–1939) was a barrister at Lincoln's Inn. He graduated from Cambridge with a bachelor's degree (1897) and was called to the Bar (1899). During the First World War he was in military intelligence at the War Office. He was also an ichthyologist and Fellow of the Zoological Society of London. His writings were varied – anything from the 6th edition of *The Trustee Acts, including a Guide for Trustees to Investments* (1903) to *Third Report on the Fishes of the Irish Atlantic Slope: Fishes of the Genus Scopelus* (1908).

C

Cadenat

Broadfoot Legskate *Cruriraja cadenati* Bigelow & Schroeder, 1962
Longfin Sawtail Catshark *Galeus cadenati* S Springer, 1966

Crocodile Shark genus *Pseudocarcharias* Cadenat, 1963
Whitefin Hammerhead *Sphyrna couardi* Cadenat, 1951
African Sawtail Catshark *Galeus polli* Cadenat, 1959
Devil Ray *Mobula rancureli* Cadenat, 1959
Violet Skate *Dipturus doutrei* Cadenat, 1960
Sicklefin Devil Ray *Mobula coilloti* Cadenat & Rancurel, 1960
Rosette Torpedo *Torpedo bauchotae* Cadenat, Capapé & Desoutter, 1978

Dr Jean Cadenat (1908–1992) of the University of Dakar, was a French researcher and ichthyologist at the Office de la Recherche Scientifique et Technique Outre-Mer (ORSTOM). He is honoured in the names of many other taxa including flatworm, fish, copepod and others. Springer named the shark after him because he had described the similar species *G Polli*. The skate was named after him 'for his work on the elasmobranchs of the west coast of Africa'.

Cailliet

Angelshark sp. *Squatina caillieti* Walsh, Ebert & Compagno, 2011

Dr Gregor Michel Cailliet (b.1943) is an ichthyologist and marine biologist who was professor at Moss Landing Marine Laboratories (1972–2009) and subsequently Professor Emeritus. He started (2001) the Pacific Shark Research Center at MLML, and is still co-director with Dr David A. Ebert. He also has been (since 1997) Consultant, Regional Water Quality Control Board, California Energy Commission, and Aspen Environmental Group. The University of California, Santa Barbara awarded both his bachelor's degree (1966) and his doctorate (1972). Among his publications is *Fishes: A Field and Laboratory Manual on Their Structure, Identification and Natural History* (1986). He has been honoured for his contributions to ichthyology, especially chondrichthyan age and growth. Gregor Caillet told us that at a lunch with Ray Troll (q.v.) and Nancy Burnett (q.v.), they decided to form a very select club called 'The Chondronomes' – as they all had cartilaginous fishes named after them!

Caira

Borneo Sand Skate *Okamejei cairae* Last, Fahmi & Ishihara, 2010

Dr Janine N. Caira (b.1957) is a Canadian parasitologist who is Professor of Ecology and Evolutionary Biology at the University of Connectictut. She co-wrote *Sharks and Rays of Borneo* (2010).

Campbell

Blackspot Skate *Dipturus campbelli* Wallace, 1967

Dr George Gordon Campbell (1893–1977) was a South African physician and naturalist who inspired the Foundation for Marine Biological Research and was its President, for which he was honoured in the name of the skate. He went to study medicine at Edinburgh University (1914) but only qualified after the end of the First World War, during which he was an early member of the Royal Flying Corps in the days when dog-fighting meant firing a revolver at someone in a nearby aeroplane! He was President of the Royal Society of South Africa (1968) and Chancellor of the University of Natal.

Cantor

Requiem Shark genus *Prionace* Cantor, 1849

Dr. Theodore Edvard Cantor (1809–1860) was a Danish amateur zoologist and Superintendent Physician of the European Asylum, Bhowanipur, Calcutta. This was part of the East India Company's Bengal Medical Service. He collected in Penang and Malaca. He was interested in tropical fishes, and (c.1840) the King of Siam gave him some bettas, commonly known as fighting fish. He published an article about them that led to 'betta fever', the popular craze in Victorian England for keeping such fish. He wrote *Catalogue of Malayan Fishes* (1850). Two mammals, eleven reptiles and a bird are named after him.

Canut

Hoary Catshark *Apristurus canutus* S Springer & Heemstra, 1979
Grey Skate *Dipturus canutus* Last, 2008

Canutus is the Latin for grey or ash-coloured.

Capapé

Tortonese's Stingray *Dasyatis tortonesei* Capapé, 1975
African Skate *Raja africana* Capapé, 1977

Ray sp. *Raja rouxi* Capapé, 1977
Rosette Torpedo *Torpedo bauchotae* Cadenat, Capapé
& Desoutter, 1978

Dr Christian Capapé (b.1940) was a professor (retired 2005) in the ichthyology department at the University of Montpellier, France (1987–1993 and 1997–2005). His master's degree in biology was awarded by the Université Pierre et Marie Curie, Paris (1973) as was his doctorate in oceanography (1974). He received a state doctorate in life sciences at the University of Montpellier 2 (1986). He taught in Tunisia, the country of his birth, as Professor of Natural Sciences at the Lycée Français de Tunis-Mutuel (1970–1981) and was attached as an expert to the Pasteur Institute in Tunis (1981–1987). Between his appointments in Montpellier, he was Professor and Chairman of the Department of Animal Biology at the Université Cheikh Anta Diop, Dakar, Sénégal. He is a member of scientific societies and editorial boards of international reviews. He is a specialist in the reproductive biology and ecology of sharks, skates and rays and has written 280 papers about these topics, including *Nouvelles Données sur la Morphologie de* Mustelus asterias *Cloquet, 1821* (1983) and as co-author *New Data on the Eggs of the Angelshark,* Squatina oculata *Bonaparte, 1840 (Pisces: Squatinidae)* (1999) and *A Case of Hermaphroditism in Tortonese's Stingray* Dasyatis tortonesei *from the Lagoon of Bizerte (Northeastern Tunisia, Central Mediterranean)* (2012). He is very critical of the film 'Jaws' for giving sharks a bad name.

Capello

Smalleyed Rabbitfish *Hydrolagus affinis* Capello, 1868

(See **Brito Capello**)

Carpenter

Carpenter's Chimaera *Chimaera lignaria* Didier, 2002

(See **Dadit**)

Carpenter, JS

Dwarf Sicklefin Chimaera *Neoharriotta carri* Bullis & JS Carpenter, 1966

Dr James S. Carpenter was a supervisory fishery biologist in the 1960s (resigned 1968) for the US Fish & Wildlife Service at Pascagoula, Mississippi where Harvey Bullis was base director. He wrote *A Review of the Gulf of Mexico Red Snapper Fishery* (1965).

Carpenter, KE

Slender Weasel Shark *Paragaleus randalli* Compagno, Krupp & KE Carpenter, 1996

Dr Kent E. Carpenter is an ichthyologist who is Professor of Biological Sciences at Old Dominion University, Norfolk, Virginia. The Florida Institute of Technology awarded his BSc in marine biology (1975) and he undertook his doctorate in zoology at the University of Hawaii (1985). His research emphasis is on the systematics and evolution of marine fishes. He has worked on marine biogeography of the Indian and west Pacific Oceans and in marine conservation with a special interest in the Philippines. He is also a long-term collaborator with the Food and Agriculture Organization of the United Nations Species Identification and Data Programme for Fisheries, producing identification guides for regions such as the western Pacific, western Atlantic and eastern Atlantic oceans. He has also carried out fieldwork in the Caribbean, West Africa and the Philippines. In addition to research and teaching responsibilities, he is also the coordinator for the IUCN Global Marine Species Assessment, completing the first global review of every marine vertebrate species, and of selected marine invertebrates and marine plants, to determine conservation status and possible extinction risk for approximately 20,000 marine species. He has written a number of books including *The Living Marine Resources of the Western Central Pacific* (2010). Two fish species are named after him.

Carr

Dwarf Sicklefin Chimaera *Neoharriotta carri* Bullis & JS Carpenter, 1966

James K. Carr (1914–1980) was Under Secretary at the US Department of the Interior (1961–1964). He was honoured in the name of the chimaera for 'his great personal interest and counsel in the [Bureau of Commercial Fisheries'] exploratory fishing programs'.

Carter

Cylindrical Lantern Shark *Etmopterus carteri* S Springer & Burgess, 1985

(See **Gilbert**)

Carvalho

Electric Ray sp. *Narcine leopard* Carvalho, 2001

Madagascar Numbfish *Narcine insolita* Carvalho, Séret & Compagno, 2002

Last's Numbfish *Narcine lasti* Carvalho & Séret, 2002

Bigeye Electric Ray *Narcine oculifera* Carvalho, Compagno & Mee, 2002

Aden Gulf Torpedo *Torpedo adenensis* Carvalho, Stehmann & Manilo, 2002

Atz's Numbfish *Narcine atzi* Carvalho & Randall, 2003

Taiwanese Blind Electric Ray *Benthobatis yangi* Carvalho, Compagno & Ebert, 2003

Groovebelly Stingray *Dasyatis hypostigma* Santos & Carvalho, 2004

Brazilian Soft Skate *Malacoraja obscura* Carvalho, Gomes & Gadig, 2005

Eastern Numbfish *Narcine nelson* Carvalho, 2008

Ornate Numbfish *Narcine ornate* Carvalho, 2008

Freshwater Ray sp. *Potamotrygon boesemani* Rosa, Carvalho & Almeida Wanderley, 2008

Roundray genus *Heliotrygon* Carvalho & Lovejoy, 2011

Gomes' Roundray *Heliotrygon gomesi* Carvalho & Lovejoy, 2011

Rosa's Roundray *Heliotrygon rosai* Carvalho & Lovejoy, 2011

Long-tailed River Stingray *Plesiotrygon nana* Carvalho & Ragno, 2011

Parnaiba River Stingray *Potamotrygon tatianae* JP da Silva & Carvalho, 2011

Tiger Ray sp. *Potamotrygon tigrina* Carvalho, Sabaj Pérez & Lovejoy, 2011

Freshwater Stingray sp. *Potamotrygon amandae* Loboda & Carvalho, 2013

Freshwater Stingray sp. *Potamotrygon pantanensis* Loboda & Carvalho, 2013

Freshwater Stingray sp. *Potamotrygon limai* Fontenelle, JP da Silva & Carvalho, 2014

Dr Marcelo Rodrigues de Carvalho works at the Departamento de Zoologia, Instituto de Biociências, Universidade de São Paulo, where he is a professor. He studied at Universidade Santa Úrsula, USU, Brazil for his first degree (1987–1991). He obtained a doctorate through a programme on evolutionary biology run jointly (1992–1999) by New York City University and AMNH, where he subsequently did post-doctoral research (1999–2001).

Castello

Electric Ray sp. *Discopyge castelloi* Menni, Rincón & Garcia, 2008

Freshwater Stingray genus *Plesiotrygon* Rosa, Castello & Thorson, 1987

Maracaibo River Stingray *Potamotrygon yepezi* Castex & Castello, 1970

White-blotched River Stingray *Potamotrygon leopoldi* Castex & Castello, 1970

[Alt. Xingu River Ray, Polka-Dot Stingray, Black Devil Stingray, Yellow-Spotted Black Amazon Stingray, Black Ray, Eclipse Ray]

Long-tailed River Stingray *Plesiotrygon iwamae* Rosa, Castello & Thorson, 1987

Dr Hugo Patricio Castello is an Argentine zoologist, marine biologist and ichthyologist who gave the holotype of the electric ray to the describers, suggesting it might be new to science. He worked at Universidad del Salvador, Buenos Aires and later as chief of the Marine Mammal Laboratory, Argentine Museum of Natural Sciences, Buenos Aires.

Castelnau

Spotback Skate *Atlantoraja castelnaui* Miranda-Ribeiro, 1907

The Maskray genus *Neotrygon* Castenau, 1873

Bigtooth River Stingray *Potamotrygon henlei* Castelnau, 1855

Smooth Back River Stingray *Potamotrygon orbignyi* Castelnau, 1855

New Caledonian Maskray *Neotrygon trigonoides* Castelnau, 1873

Francis(co) Louis Nompar de Caumont, Comte de Laporte de Castelnau (1810–1880) was a career diplomat and naturalist who was born in London, studied natural sciences in Paris, and then led a French scientific expedition to study the lakes of Canada, the USA and Mexico (1837–1841). He led the first expedition (1843–1847) to cross South America from Peru to Brazil, following the watershed between the Amazon and Río de la Plata systems. Soon after his return to France he undertook another long voyage of exploration. Following this he took several diplomatic posts. He lived in Melbourne (1864–1880), being Consul-General (1862) and then French Consul (1864–1877). Despite becoming almost completely blind when returning to Paris, he published 15 volumes on the geography, botany and fauna of South America (1850s). Five birds and a reptile are named after him.

Castex

Otorongo Ray *Potomotrygon castexi* Castello & Yagolkowski, 1969

Maracaibo River Stingray *Potamotrygon yepezi* Castex
& Castello, 1970

Large-spot River Stingray *Potamotrygon falkneri*
Castex & Maciel, 1963

Freshwater Stingray sp. *Potamotrygon labradori*
Castex, Maciel & Achenbach, 1963 [Junior Syn.
Potamotrygon motoro]

Freshwater Stingray sp. *Potamotrygon pauckei* Castex,
1963 [Junior Syn. *Potamotrygon motoro*]

White-blotched River Stingray *Potamotrygon leopoldi*
Castex & Maciel, 1970

[Alt. Xingu River Ray, Polka-Dot Stingray, Black Devil
Stingray, Yellow-Spotted Black Amazon Stingray,
Black Ray, Eclipse Ray]

Schuhmacher's stingray *Potamotrygon schuhmacheri*
Castex, 1964

Dr Mariano Narciso Antonio José Castex Ocampo
(b.1932) is an Argentine ichthyologist and physician who was Professor of Biology at the Zoology
Department, Universidad del Salvador, Buenos
Aires (1960s). He was ordained as a Jesuit priest
and received a PhD in canon law. After leaving the
priesthood, he married, practised as a psychiatrist
and was a consultant on mental problems in crimi-
nal cases. He was honoured in the name of the ray
as the person 'whose studies in the last years have
cast many and new lights on this difficult group of
elasmobranchs'.

Castro

Cryptic Horn Shark *Heterodontus sp. X* Castro

Dr José I. Castro was a major in the US Air Force (1970–
1975) and its reserve (1975–1989). He is a marine
biologist whose bachelor's degree was awarded by
the University of Miami and doctorate by Clemson
University (1988). He worked for the UN Food and
Agriculture Organisation, Rome (1985–1986) as a
shark biologist, then for the National Marine Fisheries Service (1989–1999) and subsequently as Senior
Scientist, Mote Marine Laboratory, Sarasota, Florida.
He wrote *The Sharks of North America* (2010).

Castro-Aguirre

Mexican Horn Shark *Heterodontus mexicanus* LR
Taylor & Castro-Aguirre, 1972

White-margin Fin Houndshark *Mustelus albipinnis*
Castro-Aguirre, Antuna-Mendiola, González-
Acosta & De La Cruz-Agüero, 2005

Gulf Angelshark *Squatina heteroptera* Castro-Aguirre,
Espinoza-Pérez & Huidobro-Campos, 2007

Mexican Angelshark *Squatina Mexicana* Castro-
Aguirre, Espinoza-Pérez & Huidobro-Campos,
2007

Dr José Luis Castro-Aguirre (1943–2011) was a
Mexican biologist and ichthyologist. The National
School of Biological Sciences awarded his bachelor's
degree (1967), his master's (1974) and his doctorate
(1986). Much of his career was as a research professor
at the Universidad Autónoma Metropolitana (1979–
1987), the Biological Research Center (1987–1995)
and at the Interdisciplinary Marine Science Center
(1994–2011). De La Cruz-Agüero wrote his obituary,
paying tribute to his status as the ultimate authority
on Mexican elasmobranch species.

Cavanagh

Borneo River Shark *Glyphis fowlerae* Compagno, WT
White & Cavanagh, 2010

Dr Rachel D. Cavanagh is a zoologist and marine
ecologist with the British Antarctic Survey, and the
IUCN Shark Specialist Group Programme Advisor,
Cambridge. The University of Liverpool awarded
both her bachelor's degree and her doctorate.
Amongst her co-written works is *Sharks, Rays and
Chimaeras: The Status of the Chondrichthyan Fishes*
(2002).

Cépède

Broadnose Sevengill Shark *Notorynchus cepedianus*
Péron, 1807

(See **LaCépède**)

Cervigón

West African Catshark *Scyliorhinus cervigoni* Maurin &
Bonnet, 1970

Finspot Ray *Raja cervigoni* Bigelow & Schroeder, 1964

Caribbean Roughshark *Oxynotus caribbaeus*
Cervigón, 1961

Dr Fernando Cervigón Marcos (b.1930) is a Spanish-
born Venezuelan ichthyologist, who founded (1983)
and is president of the Museo del Mar, Margarita
Island, Venezuela. The University of Barcelona
awarded his doctorate. He was Scientific Director
of the marine research station on Margarita Island
(1961–1970) and Professor at the Universidad de
Oriente (1970–1980). He wrote the five-volume
Marine Fishes of Venezuela (1991–2000). He is now the
Director of the Centre of Spanish-American studies

at the Universidad Monteavila, Caracas. He was the first (1960) to recognise the catshark as a distinct species. He was honoured in the ray name for providing the opportunity to describe Venezuelan specimens.

Chabanaud

Marbled Freshwater Whipray *Himantura krempfi* Chabanaud, 1923
Madagascar Guitarfish *Rhinobatos petiti* Chabanaud, 1929

Dr Paul Chabanaud (1876–1959) was a French ichthyologist and herpetologist. He took his first degree at Poitiers (1897). He volunteered (1915) his services to the Natural History Museum, Paris under Louis Roule, who asked him to identify herpetological specimens and sent him on a scientific expedition to French West Africa (1919). He travelled to Senegal and Guinea before walking 1,200 kilometres through southern Guinea and Liberia. He returned to France (1920) and became a preparator of fishes at the Museum with a special interest in flatfish. He took his doctorate at the Sorbonne (1936). He wrote 40 papers on herpetology (1915–1954). Three reptiles and an amphibian are named after him.

Challenger

Challenger Skate *Rajella challengeri* Last & Stehmann, 2008

This species is named after the fisheries research vessel 'Challenger' which was used to survey deep-water fish resources off Tasmania (1980s).

Chan

Blackbodied Legskate *Sinobatis melanosoma* Chan, 1965
Reticulated Swellshark *Cephaloscyllium fasciatum* Chan, 1966
Blackgill Catshark *Parmaturus melanobranchus* Chan, 1966
Comtooth Lantern Shark *Etmopterus decacuspidatus* Chan, 1966

William Lai-Yee Chan was an ichthyologist at the Fisheries Research Station, Aberdeen, Hong Kong. He wrote *A Systematic Revision of the Indo-Pacific Clupeid Fishes of the Genus Sardinella (Family Clupeidae)* (1965).

Chandler

Rounded Skate *Raja texana* Chandler, 1921

Dr Asa Crawford Chandler (1891–1958) was an American parasitologist. He died of a heart attack while on his way to a scientific meeting in Portugal.

Charvet-Almeida

Porcupine Ray *Dasyatis colarensis* Santos, Gomes & Charvet-Almeida, 2004

Dr Patricia Charvet-Almeida is a Brazilian ichthyologist and biologist at the Museu Emílio Goeldi Paraense in Belém, where she is a professor. The Pontifícia Universidade Católica do Paraná awarded her bachelor's degree (1993), the Museu Paraense Emílio Goeldi and Universidade Federal do Pará her master's (2001) and Universidade Federal da Paraíba her doctorate (2006).

Chen, JTF

Hagfish sp. *Eptatretus cheni* S-C Shen & Tao, 1975

Smalleye Whipray *Himantura microphthalma* JTF Chen, 1948
Straight-tooth Weasel Shark *Paragaleus tengi* JTF Chen, 1963
Ocellated Angelshark *Squatina tergocellatoides* JTF Chen, 1963

Dr Johnson (Jian-shan) T.F. Chen (1897–fl.1996) was a Taiwanese vertebrate zoologist formerly of the Taiwan Museum, Taipei. Amongst other works he wrote *A Review of the Sharks of Taiwan* (1963) and *Synopsis of the Vertebrates of Taiwan* (1969).

Chen, R-R

Taiwanese Wedgefish *Rhynchobatus immaculatus* Last, Ho & R-R Chen, 2013

Rou-Rong Chen is a marine biologist at the Institute of Marine Biodiversity & Evolutionary Biology, National Dong Hwa University, Taiwan.

Chidlow

Western Wobbegong *Orectolobus hutchinsi* Last, Chidlow & Compagno, 2006
Floral Banded Wobbegong *Orectolobus floridus* Last & Chidlow, 2008
Dwarf Spotted Wobbegong *Orectolobus parvimaculatus* Last & Chidlow, 2008

Justin Anthony Chidlow (b.1973) is an ichthyologist and marine biologist who worked for the Department of Fisheries, Western Australia (1997–2008). Curtin University of Technology awarded his BSc

(1993) and James Cook University his MSc (2003). He became a principal consultant (2008–2011) and since 2011 has been working as an environmental scientist for Apache Energy, an oil and gas company operating in Western Australia. He is the senior author of *Variable Growth Band Deposition Leads to Age and Growth Uncertainty in the Western Wobbegong Shark*, Orectolobus hutchinsi (2007), wrote *First Record of the Freshwater Sawfish*, Pristis microdon, *from Southwestern Australia* (2007) and is co-author of *Two New Wobbegong Sharks*, Orectolobus floridus sp. nov. *and* O. parvimaculatus sp. nov. *(Orectolobiformes: Orectolobidae) from Southwestern Australia* (2008).

Chirichigno

Peruvian Torpedo *Torpedo peruana* Chirichigno, 1963
Humpback Smooth-Hound *Mustelus whitneyi* Chirichigno, 1973
Velez Ray *Raja velezi* Chirichigno, 1973
Tumbes Round Stingray *Urobatis tumbesensis* Chirichigno & McEachran, 1979

Dr Norma Victoria Chirichigno Fonseca (b.1929) is a Peruvian ichthylogist who worked at the Instituto del Mar del Perú, Lima. She wrote *Nuevos Tiburones para la Fauna del Perú* (1963). She is now Professor Emeritus at the Faculty of Oceanography at the Universidad Nacional Federico Villarreal, Lima.

Chu

Kwangtung Skate *Dipturus kwangtungensis* Chu, 1960
Yantai Stingray *Dasyatis laevigata* Chu, 1960
South China Catshark *Apristurus sinensis* Chu & Hu, 1981
Blackfin Gulper Shark *Centrophorus isodon* Chu, Meng & JX Liu, 1981
Spotless Catshark *Bythaelurus immaculatus* Chu & Meng, 1982
Dogfish Shark sp. *Squalus acutirostris* Chu, Meng & Li, 1984
Humpback Catshark *Apristurus gibbosus* Chu, Meng & Li, 1985
Broadmouth Catshark *Apristurus macrostomus* Meng, Chu & Li, 1985
Smalldorsal Catshark *Apristurus micropterygeus* Meng, Chu & Li, 1986

Professor Yuan-Ting Chu (Zhu Yuang-ding) (1896–1986) worked at the Shanghai Fisheries Institute, Shanghai, China (1958). He wrote *Contributions to the Ichthyology of China* (1930). (See **Zhu**)

CIMAR

Chilean Roundray *Urotrygon cimar* López & Bussing, 1998

This is an abbreviation for the Centro de Investigación en Ciencias del Mar y Limnología, a research centre of the Universidad de Costa Rica (where both authors work) in honour of its twentieth anniversary.

Clark, E

Steven's Swellshark *Cephaloscyllium stevensi* E Clark & Randall, 2011

Dr Eugenie Clark (b.1922), sometimes known as the 'shark lady', is a Japanese-American ichthyologist. Hunter College awarded her bachelor's degree (1942) and New York University her master's (1946) and doctorate (1950). She is the founding director (1955–1967) of the Mote Marine Laboratory, Sarasota, Florida. She joined the faculty at the University of Maryland College Park (1968) and is now Professor Emeritus of Zoology. She wrote *Lady with a Spear* (1951).

Clark, RS

Hookskate *Dactylobatus clarkia* Bigelow & Schroeder, 1958

Maltese Skate *Leucoraja melitensis* RS Clark, 1926

Dr Robert Selbie Clark (1882–1950) was a British marine zoologist and explorer who was a member of Shackleton's Imperial Transantarctic Expedition (1914–1917). He was one of those left on Elephant Island when Shackleton sailed in an open boat to South Georgia to get help. Aberdeen University awarded his master's degree (1908). He gained a DSc degree (1925). He was zoologist to the Scottish Oceanographical Laboratory (1911–1913) and naturalist to the Plymouth Marine Biological Association (1913–1914 and 1919–1925), having served in minesweepers in the Royal Navy (1917–1919). He was Director of the Fisheries Research Laboratory, Aberdeen (1925–1934) and Superintendent of Scientific Investigations for the Fishery Board for Scotland (1934–1948). He was honoured in the name of the Hookskate for his *Rays and Skates: A Revision of the European Species* (1926).

Cléva

Broadheaded Catshark *Bythaelurus clevai* Séret, 1987

Dr Régis Cléva is a zoologist at the Muséum Nationale d'Histoire Naturelle, Paris. He specialises in the study of crustaceans and co-wrote *Report on Some Caridean Shrimps (Crustacea: Decapoda) from Mayotte, Southwest Indian Ocean* (2012). He collected the holotype.

Cloquet

Starry Smooth-Hound *Mustelus asterias* Cloquet, 1821

Dr Hippolyte Cloquet (1787–1840) was a French anatomist who qualified as a physician (1815) and was also interested in zoology. He wrote a treatise, *Poissons et Reptiles*, that appeared in Cuvier's *Dictionnaire des Sciences Naturelles* (published in tranches 1816–1830).

Coates

Coates' Shark *Carcharhinus coatesi* Whitley, 1939
[Alt. White Cheek Shark, Whitecheek Whaler,
    Widemouth Blackspot Shark]

George Coates (d.1980) of Townsville, Queensland was an artist and illustrator who collected the holotype. He supplied many fish, including the type, to Whitley who named other fish after him in addition to this shark. He wrote: *Fishing on the Barrier Reef and Inshore* (1943). A training restaurant at the Barrier Reef Institute of TAFE (Training and Further Education) and appropriately, a fishing boat are named after him.

Coillot

Sicklefin Devil Ray *Mobula coilloti* Cadenat &
    Rancurel, 1960

M. Coillot was the chief engineer of the oceanographic research vessel 'Reine Pokou'. He harpooned and captured the holotype.

Colclough

Colclough's Shark *Heteroscyllium colcloughi* Ogilby, 1908
[Alt. Bluegrey Carpetshark]

John Colclough was a friend of the describer. He was a keen fisherman, member of the Amateur Fisherman's Association of Queensland and collector. He later sent a considerable collection to Ogilby from the Aru Islands and from the Northern Territory.

Coles

Lightnose Skate *Breviraja colesi* Bigelow & Schroeder,
    1948
[Alt. Bahama Skate, Syn. *Fur macki, Fur ventralis*]

No etymology is given but we believe this to be named after Russell Jordan Coles (b.1865). He was a tobacco trader in Virginia and a keen amateur fisherman and 'shark watcher' who spent three or four months each year anywhere between Florida and Newfoundland in search of the biggest fishes, which he sometimes collected for the AMNH. Among his friends and fishing companions was Theodore Roosevelt (he gave a memorial address about him to the Explorers Club). He was known as 'Doctor' as he had studied medicine without qualifying as a physician but, at Roosevelt's suggestion, Trinity University, Hartford, Connecticut awarded him an honorary doctorate (1918). He wrote, among other scientific articles, *Natural History Notes on the Devilfish*, Manta birostris *(Walbaum) and* Mobula oflfersi *(Müller)* (1916).

Collett

Arctic Skate *Amblyraja hyperborean* Collett, 1879
Great Lantern Shark *Etmopterus princeps* Collett, 1904
Mouse Catshark *Galeus murinus* Collett, 1904
Large-eyed Rabbitfish *Hydrolagus mirabilis* Collett,
    1904

Professor Robert Collett (1842–1913) was a Norwegian zoologist and amateur photographer. He was the Director of the Museum of Natural History at Christiania (1882–1913) and Professor of Zoology (1884–1913). He also described the hooded parakeet (1898) based on a collection made by the Norwegian ornithologist Knut Dahl in northern and northwest Australia. A mammal, an amphibian, two reptiles and five birds are named after him.

Collie

Spotted Ratfish *Hydrolagus colliei* Lay & ET Bennett,
    1839

Lieutenant Dr Alexander Collie (1793–1835) was the naval surgeon and naturalist, along with Lay (q.v.), on an expedition (1825–1828) led by Captain Frederick Beechey (q.v.) on 'HMS Blossom', which made some significant zoological findings during the voyage from Chile to Alaska. Collie collected many specimens that did not survive the return journey to England in good condition, but he made some coloured drawings of taxa he thought were new and also took extensive notes. Collie also collected many live birds that went on to be exhibited at London Zoo. Dr Collie went to Perth as a colonial administrator and died there. When aboard 'HMS Sulphur', he discovered what is now the

Collie River in Western Australia. A town in Australia, a mammal, a bird and a reptile are named after him.

Compagno

Lantern Shark sp. *Etmopterus compagnoi* Fricke & Koch, 1990

[Possibly Junior Syn for *Etmopterus (Spinax) unicolor* Engelhardt, 1912]
Tigertail Skate *Leucoraja compagnoi* Stehmann, 1995

Stingrays & Relatives family *Myliobatiformes* Compagno, 1973
Megamouth Shark Family *Megachasmidae* LR Taylor, Compagno & Struhsaker, 1983

Houndshark genus *Iago* Compagno & S Springer, 1971
Harlequin Shark genus *Ctenacis* Compagno, 1973
Smoothhound Shark genus *Gollum* Compagno, 1973
Houndshark genus *Gogolia* Compagno, 1973
Crocodile Shark genus *Pseudocarchariidae* Compagno, 1973
Ground Sharks *Carcharhiniformes* Compagno, 1977
Pygmy Skate genus *Fenestraja* McEachran & Compagno, 1982
Pygmy Skate genus *Neoraja* McEachran & Compagno, 1982
Megamouth Shark genus *Megachasma* LR Taylor, Compagno & Struhsaker, 1983
White-nose Shark genus *Nasolamia* Compagno & Garrick, 1983
Weasel Shark genus *Hemigaleidae* Compagno, 1984
Finback Catshark genus *Proscylliidae* Compagno, 1984
Hammerhead Shark genus *Sphyrnidae* Compagno, 1984
Catshark genus *Bythaelurus* Compagno, 1988

Sailback Houndshark *Gogolia filewoodi* Compagno, 1973
Onefin Skate *Gurgesiella dorsalifera* McEachran & Compagno, 1980
White-rimmed Whipray *Himantura signifier* Compagno & TR Roberts, 1982
Megamouth Shark *Megachasma pelagios* LR Taylor, Compagno & Struhsaker, 1983
Pearl Stingray *Dasyatis margaritella* Compagno & TR Roberts, 1984
Dragon Stingray *Himantura draco* Compagno & Heemstra, 1984
Whitetip Weasel Shark *Paragaleus leucolomatus* Compagno & Smale, 1985
Spotted Guitarfish *Rhinobatos punctifer* Compagno & Randall, 1987

Comoro Catshark *Scyliorhinus comoroensis* Compagno, 1988
Paddle-nose Chimaera *Rhinochimaera africana* Compagno, Stehmann & Ebert, 1990
Banded Sand Catshark *Atelomycterus fasciatus* Compagno & Stevens, 1993
Slender Sawtail Catshark *Galeus gracilis* Compagno & Stevens, 1993
Deepwater Sicklefin Houndshark *Hemitriakis abdita* Compagno & Stevens, 1993
Sicklefin Houndshark *Hemitriakis falcate* Compagno & Stevens, 1993
Salalah Guitarfish *Rhinobatos salalah* Randall & Compagno, 1995
Slender Weasel Shark *Paragaleus randalli* Compagno, Krupp & KE Carpenter, 1996
Blotched Catshark *Asymbolus funebris* Compagno, Stevens & Last, 1999
Dwarf Catshark *Asymbolus parvus* Compagno, Stevens & Last, 1999
Variagated Catshark *Asymbolus submaculatus* Compagno, Stevens & Last, 1999
Madagascar Numbfish *Narcine insolita* Carvalho, Séret & Compagno, 2002
Bigeye Electric Ray *Narcine oculifera* Carvalho, Compagno & Mee, 2002
Taiwanese Blind Electric Ray *Benthobatis yangi* Carvalho, Compagno & Ebert, 2003
Bareback Shovelnose Ray *Rhinobatos nudidorsalis* Last, Compagno & Nakaya, 2004
Australian Weasel Shark *Hemigaleus australiensis* WT White, Last & Compagno, 2005
Natal Shyshark *Haploblepharus kistnasamyi* Human & Compagno, 2006
Western Wobbegong *Orectolobus hutchinsi* Last, Chidlow & Compagno, 2006
Ornate Sleeper Ray *Electrolux addisoni* Compagno & Heemstra, 2007
Western Gulper Shark *Centrophorus westraliensis* WT White, Ebert & Compagno, 2008
Southern Dogfish *Centrophorus zeehaani* WT White, Ebert & Compagno, 2008
Northern River Shark *Glyphis garricki* Compagno, WT White & Last, 2008
Eyebrow Wedgefish *Rhynchobatus palpebratus* Compagno & Last, 2008
Indonesian Houndshark *Hemitriakis indroyonoi* WT White, Compagno & Dharmadi, 2009
Southern African Frilled Shark *Chlamydoselachus africana* Ebert & Compagno, 2009
Bahamas Ghostshark *Chimaera bahamaensis* Kemper, Ebert, Didier & Compagno, 2010

Cape Chimaera *Chimaera notafricana* Kemper, Ebert, Compagno & Didier, 2010

Roughnose Legskate *Cruriraja hulleyi* Aschliman, Ebert & Compagno, 2010

Borneo River Shark *Glyphis fowlerae* Compagno, WT White & Cavanagh, 2010

Broadnose Wedgefish *Rhynchobatus springeri* Compagno & Last, 2010

Angelshark sp. *Squatina caillieti* Walsh, Ebert & Compagno, 2011

Sculpted Lantern Shark *Etmopterus sculptus* Ebert, Compagno & De Vries, 2011

Dr Leonard Joseph Victor Compagno (b.1943) is an American ichthyologist and internationally recognised authority on shark taxonomy. His early career was in the US as a research assistant in chemistry (1964) and ornithology (1965–1966) before becoming a curatorial assistant of fishes at Stanford (1966–1967). He worked in various jobs within and without the University while studying for his PhD, awarded by Stanford (1979). Since 1977 he has been research consultant and contract writer for the Fisheries Division of the Food and Agriculture Organization of the United Nations and the Save Our Seas Foundation, Geneva. He was Adjunct Professor at San Francisco State University (1979–1985) and Curator of Fishes, Iziko Museum, Cape Town, South Africa, becoming (1987) Director of the Shark Research Centre there until retiring (2008). He is currently (2014) Director of the International Shark Research Institute at Princeton and Extraordinary Professor, Department of Zoology & Botany, Stellenbosch University, South Africa. He has written over 1,000 scientific papers and longer works including *Sharks of the World* (2002) and *A Field Guide to the Sharks of the World* (2005). He has described or co-described 52 species, genera or families of elasmobranch. A claim to fame is that he is mentioned in the credits for the film 'Jaws' (1975). He has awards and honours too numerous to mention and belongs to many societies and organisations, including in particular the American Society of Ichthyologists and Herpetologists, the American Elasmobranch Society, the Royal Society of South Africa and the Oceania Chondrichthyan Society, Australia. He divides his time between South Africa, New Jersey and Oregon. He was honoured in the lantern shark name for his research on South African sharks.

Cook

Cook's Swellshark *Cephaloscyllium cooki* Last, Séret & WT White, 2008

Sidney 'Sid' F. Cook (1953–1997) was a shark fishery conservationist and biologist, in the describers words, 'whose energy, dedication and contribution to shark conservation is sadly missed'. He wrote *Cook's Book: A Guide to the Handling & Eating of Sharks and Skates* (1985).

Cooke

Prickly Shark *Echinorhinus cookei* Pietschmann, 1928

Dr Charles Montague Cooke Jr (1874–1948) was a malacologist at the Bishop Museum, Hawaii and Curator of the snail collection (1902–1948). He was born into a wealthy family in Honolulu and his resources enabled him to acquire some extensive collections. Yale awarded his bachelor's degree (1897) and his doctorate (1901). He took part in a number of expeditions to the South Pacific and led the Museum's Mangarevan Expedition (1934). The describer honoured him for his 'helpful assistance'.

Cooper

Haller's Roundray *Urobatis halleri* Cooper, 1863

California Butterfly Ray *Gymnura marmorata* Cooper, 1864

James Graham Cooper (1830–1902) was a US ornithologist, ichthyologist, malacologist and surgeon. Spencer F. Baird, Assistant Secretary of the Smithsonian, helped get Cooper work with the Pacific railroad survey parties in the Washington territory. Along with Baird he wrote a book on the birds of California, *Ornithology, Volume I, Land Birds* (1870). He also wrote *Botanical Report: Explorations and Surveys for a Railroad Route from the Mississippi River to the Pacific Ocean* (1860). A mammal and two birds are named after him.

Cope

Shorttail Fanskate *Sympterygia brevicaudata* Cope, 1877

Speckled Smooth-Hound *Mustelus mento* Cope, 1877

Edward Drinker Cope (1840–1897) was an American palaeontologist, anatomist, herpetologist and ichthyologist. He studied under Baird at the Smithsonian (1859), at the British Museum, London, and the Jardin des Plantes, Paris (1863–1867). He was Professor of Comparative Zoology and Botany, Haverford College, Pennsylvania (1864–1867), and was appointed Curator, Philadelphia Academy of Natural Sciences (1865). He was the palaeontologist on the Wheeler Survey (1874–1877) west of the

100th meridian in New Mexico, Oregon, Texas and Montana. He was a professor at the University of Pennsylvania of Geology and Mineralogy (1889–1895) and Zoology and Comparative Anatomy (1895–1897). He was Senior Naturalist (1878) on the periodical *American Naturalist*, which he co-owned. He wrote many articles including *On the Fishes of the Ambylacu River* (1872). In his will he donated his body to science. His cause of death was listed as uremic poisoning. It was rumoured that he died of syphilis but when (1995) permission was granted for his skeleton to be medically examined, no evidence of bony syphilis was found. Fifty-nine reptiles and 19 amphibians are named after him.

Cortez

Cortez' Ray *Raja cortezensis* McEachran & Miyake, 1988

This species is named after the Sea of Cortez where the holotype was caught.

Costa

Iberian Pygmy Skate *Neoraja iberica* Stehmann, Séret, Costa & Baro, 2008

Dr Maria Esmeralda de Sá Leite Correia da Costa (b.1970) is an ichthyologist at the Universidade do Algarve, Faro, Portugal, which awarded her licentiate degree (1994). She has worked as a research assistant at the University of the Algarve (1996–2007), which awarded her doctorate in marine, earth and environmental sciences (2014) for a thesis, *Bycatch and Discards of Commercial Trawl Fisheries in the South Coast of Portugal*. She has written *Cases of Abnormal Hermaphroditism in Velvet Belly* Etmopterus spinax *(Chondrichthyes: Etmopteridae) from the Southern Coast of Portugal* (2013) and was the lead author of *Reproductive Biology of the Blackmouth Catshark,* Galeus melastomus *(Chondrichthyes: Scyliorhinidae) off the South Coast of Portugal* (2005).

Couard

Whitefin Hammerhead *Sphyrna couardi* Cadenat, 1951

Monsieur Couard was Director of shark fisheries off the coast of Senegal, the type's locality.

Couch

Sandy Skate *Leucoraja circularis* Couch, 1838

Dr Richard Quiller Couch (1816–1863) was a British physician, zoologist and geologist. He practised medicine in Penzance, Cornwall (1845–1863). He

was also interested in occupational diseases and did much work on those suffered by Cornish miners. He wrote *An Essay on the Zoophytes of Cornwall* (1841) and *On the Mortality of the Cornish Miners in the District of Marazion* (1860).

Cousseau

Cousseau's Skate *Bathyraja cousseauae* Díaz de Astarloa & Mabragaña, 2004
[Alt. Joined-fins Skate]

Dr Maria Berta Cousseau is an Argentine ichthyologist who is Professor Emeritus at the Department of Sciences, Universidad Nacional de Mar del Plata, Argentina. She wrote *Peces Marinos de Argentina: Biología, Distribución, Pesca* (2000) and co-wrote *Ictiología. La Vida de los Peces Sudamericanos* (2010). She was honoured for her contribution to the marine fishes of Argentina.

Cozzi

Springer's Sawtail Catshark *Galeus springeri* Konstantinou & Cozzi, 1998

Dr Joseph R. 'Coz' Cozzi is a marine archaeologist who studied at Texas A & M University (1990–2000). Originally a commercial diver, he worked at the Mote Marine Laboratory, Sarasota, Florida as a marine archaeologist (1997–2011). He is now Senior Marine Archaeologist at Fathom Research, New Bedford, Massachusetts. Konstantinou (q.v.), the senior author is his wife, with whom since about 2001 he has been training prize-winning Labrador retriever dogs in agility and taking them to shows all over the USA.

Creaser

Brook Lamprey sub-genus *Lethenteron* Creaser & CL Hubbs, 1922

Dr Charles William Creaser (1897–1965) was an ichthyologist at the Museum of Zoology, University of Michigan (1923), where he was awarded his doctorate (1926). He was a professor at Wayne University, Michigan (1940). He wrote *The Skate,* Raja erinacea Mitchill, *a Laboratory Manual* (1927).

Crook

Butterfly Ray sp. *Gymnura crooki* Fowler, 1934

Alfred Herbert Crook (b.1873) was a biologist and Headmaster of Queen's College, Hong Kong (1925–1930), who also lectured on biology at the Hong

Kong College of Medicine. He was an amateur naturalist who contributed articles to 'The Hong Kong Naturalist', of which he was co-editor. He wrote *The Flowering Plants of Hong Kong: Ranunculaceae to Meliaceae: with Thirty Diagrams, Comprising Drawings of Parts of Over 100 Different Species* (1959).

Crosnier

Madagascar Skate *Dipturus crosnieri* Séret, 1989

Dr Alain Georges Paul Crosnier (b.1930) is a French biologist and carcinologist at the Museum of Natural History, Paris, who has published much on crustaceans. He initiated the deep trawling surveys off Madagascar (1970s) and 'entrusted Séret with his valuable collection of Madagascar skates'.

Cubelio

Smoothhound sp. *Mustelus mangalorensis* Cubelio, Remya & Kurup, 2011

Dr Sherine Sonia Cubelio (b.1972) is an Indian ichthyologist who since 2011 has worked at the Centre for Marine Living Resources & Ecology (CMLRE), Cochin, India. Her bachelor's and master's degrees were both awarded by Kerala Agricultural University and her doctorate in marine bioscience by Tokyo University (2007). Previously she was Senior Research Fellow at the Central Institute of Fisheries Technology, ICAR, Cochin (1998–2002), Research Associate at the School of Industrial Fisheries, Cochin (2008–2010) and at the National Bureau of Fish Genetic Resources, Cochin Unit (2010–2011). Her research to identify new crab species from hydrothermal vents for her doctorate involved many trips in a manned submersible to depths of 4,000 metres. She has written *New Record of Deep Sea Cusk Eel* Bassozetus robustus *Smith and Radcliff (1913)* (ophidiiformes; ophidiidae) *from Indian EEZ with a Redescription* (2009) and *New Species of a* Traikidae *Shark from Indian EEZ* (2011).

Cummins

Northern Brook Lamprey *Ichthyomyzon fossor* Reighard & Cummins, 1916

Dr Harold Cummins (1894–1976) was an American dermatoglyphics (fingerprint) specialist – in fact he coined the term. He was at the University of Michigan and Vanderbilt University before attending Tulane, which awarded his PhD. He became

Professor of Anatomy at Tulane University School of Medicine, New Orleans until retirement (1964), after which he continued as Professor Emeritus. His most famous publication was *Finger-prints, Palms and Soles* (1943).

Cuvier

Tiger Shark *Galeocerdo cuvier* Péron & Lesueur, 1822

Mackerel Shark genus *Lamna* Cuvier, 1816
Eagle Ray genus *Myliobatis* Cuvier, 1816

Yellow Stingray *Urobatis jamaicensis* Cuvier, 1816
Winghead Shark *Eusphyra blochii* Cuvier, 1817
Granulated Guitarfish *Glaucostegus granulatus* Cuvier, 1829
Sawback Angelshark *Squatina aculeate* Cuvier, 1829

Georges Léopold Chrétien Frédéric Dagobert Baron Cuvier (1769–1832), better known by his pen name Georges Cuvier, was a French naturalist and one of the scientific giants of his age. He supported the geological school of thought termed 'catastrophism', according to which paleontological discontinuities are evidence of sudden and widespread catastrophes and extinctions which took place suddenly. The harshness of his criticism towards scientific opponents, and the strength of his reputation, discouraged other naturalists from speculating about the gradual transmutation of species, right up until Darwin's time. Cuvier is also famed for having stayed in a top government post, as permanent secretary at the Academy of Sciences, through three regimes, including Napoleon's. Cuvier's research on fish was begun in 1801, and culminated in a work co-written with Achille Valenciennes, *Histoire Naturelle des Poissons* (1828–1831), which described c.5,000 species of fishes. He described six species or genera of *Elasmobranchii*. Seven birds, six reptiles, three mammals and an amphibian are named after him.

Cyrano

Cyrano Spurdog *Squalus rancureli* Fourmanoir & Rivaton, 1979

Hercule-Savinien de Cyrano de Bergerac (1619–1655) was a French dramatist, poet and duelist. A play, 'Cyrano de Bergerac', based loosely on his life, was written by Edmond Rostand (1897). It revolved around the fact that Cyrano had a very prominent nose, as does this shark.

D

D'Aubrey

Speckled Catshark *Halaelurus boesemani* S Springer &
D'Aubrey, 1972
Lined Catshark *Halaelurus lineatus* Bass, D'Aubrey &
Kistnasamy, 1975
Thorny Lantern Shark *Etmopterus sentosus* Bass,
D'Aubrey & Kistnasamy, 1976

Jeanette Dymock D'Aubrey (b.1939) is an ichthyologist at the Oceanographic Research Institute,
Durban, who undertook (1959–1976) a taxonomic
and biological survey of the sharks of the east coast
of southern Africa. She wrote *Preliminary Guide to the
Sharks Found off the East Coast of South Africa* (1964).

Da Silva, JP

Parnaiba River Stingray *Potamotrygon tatianae* JP da
Silva & Carvalho, 2011
Freshwater Stingray sp. *Potamotrygon limai*
Fontenelle, JP da Silva & Carvalho, 2014

João Paulo Capretz Batista da Silva is a Brazilian
zoologist and ichthyologist. His bachelor's degree
(2006) and master's (2010) were both awarded by the
University of São Paulo, Brazil, where he is currently
studying and researching for his doctorate.

Da Silva, KG

Hidden Angelshark *Squatina occulta* Vooren & KG da
Silva, 1992

Dr Kleber Grübel da Silva is an oceanographer
and Director of the Centre for Education and Environmental Monitoring in Brazil. He co-wrote *The
Distribution and Occurrence of the Marine Manatee* (Tri-
chechus manatus) *in the Estuary of the Mamanguape
River, Paraíba, Brazil* (2011).

Dadit

Carpenter's Chimaera *Chimaera lignaria* Didier, 2002

Kevin J. Dadit is the son of the describer and a
woodworker, carpenter and 'supporter of research
on chimaeroid fishes in his spare time'. The second
part of the binomial is the sort of play on words
much loved by taxonomists: 'lignaria' means 'of or
belonging to wood' thus referring to Dadit's occupation.

Danford

Carpathian Lamprey *Eudontomyzon danfordi* Regan,
1911

Charles George Danford (1843–1928) was a geologist, paleontologist, zoologist, artist, traveler and
explorer. He was in Asia Minor (Turkey) (1875–1876
and 1879). The Danford iris was named after his
wife, who introduced it to England. A reptile and a
bird are also named after him.

Davies

Davies' Stingaree *Plesiobatis daviesi* Wallace, 1967

Dr David Herbert Davies (1922–1965) was a South
African ichthyologist and oceanographer. The University of Cape Town awarded his BSc (1942). He
then served in the South African Air Force rising
to lieutenant with active service in Italy and North
Africa. After the war he became Senior Demonstrator
in Zoology at Cape Town University (1945) and then
(1946) as a researcher with the Division of Sea Fisheries, quickly rising to be its Chief Biologist (during
this time he also received his MSc and PhD). He took
part in a Danish circumnavigation (1951). He left
Cape Town (1957), joining the staff of the Institute of
Marine Resources in California. He then became the
first director (1958) of the Oceanographic Research
Institute, a division of the South African Association
for Marine Biological Research. Its research vessel is
also named after him, as is its library. He was also
Research Professor at the University of Natal (1960).
He wrote many papers (1947–1965) but was probably best known for *About Sharks and Shark Attack*
(1964). He was killed in an accident whilst attending
a scientific meeting. He was honoured as the person
'who was responsible for the initiation of research on
the batoid fishes of the east coast of southern Africa.'

Dawson

New Zealand Catshark *Bythaelurus dawsoni*
S Springer, 1971

Elliot Watson Dawson (b.1930) was a biologist with
the Oceanographic Institute of the New Zealand
Department of Scientific & Industrial Research
(1955–1990), principally involved in carrying out
surveys of the marine benthic fauna of the New
Zealand sub-Antarctic region. He led a joint New
Zealand/United States expedition to the Ross Sea
(1965), making the first bathymetric charts of the
Balleny Islands. The highlight of his career was
leading the multidiscipline Royal Society Cook

Bicentenary Expedition to the South Pacific (1969). He is now an honorary research associate of the Museum of New Zealand Te Papa Tongarewa, Wellington, where he works on decapod crustaceans of the southwest Pacific islands. As well as various papers on crustaceans, molluscs, brachiopods and sub-fossil birds, his publications include a comprehensive bibliography of king crabs of the world and their fisheries. He has also written accounts of the little-known German Transit of Venus expedition to the Auckland Islands (1874) and a comprehensive bibliography of the human and natural history of these sub-Antarctic islands. He is a graduate of the University of New Zealand and of Cambridge University, where he was the first holder of the John Stanley Gardiner Studentship in zoology. Springer named this species after him as he had brought the first specimens to Springer's attention.

Day

Indian Sand Tiger *Carcharias tricuspidatus* Day, 1878

Francis Day CIE (1829–1889) was Inspector-General of Fisheries in India (1871–1877) and Burma and an ichthyologist. He became the medical officer in the Madras Presidency East India Company services (1852). He wrote the two volumes on fish in *The Fauna of British India, including Ceylon & Burma* (1875–1878) in which he described 1,400 species. He also wrote *Fishes of Malabar* (1865). Later Günther questioned much of his work. He retired (1877) to Cheltenham where he died of stomach cancer.

Dean

Shark genus *Deania* Jordan & Snyder, 1902

Black Hagfish *Eptatretus deani* Evermann & Goldsborough, 1907
Philippine Chimaera *Hydrolagus deani* HM Smith & Radcliffe, 1912

Hagfish genus *Paramyxine* Dean, 1908

Brown Hagfish *Eptatretus atami* Dean, 1904
Hagfish sp. *Eptatretus okinoseanus* Dean, 1904

Dr Bashford Dean (1857–1928) was an American zoologist, ichthyologist and acknowledged expert on mediaeval armour. The College of the City of New York awarded his bachelor's degree (1886) and Columbia University, where he was later Professor of Zoology, his doctorate (1890). He is the only person to have held concurrent positions at two of New York's great museums, the Metropolitan Museum of Art, where he was Honorary Curator of Arms and Armour, and the American Museum of Natural History. He wrote *Bibliography of Fishes* (1916) and *Catalogue of European Court Swords and Hunting Swords Including the Ellis, De Dino and Reubell Collections* (1929). He was honoured in the name of the hagfish for his work on the embryology of *E. stoutii*.

De Buen

Hagfish sp. *Myxine debueni* Wisner & CB McMillan, 1995

Primitive Sharks *Hexanchiformes* de Buen, 1926
Electric Rays *Torpediniformes* de Buen, 1926
Dogfish genus *Aculeola* de Buen, 1959
Skate genus *Gurgesiella* de Buen, 1959

Largenose Catshark *Apristurus nasutus* de Buen, 1959
Hooktooth Dogfish *Aculeola nigra* de Buen, 1959
Slimtail Skate *Bathyraja longicauda* de Buen, 1959
Dusky Finless Skate *Gurgesiella furvescens* de Buen, 1959
Ghost Shark sp. *Hydrolagus macrophthalmus* de Buen, 1959
Chilean Torpedo *Torpedo tremens* de Buen, 1959
Blackish Skate *Rajella nigerrima* de Buen, 1960

Don Fernando de Buen y Lozano (1895–1962) was a Spanish ichthyologist and oceanographer. He and his brother Rafael were the first curators of the ichthyological collection at the Malaga laboratory (1915–1926). They enhanced the collection with many specimens collected by Spanish trawlermen off Morocco, which Fernando catalogued. He worked and lived in Mexico, Uruguay and Chile. He was Director of the Department of Oceanography and Fisheries Service in Uruguay and Professor of Hydrobiology and Protozoology in the Faculty of Arts and Sciences. He was honoured in the name of the hagfish for his 'extensive work' on South American fishes. (See **Buen**)

De Forges

Chesterfield Island Stingaree *Urolophus deforgesi* Séret & Last, 2003

Dr Bertrand Richer-de-Forges (b.1948) is a French carcinologist and marine biologist who (since 1984) has been based at ORSTOM, Noumea, New Caledonia. He studied at the Pierre-et-Marie-Curie University and at the Muséum National d'Histoire Naturelle, Paris. He has organised many dredging expeditions in New Caledonia and the southwest Pacific Ocean and has collected numerous novelties in every group of marine animals. He co-wrote *Lagons et Récifs de Nouvelle-Calédonie* (2004). A large part of the zoological results from his expeditions are published in a series of the Muséum National d'Histoire Naturelle, *Tropical Deep Sea Benthos*, of which the 27th volume appeared in 2013. He was honoured for promoting the exploration of the bathyal fauna off New Caledonia and for collecting valuable fish specimens from cruise surveys.

De La Cruz-Agüero

White-margin Fin Houndshark *Mustelus albipinnis* Castro-Aguirre, Antuna-Mendiola, González-Acosta & De La Cruz-Agüero, 2005

Dr José De La Cruz-Agüero (b.1960) is an ichthyologist. The Autonomous University of Baja California Sur awarded his bachelor's degree and the National Polytechnic Institute La Paz, Baja California, Mexico, where he is now a research professor, his master's and doctorate. He co-wrote *Morphometric and Molecular Data on Two Mitochondrial Genes of a Newly Discovered Chimaeran Fish* (Hydrolagus melanophasma, Chondrichthyes) (2012) and *Mass Stranding of Fish in the Cape Region of the Gulf of California* (2012).

De Vis

Ornate Wobbegong *Orectolobus ornatus* De Vis, 1883

Charles Walter De Vis (1829–1915) was a zoologist and clergyman. Magdalene College, Cambridge awarded his BA (1849) and he became a deacon (1852) and then a rector in Somerset (1855–1859). He gave up his work as a clergyman (1862) to become Curator, Queens Park Museum, Manchester. He emigrated from England to Australia (1870), where he became Librarian, School of Arts, Rockhampton, Queensland. He published many popular articles under the pen name of 'Thickthorn', which brought him to the attention of the Trustees of the Queensland Museum, who recruited him to be its first director (1882–1905); he remained a consultant until 1912. He was a founding member of the Royal Society of Queensland (1884) and its President (1888–1889), and a founder and first vice-president of the Australasian Ornithologists' Union (1901). He wrote around 50 papers on herpetological subjects (1881–1911). Two mammals, five reptiles and two birds are named after him.

De Vries

Sculpted Lanternshark *Etmopterus sculptus* Ebert, Compagno & De Vries, 2011

Marlee J. De Vries (d.2010) was an ichthyologist at the Department of Biodiversity and Conservation Biology, University of the Western Cape, South Africa. She died before this description could be published and her co-authors dedicated the paper in her memory.

DeFino

Burmese Bamboo Shark *Chiloscyllium burmensis* Dingerkus & DeFino, 1983

Terry C. DeFino wrote a number of papers, including the co-written *A Revision of the Orectolobiform Shark Family* Hemiscyllidae *(Chondrichthyes, Selachii)* (1983). He was probably employed by the AMNH but we cannot find any more about him.

DeKay

Stingray sp. *Dasyatis hastate* DeKay, 1842
American Brook Lamprey *Lethenteron appendix* DeKay, 1842

Dr James Ellsworth De Kay (1792–1851) was an American zoologist. He studied at Yale (1807–1812) but did not graduate. He studied medicine at Edinburgh, qualifying in 1819. He returned to the USA, married, and then travelled with his father-in-law to Turkey as a ship's physician (1831–1832). He returned to America, forsaking medicine to study natural history, and was on the Geological Survey of New York (1835). He wrote *The Zoology of New York* (1842–1849). A reptile is named after him.

Delalande

Brazilian Sharpnose Shark *Rhizoprionodon lalandii* Müller & Henle, 1839

Pierre Antoine Delalande (1787–1823) worked for Muséum National d'Histoire Naturelle, Paris. He collected in the region around Rio de Janeiro (1816) with Auguste de Saint-Hilaire, and in the African Cape with his nephew Jules Verreaux, and Andrew Smith (1818). Later Geoffroy Saint-Hilaire employed him as a taxidermist. He collected the type. Three birds and three reptiles are named after him.

Delaroche

Mediterranean Starry Ray *Raja asterias* Delaroche, 1809
Rough Ray *Raja radula* Delaroche, 1809

François Étienne Delaroche (1743–1812) was a Swiss physician and naturalist. He studied medicine in

Edinburgh and Geneva, where he practised for some years before moving to Paris as private physician to the Duke of Orleans and later at the hospital in Faubourg Saint-Martin. He wrote *Observations sur des Poissons Recueillis dans un Voyage aux Îles Baléares* (1809).

Delpiani

Shortnose Eagle Ray *Myliobatis ridens* Ruocco, Lucifora, Díaz de Astarloa, Mabragaña & Delpiani, 2012

Sergio Matías Delpiani is a marine biologist who graduated at Universidad de Mar del Plata, Argentina (2009) and was employed there (2012) in the Faculty of Exact and Natural Sciences. He is the lead author of *The use of Otoliths and Bones of the Head for the Identification of Two Species of the Genus* Merluccius, *in Studies of Prey-predator* (2012).

Delsman

Smalleyed Round Stingray *Urotrygon microphthalmum* Delsman, 1941

Hendricus Christoffel Delsman (1886–1969) was a fisheries biologist. He was Director of the fishery station in Batavia (Jakarta), Dutch East Indies (Indonesia). He co-wrote *De Indische Zeevisschen en Zeevisscherij* (1934).

Deng

East China Legskate *Anacanthobatis donghaiensis* Deng, Xiong & Zhan, 1983
Fat Catshark *Apristurus pinguis* Deng, Xiong & Zhan, 1983
Gulper Shark sp. *Centrophorus robustus* Deng, Xiong & Zhan, 1985
Shortnose Demon Catshark *Apristurus internatus* Deng, Xiong & Zhan, 1988

Deng Si-Ming is an ichthyologist at East China Sea Fisheries Institute, National Bureau of Aquatic Products of China. He co-wrote *On Three New Species of Sharks of the Genus* Carcharhinus *from China* (1981).

Desoutter

Rosette Torpedo *Torpedo bauchotae* Cadenat, Capapé & Desoutter, 1978

Martine Desoutter-Méniger has spent her entire career at the Muséum National d'Histoire Nationale, Paris, where she studied the collections and made a special study of Acanthomorphs.

Devicenzi

Freshwater Stingray sp. *Potamotrygon brumi* Devincenzi & Teague, 1942
[JS. *Potamotrygon brachyura*]

Dr Garibaldi José Devincenzi (1882–1943) was a Uruguayan naturalist and ichthyologist, who was qualified as a physician (1909). He was Director, Uruguay Museum, Montevideo (1912–1942). He produced the first systematic catalogue of Uruguayan fishes, reptiles, mammals and birds, and wrote *Peces del Uruguay* (1924). An amphibian is named after him.

Deynat

Freshwater Stingray sp. *Potamotrygon marinae* Deynat, 2006

Dr Pascal Pierre Deynat (b.1965) is a French ichthyologist and naturalist. University Paris 7 awarded his doctorate (1996). He spent time at both the AMNH and the Smithsonian (1994). After refusing a job in fisheries in Kergeuelun (1997) he was forced to change tack and worked for the bookseller Gibert Joseph (1997–2010) whilst continuing his researches into odontodes, using a database that he had developed to enable exact recognition of the skin of cartilaginous fishes and even to identify individuals.

Dharmadi

Bali Catshark *Atelomycterus baliensis* WT White, Last & Dharmadi, 2005
Indonesian Houndshark *Hemitriakis indroyonoi* WT White, Compagno & Dharmadi, 2009

Dharmadi (b.1957) is a researcher at Indonesia's Research Centre for Fisheries Conservation & Management where he has been co-ordinator of Fish Resources Assessment since 2009. The National University of Indonesia awarded both his bachelor's degree (1983) and his doctorate (1986). He wrote *New Species of Shark and Ray* (2010) and co-wrote *Economically Important Sharks and Rays of Indonesia* (2006).

Díaz de Astarloa

Joined-fins Skate *Bathyraja cousseauae* Díaz de Astarloa & Mabragaña, 2004

Argentine Skate *Dipturus argentinensis* Díaz de Astarloa, Mabragaña, Hanner & Figueroa, 2008
Shortnose Eagle Ray *Myliobatis ridens* Ruocco, Lucifora, Díaz de Astarloa, Mabragaña & Delpiani, 2012

Dr Juan Martin Díaz de Astarloa is an Argentine ichthyologist who is an independent researcher and co-ordinator of the Oceanographic Commission of the National Research Council of Argentina (CONICET). He is also a member of the board, Committee of the Marine and Coastal Research Institute, Faculty of Natural Sciences, University de Mar del Plata where he is a full professor. Among his publications, often as co-author, are *DNA Barcoding Identifies Argentine Fishes from Marine and Brackish Waters* (2011) and *From Coexistence to Competitive Exclusion: Can Overfishing Change the Outcome of Competition in Skates* (Chondrichthyes, Rajidae)? (2011). He led expeditions to the Antarctic (2011–2012 and 2013).

Didier

Dwarf Chimaera *Neoharriotta pumila* Didier & Stehmann, 1996
Leopard Chimaera *Chimaera panther* Didier, 1998
Carpenter's Chimaera *Chimaera lignaria* Didier, 2002
Pale Ghost Shark *Hydrolagus bemisi* Didier, 2002
Pointy-nosed Blue Chimaera *Hydrolagus trolli* Didier & Séret, 2002
White-spot Chimaera *Hydrolagus alphus* Quaranta, Didier, Long & Ebert, 2006
Galápagos Ghostshark *Hydrolagus mccoskeri* Barnett, Didier, Long & Ebert, 2006
Southern Chimaera *Chimaera fulva* Didier, Last & WT White, 2008
Longspine Chimaera *Chimaera macrospina* Didier, Last & WT White, 2008
Shortspine Chimaera *Chimaera obscura* Didier, Last & WT White, 2008
Black Ghostshark *Hydrolagus homonycteris* Didier, 2008
Marbled Ghostshark *Hydrolagus marmoratus* Didier, 2008
Eastern Pacific Black Ghostshark *Hydrolagus melanophasma* James, Ebert, Long & Didier, 2009
Bahamas Ghost Shark *Chimaera bahamaensis* Kemper, Ebert, Didier & Compagno, 2010
Cape Chimaera *Chimaera notafricana* Kemper, Ebert, Compagno & Didier, 2010

Dr Dominique Ann Didier (b.1965) is a zoologist, marine biologist and ichthyologist. The Illinois Wesleyan University awarded her bachelor's degree in biology (1987) and University of Massachusetts, Amherst her doctorate (1991). She is now an associate professor of biology at Millersville University of Pennsylvania. Her research interests and expertise are the systematics and evolution of Chondrichthyan fishes with a particular focus on the chimaeroid fishes. Among her notable publications is a monograph on the *Phylogenetic Systematics of Extant Chimaeroid Fishes (Holocephali, Chimaeroidei)*. More recently her publications have focused on taxonomic descriptions of new species of chimaeroid fishes and she and her colleagues have recently synthesized much of this work in the second edition of *Biology of Sharks and their Relatives* (*Phylogeny, Biology and Classification of Extant Holocephalans* by Didier, Kemper and Ebert) (2012). Her most recent publication was a collaborative effort in which the fossil shark, *Helicoprion* Tapanila *et al*. 2013, was described (2013).

Diehl

Thorny-tail Skate *Dipturus diehli* Soto, Jules & Mincarone, 2001

Professor Fernando Luiz Diehl (b.1959) is a Brazilian oceanographer. He received his bachelor's degree in oceanography (1984) and his master's in geography (1997). He was Director of the School of Marine Sciences of the Universidade do Vale do Itajaí – UNIVALI (1992–2004), and was President of the Brazilian Association of Oceanography – AOCEANO six times (1990–2009) and President of the Latin American Association of Researchers in Marine Sciences – ALICMAR (2007–2009 and 2011–2013). He was honoured in recognition of his 'extensive work and tireless dedication to oceanography in Brazil'.

Dingerkus

Nurse Shark genus *Pseudoginglymostoma* Dingerkus, 1986

Burmese Bamboo Shark *Chiloscyllium burmensis* Dingerkus & DeFino, 1983

Guido Dingerkus (1953–2004) was an ichthyologist whose bachelor's degree (1975) and master's (1977) were both awarded by Cornell University. He worked in the Ichthyology Department of AMNH and later was Director, Natural History Consultants Inc. in New York. He wrote *The Shark Watcher's Guide* (1985). He died from the effects of exposure to radiation.

Dixon

Dixon's Stingaree *Urolophus paucimaculatus* Dixon, 1969
[Alt. Sparsely-spotted/White-spotted Stingaree]

Joan Maureen Dixon (b.1937) was Curator of Vertebrates at the National Museum of Victoria and since retirement has been Curator Emeritus. She has written widely on many vertebrates from mammals to reptiles. She described the ray in *The Victorian Naturalist* published by the Museum. A reptile is named after her.

Doello-Jurado

Southern Thorny Skate *Amblyraja doellojuradoi* Pozzi, 1935

Professor Martin Doello-Jurado (1884–1948) was a marine biologist and malacologist with an interest in palaeontology who directed several oceanic explorations of the Magellan Region and elsewhere (1914–1915 and 1921) aboard the 'A.R.A. Patria' while teaching at the University of Buenos Aires. He became Director of the Bernardino Rivadavia Natural Sciences Museum in Buenos Aires (1923) and was dismissed by President Juan Perón (1946).

Dolganov

Pocket Shark genus *Mollisquama* Dolganov, 1984

Bottom Skate *Bathyraja bergi* Dolganov, 1983
Mud Skate *Bathyraja taranetzi* Dolganov, 1983
Pocket Shark *Mollisquama parini* Dolganov, 1984
Fedorov's Catshark *Apristurus fedorovi* Dolganov, 1985
Little-eyed Skate *Bathyraja andriashevi* Dolganov, 1985
Fedorov's Skate *Bathyraja fedorovi* Dolganov, 1985
[Alt. Cinnamon Skate]
Longnose Deep-sea Skate *Bathyraja shuntovi*
    Dolganov, 1985
Creamback Skate *Bathyraja tzinovskii* Dolganov, 1985
Mid-Atlantic Skate *Rajella kukujevi* Dolganov, 1985
Lanternshark sp. *Etmopterus schmidti* Dolganov, 1986
Rasptooth Dogfish *Miroscyllium sheikoi* Dolganov, 1986

Dr Vladimir Nikolaevich Dolganov (b.1949) is a Russian ichthyologist. His doctorate was awarded (2003). He was at the Institute of Marine Biology, Russian Academy of Sciences, Vladivostok (2008). He wrote *Slopes of Suborder RAJOIDEI of the World Ocean (the Parentage, Evolution and Moving)* (2003).

Donovan

Starry Skate *Amblyraja radiate* Donovan, 1808

Edward Donovan (1768–1837) was an Anglo-Irish writer, natural history illustrator and amateur zoologist. He collected specimens, buying them at auction as well as directly from returning voyages of exploration. He founded the London Museum and Institute of Natural History (1807) exhibiting birds,

fish and other taxa, coral and botanical specimens. He was a successful author of natural history books writing, among others, *Natural History of British Fishes* (1802–1808) and *Natural History of British Birds* (1792–1797). He wrote articles on conchology, entomology etc. and compiled albums of his own artwork. In decline his museum closed (1817) and the collection was auctioned off (1818). He eventually died penniless leaving a large family destitute.

Doutre

Violet Skate *Dipturus doutrei* Cadenat, 1960
[Alt. Javelin Skate]

M. P. Doutre was a French veterinary surgeon at the Laboratoire National de l'Élevage et de Recherches Vétérinaires, Dakar, Sénégal off whose coast this skate occurs. He wrote *Les merlus de Sénégal. Mise en evidence d'une nouvelle espèce* (1960), naming it *Merluccius polli cadenati* after Cadenat (1960).

Driggers

Carolina Hammerhead *Sphyrna gilberti* Quattro, Driggers, Grady, Ulrich & MA Roberts, 2013

Dr William B. Driggers III (b.1968) is an American ichthyologist and biologist who works as a research fisheries biologist at the National Oceanic and Atmospheric Administration, National Marine Fisheries Service, Pascagoula, Mississippi. Clemson University, South Carolina, awarded his bachelor's degree (1992) and the University of South Carolina his master's (1998) and his doctorate in marine science (2001). He is also a member of the faculty of the University of New England. He co-wrote *Variability in the Reproductive Cycle of Finetooth Sharks,* Carcharhinus isodon, *in the Northern Gulf of Mexico* (2009).

Dudgeon

Milne Epaulette Carpetshark *Hemiscyllium michaeli* Allen & Dudgeon, 2010

Dr Christine 'Chris' L. Dudgeon (b. 1977) is a marine scientist whose bachelor's degree in marine biology and zoology was awarded by James Cook University (1998). The School of Integrative Biology, University of Queensland awarded her doctorate (2009) and since 2011 she has been a post-doctoral fellow at the School of Veterinary Science, University of Queensland. Her research interests include contemporary and historical population dynamics, speciation and conservation biology of marine organisms, with an emphasis on chondrichthyan fishes. She co-wrote (both 2012) *First Record of Potential Batesian Mimicry in an Elasmobranch: Juvenile Zebra Sharks Mimic Banded Sea Snakes?* and *A Review of the Application of Molecular Genetics for Fisheries Management and Conservation of Sharks and Rays.*

Duffy

Kermadec Spiny Dogfish *Squalus raoulensis* Duffy & Last, 2007

Clinton A. J. Duffy (b.1966) is a New Zealand marine scientist. The University of Canterbury awarded both his BSc and MSc. He works for the marine ecosystems team of the Department of Conservation. He is an expert on sharks and has worked on their conservation, age and growth, reproduction and phylogeography, as well as their taxonomy. He is a marine research associate of the Auckland Museum, a member of the IUCN Shark Specialist Group (Australia and Oceania) and PhD student (part time) at the University of Auckland. His PhD is focused on tracking the movements and estimating the population size of New Zealand great white sharks. He has performed a public dissection of a great white at the Auckland Museum. His current taxonomic projects include the description of new spiny dogfishes and spotted smoothhounds from the southwest Pacific. He has published a number of papers including the co-written *Age, Growth, Longevity and Natural Mortality of the Shortfin Mako Shark* (Isurus oxyrinchus) *in New Zealand Waters* (2006).

Duméril, AHA

Electric Ray genus *Hypnos* AHA Duméril, 1852
Ray genus *Paratrygon* AHA Duméril, 1865

Coffin Ray *Hypnos monopterygius* AHA Duméril, 1852
Australian Swellshark *Cephaloscyllium laticeps* AHA Duméril, 1853
Necklace Carpetshark *Parascyllium variolatum* AHA Duméril, 1853
Cape Elephantfish *Callorhinchus capensis* AHA Duméril, 1865
Magdalena River Stingray *Potamotrygon magdalenae* AHA Duméril, 1865

Auguste Henri André Duméril (1812–1870) was a physician and zoologist like his father, and followed much in his father's footsteps becoming more and more important (1850s). He produced the two-volume *Histoire Naturelle des Poissons, ou Ichtyologie Générale* (1865–1870). (See **Duméril, AMC** below)

Duméril, AMC

Sand Devil *Squatina dumeril* Lesueur, 1818

Angel Shark genus *Squatina* AMC Duméril, 1806

Dr André Marie Constant Duméril (1774–1860) was a zoologist and herpetologist who qualified as a physician (1793). He was Professor of Anatomy, Muséum National d'Histoire Naturelle, Paris (1801–1812), changing to Professor of Herpetology and Ichthyology (1813–1857). He built up the largest herpetological collection of the time. Toward the end of his career his son assisted him. He retired entirely (1857) and his son took over his professorship. A very large number of taxa are named after them including 6 amphibians, 2 birds and 21 reptiles. He was honoured in the sand devil name 'in testimony of [Lesueur's] remembrance and esteem'. (See **Duméril, AHA** above)

Dunn

Ray family *Urotrygonidae* McEachran, Dunn & Miyake, 1996

Dr Katherine A. Dunn is an ichthyologist, geneticist and microbiologist who is a research associate with the Bielawski Research Group at the Department of Biology, Dalhousie University, Halifax, Nova Scotia. Long Island University, Southampton awarded her bachelor's degree (1990), Hofstra University her master's (1994) and Texas A & M University her doctorate in wildlife and fisheries sciences (2001).

Durand

Fine-spotted Leopard Whipray *Himantura tutul* Borsa, Durand, K-N Shen, Arlyza, Solihin & Berrebi, 2013

Dr Jean-Dominique Durand (b.1972) is a marine ecologist and taxonomist, mainly involved in the evolutionary ecology of tropical fishes. He is a member of the faculty at the Département des Ressources Vivantes, Institut de Recherche pour le Développement, Montpellier, which he joined (2001) and where he has been since 2011 a senior research scientist. Claude Bernard Lyon 1 University awarded his doctorate (1999). Much of his work has been about the *Mugilidae* family including being co-author of *Genus-level Taxonomic Changes Imposed by the Mitochondrial Phylogeny of Grey Mullets (Teleostei: Mugilidae)* (2012).

Dussumier

Whitecheek Shark *Carcharhinus dussumieri* Müller &
Henle, 1839

Jean-Jacques Dussumier (1792–1883) was a French
merchant, collector, traveller and ship owner. He
was most active (1816–1840) in southeastern Asia
and around the Indian Ocean. He collected molluscs
and fishes, a large number being named after him,
including a whole genus of herring, *Dussumieria*. He
collected the type. A mammal, four birds and five
reptiles are also named after him.

Dybowski

Far Eastern Brook Lamprey *Lethenteron reissneri*
Dybowski, 1869

Benedykt (Benoit) Dybowski (1833–1930) was a
Polish biologist who was born in Belarus. He was an
ardent proponent of Darwin's theory of evolution.
He was appointed adjunct Professor of Zoology in
Warsaw (1862), but after the failure of the Uprising (January 1863 against the Russian Empire) he
was banished and spent time as a political exile in
Siberia. There support from the Zoological Cabinet
at Warsaw allowed him to undertake investigations into the natural history of Lake Baikal and
other parts of the Russian Far East. He was par-
doned (1877) and went to Kamchatka as a physician
(1878). He was appointed (1883–1906) to the Chair
of Zoology, University of Lemburg, Poland (Lwow,
Ukraine) until his retirement. A mammal, three birds
and two amphibians are named after him.

E

Eastman

Gecko Catshark *Galeus eastmani* Jordan & Snyder, 1904

Dr Charles Rochester Eastman (1868–1918) was an American palaeontologist specialising in fossil fish. He graduated from Harvard (1891) and was awarded his MSc by Johns Hopkins University (1892), after which he obtained his PhD at the University of Munich (1894). He taught geology and palaeontology at Harvard and Radcliffe while further studying fossil fish under Agassiz (q.v.). He became Curator of Vertebrate Palaeontology at the Museum of Comparative Zoology and then Curator at Carnegie Museum Pittsburgh (1910). He was also an assistant geologist for the US Geological Survey's New England Division. He wrote more than 100 scientific papers. Moreover he was the foremost authority on the literature of fishes and produced a massive bibliography. His body was found drowned at Long Beach where he had been walking while recuperating from Spanish flu; it was presumed that he had collapsed and fallen in.

Eaton

Eaton's Skate *Bathyraja eatonii* Günther, 1876

Reverend Alfred Edmund Eaton (1845–1929) was an English explorer, entomologist and naturalist who published many scientific papers (1860s–1920). He collected on Kerguelen Island with the Transit of Venus Expedition (1874–1875) and collected the holotype of the skate off Madeira and Tenerife (1902). Two birds are named after him.

Ebert

Paddle-nose Chimaera *Rhinochimaera africana* Compagno, Stehmann & Ebert, 1990

Taiwanese Blind Electric Ray *Benthobatis yangi* Carvalho, Compagno & Ebert, 2003

Broad-snout Lanternshark *Etmopterus burgessi* Schaaf-Da Silva & Ebert, 2006

White-spot Chimaera *Hydrolagus alphus* Quaranta, Didier, Long & Ebert, 2006

Galápagos Ghostshark *Hydrolagus mccoskeri* Barnett, Didier, Long & Ebert, 2006

Taiwan Torpedo *Torpedo formosa* Haas & Ebert, 2006

Western Gulper Shark *Centrophorus westraliensis* WT White, Ebert & Compagno, 2008

Southern Dogfish *Centrophorus zeehaani* WT White, Ebert & Compagno, 2008

Australian Reticulate Swellshark *Cephaloscyllium hiscosellum* WT White & Ebert, 2008

Spotted Swellshark *Cephaloscyllium maculatum* Schaaf-Da Silva & Ebert, 2008

Leopard-spotted Swellshark *Cephaloscyllium pardelotum* Schaaf-Da Silva & Ebert, 2008

Southern African Frilled Shark *Chlamydoselachus africana* Ebert & Compagno, 2009

Eastern Pacific Black Ghostshark *Hydrolagus melanophasma* James, Ebert, Long & Didier, 2009

Bahamas Ghostshark *Chimaera bahamaensis* Kemper, Ebert, Didier & Compagno, 2010

Cape Chimaera *Chimaera notafricana* Kemper, Ebert, Compagno & Didier, 2010

Roughnose Legskate *Cruriraja hulleyi* Aschliman, Ebert & Compagno, 2010

Lantern Shark sp. *Etmopterus joungi* Knuckey, Ebert & Burgess, 2011

African Dwarf Sawshark *Pristiophorus nancyae* Ebert & Cailliet, 2011

Sculpted Lanternshark *Etmopterus sculptus* Ebert, Compagno & De Vries, 2011

Shortfin Smooth Lanternshark *Etmopterus joungi* Knuckey, Ebert & Burgess, 2011

Angelshark sp. *Squatina caillieti* Walsh, Ebert & Compagno, 2011

Lana's Sawshark *Pristiophorus lanae* Ebert & Wilms, 2013

Dr David A. Ebert is Program Manager at the Pacific Shark Research Center (PSRC). He has collaborated with many of the world's leading shark experts. He has a wide ichthyological interest covering the biology, ecology, biogeography, fisheries and systematics of chondrichthyan fishes (sharks, ray, chimaeras). He followed his father (Earl) into marine biology. He was awarded a bachelor's degree in zoology at Humboldt State University (1980) and began formal research training at the Moss Landing Marine Laboratories under the supervision of Gregor Cailliet (q.v.). He gained his MSc (1984) and went on to do his PhD (1990) at the Department of Ichthyology and Fisheries Science at Rhodes University, South Africa under the supervision of Dr Compagno (q.v.). He did postdoctoral research at the Shark Research Center, South African Museum then returned to California to start up an abalone farming business. He wrote *Sharks, Rays, and Chimaeras of California* (2003) following the earlier *Guide to the Sharks and Rays of Southern Africa* (1989).

Economidis

Macedonia Brook Lamprey *Eudontomyzon hellenicus*
Vladykov, Renaud, Kott & Economidis, 1982

Epirus Brook Lamprey *Eudontomyzon graecus* Renaud
& Economidis, 2010

Professor Dr Panos Stavros Economidis was formerly Director of the Ichthyology and Zoology
Laboratories in the School of Biology, Aristotle
University of Thessaloniki and was the main contributor to its fish collection. He is now Emeritus
Professor. He is an expert in fish taxonomy, distribution and conservation and in fisheries management
with more than four decades of experience. He has
written or co-written many papers including *Fish
Fauna of Greece in a Protected Greek Lake: Fish Biodiversity, Impact of Introduced Fish Species on the Ecosystem*
(2008).

Edmunds

Edmund's Spurdog *Squalus edmundsi* Last, WT White
& Stevens, 2007

Fatspine Spurdog *Squalus crassispinus* Last, Edmunds
& Yearsley, 2007

Dr Matthew J. 'Matt' Edmunds is an Australian
ecologist. His bachelor's degree (1990) and his doc-
torate in zoology (1995) were both awarded by the
University of Tasmania, Hobart. He was employed
as a marine biologist by Consulting Environmental
Engineers Pty Ltd (1990–1999) and became (1999)
Director, Principal Ecologist at Australian Marine
Ecology Pty Ltd, a company that provides scien-
tific consulting services. He was honoured for his
'high-quality preliminary research' on Australian
Squalus during a summer vacation scholarship at
CSIRO Marine Laboratories in the early 1990s.

Edwards

Puffadder Shyshark *Haploblepharus edwardsii* Schinz,
1822

George Edwards (1694–1773) was an illustrator, naturalist and ornithologist. He was Librarian, Royal
College of Physicians, London (1733–1764) and
corresponded regularly with Linnaeus. Edwards
wrote the four-volume *A Natural History of Birds*
(1743–1751). The first known reference to the puffadder shyshark was in some drawings by Edwards
of specimens that were later lost. Cuvier described
the species (1817) as 'Scyllium d'Edwards' based on
Edwards' drawings. Although he was not assign-

ing a scientific name, it was translated by Schinz
(1822) as *Scyllium edwardsii*. Four birds, a reptile and
mammal are named after him.

Ehrenbaum

African Wedgefish *Rhynchobatus luebberti*
Ehrenbaum, 1915

Professor Dr Ernst M. E. Ehrenbaum (1861–1942)
was a German ichthyologist and oceanographer. He
studied (1879–1884) zoology in Berlin, Würzburg
and Kiel. He became head of a zoological station
on the North Sea (1888–1892) then Assistant and
Research Assistant for fisheries at the Heligoland
Biological Institute (1892–1900). He became a professor (1900–1910) before moving to the Hamburg
Natural History Museum (1910–1920). He became
(1920) a member of the German Commission for
Marine Research until retirement (1931). He wrote
*Naturgeschichte und Wirtschaftliche Bedeutung der Seef-
ische Nordeuropas* (1936).

Ehrenberg

Arabian Smooth-Hound *Mustelus mosis* Hemprich &
Ehrenberg, 1899

Professor Christian Gottfried Ehrenberg (1795–1876)
was a German naturalist. He started studying theology at Leipzig (1815) but changed direction (1817) and
went to Berlin to study medicine. He was working
on fungi and was a lecturer at the University of
Berlin (1820), where he became a professor of medicine (1827). He travelled extensively (1820–1825),
mainly in the company of his friend Hemprich. They
travelled widely in northeast Africa and the Middle
East, from Lebanon to Sinai and from the Nile to
Abyssinia (Ethiopia). Furthermore, he travelled with
Humboldt to Asia (1829) and visited England to
meet Darwin at Oxford (1847). He is regarded as the
founder of micropalaeontology. He bequeathed his
collection to the Museum für Naturkunde in Berlin.
He published a great many articles and books, especially on fungi and corals. He was the first person to
establish that phosphorescence in the sea is caused
by plankton-like microorganisms. Two birds, a
mammal and an amphibian are named after him.

Eisenhardt

Galapagos Gray Skate *Rajella eisenhardti* Long &
McCosker, 1999

Emil Roy Eisenhardt (b.1939) is Director Emeritus
of the California Academy of Science. He attended

Dartmouth College and served in the US Marine Corps before going to law school. He practised law for 12 years (1966–1978) in San Francisco and taught at the University of California, Berkeley's Boalt Hall School of Law, from which he had graduated (1965). He was President of Oakland Athletics (1982–1989). He was an executive director of the California Academy of Science (1989–1995) since when he has been in private consultancy. He was honoured in the skate name for his generous assistance to the authors and their colleagues.

Endeavour

The Endeavour Skate *Dipturus endeavouri* Last, 2008

This species is named after the fisheries investigation steamship 'Endeavour' that was lost with all hands off the coast of Australia (1914). It was responsible for collecting the first specimens of this species and many of Australia's continental shelf fish species in the early years of the twentieth century.

Engelhardt

Brown Lantern Shark *Etmopterus unicolor* Engelhardt, 1912
Red-blotched River Stingray *Potamotrygon ocellata* Engelhardt, 1912
Baluchistan Torpedo *Torpedo zugmayeri* Engelhardt, 1912

Robert Engelhardt was an ichthyologist at the Zoologischen Staatssammlung München and a member of Bayerische Akademie der Wissenschaften. He wrote *Über einige neue Selachier-Formen* (1912).

Eos

Hagfish sp. *Eptatretus eos* Fernholm, 1991

Eos in Greek mythology was the goddess of the dawn. In this case it refers to the pink colouring of the hagfish.

Erdmann

Walking Shark *Hemiscyllium galei* Allen & Erdmann, 2008
Bamboo Shark sp. *Hemiscyllium henryi* Allen & Erdmann, 2008

Dr Mark van Nydeck Erdmann (b.1968) is an American reef fish expert and marine senior advisor with Conservation International Indonesia. Duke University awarded his bachelor's degree in biology (1990) and the University of California, Berkeley his doc-

torate (1997). He wrote *The Ecology, Distribution and Bioindicator Potential of Indonesian Coral Reef Stomatopod Communities* (1997).

Espinoza-Pérez

Gulf Angelshark *Squatina heteroptera* Castro-Aguirre, Espinoza-Pérez & Huidobro-Campos, 2007
Mexican Angelshark *Squatina mexicana* Castro-Aguirre, Espinoza-Pérez & Huidobro-Campos, 2007

Dr Héctor Salvador Espinosa-Pérez (b.1954) is an ichthyologist who is a professor and curator of fishes at the Instituto de Biología, Universidad Nacional Autónoma de México, which university awarded his bachelor's, master's and doctorate and where he joined the faculty in 1983. He specialises in the systematics and taxonomy of both extant and fossil fishes. Among his many publications he is co-author of *Common and Scientific Names of Fishes* (7th edition 2013).

Euphrasén

Spotted Eagle Ray *Aetobatus narinari* Euphrasén, 1790

Bengt Anders Euphrasén (1756–1796) was a Swedish botanist. Uppsala University awarded his degree (1784). He undertook a collecting trip in the Antilles under the sponsorship of the Royal Swedish Academy of Science, including to Saint Barthélemy and Saint Christopher Islands (St Kitts).

Evans

Blackmouth Lantern Shark *Etmopterus evansi* Last, Burgess & Séret, 2002

David Evans is an Australian fisheries scientist who worked for CSIRO at the Marmion Marine Laboratories, Western Australia. The etymology honours 'David Evans, who over the last decade has meticulously selected and donated valuable taxonomic specimens (including several excellent Etmopterus specimens of which two have been designated as holotypes) collected by commercial trawlers from the tropical deepwater of Western Australia.' He co-wrote *Crustaceans from the Deepwater Trawl Fisheries of Western Australia* (1991).

Evermann

Requiem Sharks genus *Carcharhinidae* Jordan & Evermann, 1896

Pacifric Cownose Ray *Rhinoptera steindachneri* Evermann & OP Jenkins, 1891

Southern Banded Guitarfish *Zapteryx xyster* Jordan &
  Evermann, 1896
Black Hagfish *Eptatretus deani* Evermann &
  Goldsborough, 1907

Dr Barton Warren Evermann (1853–1932) was a schoolteacher (1876–1886) and a student at Indiana University, where he was awarded his bachelor's degree (1886), master's (1888) and doctorate (1891). He worked for the Bureau of Fishes in Washington (1891–1914) in various roles that he combined with lecturing on zoology at Cornell (1900–1903), Yale (1903–1906), and later Stanford, after he became Director of the Museum, California Academy of Sciences (1914). A bird and a reptile are also named after him.

F

Fabricius

Black Dogfish *Centroscyllium fabricii* Reinhardt, 1825

Bishop Otto Fabricius (1744–1822) was a Danish missionary, explorer and naturalist in Greenland (1768–1773) and the first to study this species as well as many other Greenland fishes. He wrote *Fauna Groenlandica* (1780). A bird is named after him.

Fahmi

Jimbaran Shovelnose Ray *Rhinobatos jimbaranensis* Last, WT White & Fahmi, 2006
Indonesian Shovelnose Ray *Rhinobatos penggali* Last, WT White & Fahmi, 2006
Borneo Sand Skate *Okamejei cairae* Last, Fahmi & Ishihara, 2010
Starrynose Stingray *Pastinachus stellurostris* Last, Fahmi & Naylor, 2010

Mr Fahmi (b.1974), who holds a master's degree, is a marine biology researcher at the Research Centre for Oceanography, Indonesian Institute of Sciences, Jakarta. He focuses his research on the taxonomy, biology, ecology, fisheries and conservation of elasmobranch fishes. He led the Natuna Expedition (2010) to investigate the decline in fish stocks. He wrote *Sharks and Rays in Indonesia* (2012) and he tells us that he has contributed to some field guides on sharks and rays in Borneo (2010) and in southeast Asia (2013).

Fairchild

New Zealand Torpedo *Torpedo fairchildi* Hutton, 1872

Captain John Fairchild (1835–1898) was a New Zealander who was master on government steamers, and supply vessels to sub-Antarctic islands. He was in command of the 'Luna' when he discovered the holotype stranded on the mud inside Napier harbour (1868). Fairchild died in an accident: whilst superintending the loading of some iron rails he was struck on the head by a chain. A bird is named after him.

Falkner

Large-spot River Stingray *Potamotrygon falkneri* Castex & Maciel, 1963

Thomas Falkner (sometimes Falconer) (1702–1784) was born in Manchester, England. He studied medicine and natural sciences in London, as a student of Mead and Newton. The Royal Society of London commissioned Falkner to go to South America to study the medicinal properties of herbs. His passage to Buenos Aires was as a doctor on a slave ship. Soon after arriving he became very ill. He was aided in his quest by a Jesuit, converted to Catholicism and (1732) entered the Society of Jesus, spending the next ten years studying in Córdoba (Argentina). He was sent (1742) as a missionary explorer along the Salado and then south. He helped found a mission (1746) and studied the local flora and fauna until a general uprising of the Indians destroyed the missions. Sent to another mission in Areco, he continued his studies before returning to Córdoba (1752), where he founded the Department of Mathematics. He was arrested as a spy and with 40 other Jesuits was deported from Argentina to Spain (1776) and returned to England. Castex (also a priest) honoured him for his apostolic and scientific work in eighteenth-century Argentina.

Fechhelm

East African Skate *Okamejei heemstrai* McEachran & Fechhelm, 1982
Allen's Skate *Pavoraja alleni* McEachran & Fechhelm, 1982
Colombian Electric Ray *Diplobatis colombiensis* Fechhelm & McEachran, 1984

Mrs Janice Drukten Fechhelm (b.1954) has a master's degree in wildlife and fisheries science. She was a research assistant and scientific illustrator at Texas A & M University (1998) and is now a science teacher at Cypress Grove Intermediate School at College Grove, Texas (2012). She and McEachran are co-authors of the two-volume *Fishes of the Gulf of Mexico* (1998 and 2005).

Fedorov

Fedorov's Catshark *Apristurus fedorovi* Dolganov, 1985
Fedorov's Skate *Bathyraja fedorovi* Dolganov, 1985 [Alt. Cinnamon Skate]

Dr Vladimir Vladimirovich Fedorov (1939–2011) was a Russian ichthyologist. His doctorate was awarded by Leningrad State University (1979). He worked at the Pacific Research Institute of Fisheries and Oceanography, Valdivostock (1961–1975) and at the Zoological Institute, Academy of Sciences,

Leningrad (later St Petersburg) (1975). He collected the holotype of the skate.

Fehlmann

Harlequin Catshark *Ctenacis fehlmanni* S Springer, 1968

Dr Herman Adair Fehlmann (1917–2005) of George Vanderbilt Foundation, Bangkok and the Smithsonian was primarily an ichthyologist. His herpetological specimens were deposited in the Stanford University Museum. He wrote *Ecological Distribution of Fishes in a Stream Drainage in the Palau Islands* (1960). He was honoured for 'setting high standards for field treatment of shark specimens collected for study'. A reptile is named after him.

Feld

Jespersen's Hagfish *Myxine jespersenae* Møller, Feld, Poulsen, Thomsen & Thormar, 2005

Mrs Thea Kristine Feld Nielsen (b.1980) graduated in biology and physical education at the University of Copenhagen (2006) and since 2007 has taught biology and sports at Helsingør Gymnasium (High School).

Fernández-Yépez

Rosette River Stingray *Potamotrygon schroederi* Fernández-Yépez, 1958
[Alt. Flower Ray]

Dr Augustín Antonio Fernández-Yépez (1916–1977) was a Venezuelan architect and ichthyologist who was a curator at the Museo de Ciencias Naturales de Caracas. His brother Dr Francisco José Fernández Yépez (1923–1986) was an agronomist and entomologist and has several odonata and an earwig named after him. (See **Yépez**)

Fernholm

Hagfish sp. *Paramyxine fernholmi* Kuo, Huang & Mok, 1994
Hagfish sp. *Myxine fernholmi* Wisner & CB McMillan, 1995
Hagfish sp. *Eptatretus fernholmi* CB McMillan & Wisner, 2004

Hagfish sp. *Eptatretus minor* Fernholm & CL Hubbs, 1981
Hagfish sp. *Eptatretus multidens* Fernholm & CL Hubbs, 1981
White-headed Hagfish *Myxine ios* Fernholm, 1981
Hagfish sp. *Eptatretus caribbeaus* Fernholm, 1982
Hagfish sp. *Eptatretus eos* Fernholm, 1991

Hagfish sp. *Eptatretus lopheliae* Fernholm & Quattrini, 2008
Australian Hagfish sp. *Eptatretus alastairi* Mincarone & Fernholm, 2010
Hagfish sp. *Eptatretus astrolabium* Fernholm & Mincarone, 2010
Hagfish sp. *Eptatretus gomoni* Mincarone & Fernholm, 2010

Bo Fernholm (b.1941) is a Swedish zoologist. He is Professor Emeritus of Vertebrate Zoology at the Swedish Museum of Natural History. His specialist area is hagfish systematics. He also served 20 years as Swedish commissioner to the International Whaling Commissioner and as Swedish scientific representative to the Convention for the Conservation of Antarctic Marine Living Resources (CCAMLR). He was honoured for his many contributions to hagfish knowledge, including the discovery of the eponymous species.

Figaro

Catshark genus *Figaro* Whitley, 1928

Figaro is the eponymous *Barber of Seville* (opera by Rossini) and of Mozart's *The Marriage of Figaro*.

Figueiredo

Ghost Shark sp. *Hydrolagus lusitanicus* Moura, Figueiredo, Bordalo-Machado, Almeida & Gordo, 2005

Professor Dr Ivone Maria Ribeiro Figueiredo da Silva Rosa is a Portuguese marine biologist and professor at the Faculdade de Ciências da Universidade de Lisboa. Among her published papers is the co-written *Deep-water Sharks Fisheries off the Portuguese Continental Coast* (2005).

Figueroa

Argentine Skate *Dipturus argentinensis* Díaz de Astarloa, Mabragaña, Hanner & Figueroa, 2008

Dr Daniel Enrique Figueroa is (2012) an ichthyologist at Universidad Nacional de Mar del Plata, Argentina, where he took his doctorate (1999). He wrote *Estudio Sobre la Anatomía y Algunos Aspectos de la Biología de los Congrios que Habitan el Mar Argentino y Adyacencias* (1999).

Filewood

Sailback Houndshark *Gogolia filewoodi* Compagno, 1973

Lionel Winston Charles Filewood (b.1936) is an

Australian biologist who worked (1960s, 1970s and 1980s) for the Department of Agriculture, Stock and Fisheries, Papua New Guinea. The University of Sydney awarded his bachelor's degree (1958). He co-wrote *Scientific Names used in Birds of New Guinea and Tropical Australia* (1976). He was honoured 'for his work on the poorly known elasmobranch fauna of New Guinea'. An Australian court sentenced him to nine months' imprisonment for child pornography offences (2010). A bird is named after him.

Fischer

Fanfin Skate *Pseudoraja fischeri* Bigelow & Schroeder, 1954

E. N. Fischer made skilful drawings for the describers' publications on fishes from the Gulf of Maine so this species was named after him. We have been unable to find out anything more about him.

Fitch

Snouted Eagle Ray *Myliobatis longirostris* Applegate & Fitch, 1964

John Edgar Fitch (1918–1982) was a marine biologist and ichthyologist who was Director of the US Fisheries Laboratory. He wrote *Offshore Fishes of California* (1958).

Fitzroy

Creek Whaler *Carcharhinus fitzroyensis* Whitley, 1943

The holotype was collected in the estuary of the Fitzroy River.

Flinders

Pygmy Thornback Skate *Dipturus flindersi* Last & Gledhill, 2008

This species is named for the Flindersian Province, which is the name of the western warm temperate biogeographic region of Australia and is this skate's distributional range.

Follett

Modoc Brook Lamprey *Entosphenus folletti* Vladykov & Kott, 1976

Salmon Shark *Lamna ditropis* CL Hubbs & Follett, 1947
Western Brook Lamprey *Lampetra richardsoni* Vladykov & Follett, 1965

William Irving 'Bill' Follett (1901–1992) was a lawyer by profession but an ichthyologist by choice. He was Curator, Department of Ichthyology, California Academy of Sciences (1947–1969).

Fontenelle

Freshwater Stingray sp. *Potamotrygon limai* Fontenelle, JP da Silva & Carvalho, 2014

João Pedro Fontenelle (b.1987) is a Brazilian ichthyologist at the Department of Zoology, Institute of Biosciences, University of São Paulo, which awarded both his bachelor's (2010) and master's degrees (2013) and where he is a graduate student.

Forsskål

Halavi Ray *Glaucostegus halavi* Forsskål, 1775
Feathertail Stingray *Pastinachus sephen* Forsskål, 1775
Giant Guitarfish *Rhynchobatus djiddensis* Forsskål, 1775
Bluespotted Ribbontail Ray *Taeniura lymma* Forsskål, 1775

Dr Peter Forsskål (1732–1763) was a Swedish explorer, orientalist, naturalist and student of Linnaeus. After a chequered academic career he eventually graduated from Uppsala (1751). He studied oriental languages and philosophy at Göttingen (1753–1756), achieving his doctorate. He returned to Uppsala and published his dissertation on civil freedom for which he received a warning from the crown; the government censored it. However, he was appointed (1760) by King Frederick V of Denmark to join an expedition to Egypt (1761) and on to Arabia (1762). He collected both botanical and zoological specimens but fell ill with malaria and died. A bird and a reptile are named after him.

Forster, GR

Pale Ray *Bathyraja pallida* GR Forster, 1967

George R. 'Bob' Forster (b.1927) is a zoologist who graduated from Aberdeen University (1949) and received a research grant (1949–1951) at the Plymouth Marine Laboratory of the Marine Biological Association where he was an early pioneer of subaqua diving (1951) and where he eventually was on the staff (1951–1985). He was a full time researcher and worked on a number of subjects including sponges, crustaceans and deep line fishing. He wrote a considerable number of scientific papers including *A Note on Two Rays Lacking Part of the Snout* (1967) and *The Ecology of* Latimeria chalumnae *Smith: Results of Field Studies from the Grande Comore* (1974). He carried out a lot of his work on board the 'Sarsia' – a vessel owned by the Association (1953–1979).

Forster, JR

Broadgilled Hagfish *Eptatretus cirrhatus* JR Forster, 1801

Johann Reinhold Forster (1729–1798) was a clergyman in Danzig. He became a naturalist and accompanied James Cook on his second voyage around the world (1772–1773) that extended further into Antarctic waters than any previous attempt. He discovered five new species of penguin. However, he gained a reputation as a constant complainer and troublemaker. His complaints about Cook continued after his return and became public, destroying Forster's career in England. He went to Germany and became a professor of history and mineralogy. Unpleasant and troublesome to the end, Forster refused to relinquish his notes of the voyage. They were not found and published until c.50 years after his death, which explains how Forster could name species long after his death. His son Johann George Adam was also on Cook's voyage as an artist. A mammal and eleven birds are named after him.

Fourmanoir

Longnose Houndshark *Iago garricki* Fourmanoir & Rivaton, 1979
Blacktailed Spurdog *Squalus melanurus* Fourmanoir & Rivaton, 1979
Cyrano Spurdog *Squalus rancureli* Fourmanoir & Rivaton, 1979

Pierre Fourmanoir (1924–2007) was a French ichthyologist who worked in New Caledonia. He described many other fish species. He co-wrote *Poissons de Nouvelle Calédonie et des Nouvelles Hebrides* (1976) and *Poissons de la Pente Récifale Externe de Nouvelle-Calédonie et des Nouvelles-Hébrides* (1979), where the above species are described.

Fowler, HW

Dogfish Shark genus *Zameus* Jordan & HW Fowler, 1903
Stingray genus *Pteroplatytrygon* HW Fowler, 1910
Panray genus *Zanobatidae* HW Fowler, 1928
Catshark genus *Aulohalaelurus* HW Fowler, 1934
Lanternshark genus *Etmopteridae* HW Fowler, 1934
Butterfly Ray genus *Gymnuridae* HW Fowler, 1934
Pygmy Shark genus *Heteroscymnoides* HW Fowler, 1934
Catshark genus *Holohalaelurus* HW Fowler, 1934
Dogfish genus *Proscymnodon* HW Fowler, 1934

Whitefin Dogfish *Centroscyllium ritteri* Jordan & HW Fowler, 1903

Acutenose Skate *Dipturus tengu* Jordan & HW Fowler, 1903
Ratfish sp. *Hydrolagus waitei* HW Fowler, 1907
Broadnose Skate *Bathyraja brachyurops* HW Fowler, 1910
Spotted Ray *Raja montagui* HW Fowler, 1910
Dark Ghost Shark *Hydrolagus novaezealandiae* HW Fowler, 1911
Longfin Catshark *Apristurus herklotsi* HW Fowler, 1934
Butterfly Ray sp. *Gymnura crooki* HW Fowler, 1934
Longnose Pygmy Shark *Heteroscymnoides marleyi* HW Fowler, 1934
Brownspotted Catshark *Scyliorhinus garmani* HW Fowler, 1934

Henry Weed Fowler (1878–1965) was an American zoologist. He studied at Stanford under D. S. Jordan (q.v.). He joined the Academy of Natural Sciences in Philadelphia and worked as an assistant (1903–1922), Associate Curator of vertebrates (1922–1934), Curator of fish and reptiles (1934–1940) and finally Curator of fish (1940–1965). He published very widely across many taxa. He was a co-founder of the American Society of Ichthyologists and Herpetologists (1927), being its first treasurer.

Fowler, SL

Borneo River Shark *Glyphis fowlerae* Compagno, WT White & Cavanagh, 2010

Dr Sarah Louise Fowler (b.1958) joined Naturebureau (1989) as Managing Director. She is still (2014) associated with it as a non-executive director and consultant. She has spent her lifetime working for the conservation of sharks. Her bachelor's degree was awarded by the University College of North Wales, Bangor (1979) and her master's by University College, London (1981). She is a founding trustee of the Shark Trust. She also led the study of the family found off Sabah (1996) leading to the discovery of three new species for which she is honoured in the name. She co-wrote *Collins Field Guide to Sharks of the World* (2005).

Frade

Sailfin Roughshark *Oxynotus paradoxus* Frade, 1929

Professor Fernando Frade Viegas da Costa (1898–1983) was a Portuguese zoologist. He joined the Museu Bocage, Lisbon (1924). He led an expedition to Portuguese Guinea (Guinea Bissau) (1945–1946). He published very many articles and monographs on almost every branch of zoology (1922–1980). His library was given to the Museu Bocage after his death.

Francis

Horn Shark *Heterodontus francisci* Girard, 1855

The original description has no etymology but the specific name probably derives from the city of San Francisco as the holotype was collected near there.

Franke

Gorgona Guitarfish *Rhinobatos prahli* Acero & Franke, 1995

Dr Rebeca Franke Ante is a biologist, zoologist and researcher at Parques Nacionales Naturales de Colombia–Territorial Caribe in Santa Marta and at Universidad del Valle, Cali, Colombia. Universidad Nacional de Colombia, Bogotá awarded her master's degree (2002). She co-wrote *Peces Óseos Comerciales del Parque Gorgona, Pacífico Colombiano (Osteichthyes: Muraenesocidae, Hemiramphidae, Belonidae, Scorpaenidae, Triglidae, Malacanthidae, Gerreidae, Sparidae, Kyphosidae, Sphyraenidae e Istiophorida)* (1996).

Freminville

Bullnose Eagle Ray *Myliobatis freminvillii* Lesueur, 1824

Galapagos Bullhead Shark *Heterodontus quoyi* Fréminville, 1840

Christophe/Chrétien-Paulin de la Poix Chevalier de Fréminville (1787–1848) was a French naval officer and naturalist. He was on board 'La Syrène', which attempted to discover the Northwest Passage (1806). He was an expert of the history and archaeology of the late Middle Ages and on the history of Brittany, and of the Templars in particular. Toward the end of his life an old episode affected him and he became deranged. He was in command of the French frigate 'La Néréide' in the West Indies (1822). He fell from some rocks and was lucky enough to be rescued from drowning and nursed back to health by a beautiful local girl, Caroline. He had to sail to Martinique but when he returned, he found she had drowned herself, thinking that he had deserted her. He took away some of her dresses as keepsakes (1842) and spent the last six years of his life wearing her old clothes. A reptile is named after him.

Frerichs

Thickbody Skate *Amblyraja frerichsi* G Krefft, 1968

Thomas Frerichs was captain of the research vessel 'Walther Herwig' from which the holotype was collected. He was honoured for 'his keen interest in deepsea catches, which assured us many precious discoveries'.

Freycinet

Indian Speckled Carpetshark *Hemiscyllium freycineti* Quoy & Gaimard, 1824

Captain Louis Claude Desaules de Freycinet (1779–1842) was a French navigator who was involved, with Baudin, on board the 'Casuarina' and 'Le Géographe', in mapping the Western Australian coast north of Perth (1803). He explored in the Pacific (1817–1820) on 'L'Uranie' and, after she was wrecked, he bought 'La Physicienne', from which he collected the holotype. An amphibian and two birds are named after him.

Fricke

Lantern Shark sp. *Etmopterus compagnoi* Fricke & Koch, 1990
Pita Skate *Okamejei pita* Fricke & Al-Hassan, 1995

Dr Ronald Fricke (b.1959) is a fish taxonomist. He was at the Technische Universität Braunschweig (1980–1986) and in the same period acted as Curator of ichthyology and herpetology at the Staatliches Naturhistorisches Museum, Braunschweig, with an interim period (1983–1984) at King's College, London. He was at the University of Hamburg (1986–1987) and at Freiburg (1988), where his doctorate was awarded (1989). He has been Curator of Ichthyology, Staatliches Museum für Naturkunde, Stuttgart, Germany since 1988 and since 2000 has been an independent expert for the EU Commission (diversity, distribution and conservation of NATURA 2000 species in the Atlantic, Continental and Mediterranean regions). He co-wrote *Raja Pita, a New Species of Skate from the Arabian/Persian Gulf (Elasmobranchii: Rajiformes)* (1995) and *The Coastal Fishes of Madeira Island – New Records and an Annotated Checklist* (2008).

Fries

Sailray *Dipturus linteus* Fries, 1838

Bengt Fredrik Fries (1799–1839) was a Swedish zoologist, entomologist and ichthyologist who studied at the University of Lund. He co-wrote *Skandinaviens Fiskar, Malade efter Lefvande Exemplar och Ritade l'a Sten* (1836).

Fritz (Frithjof)

Guadaloupe Hagfish *Eptatretus fritzi* Wisner & CB McMillan, 1990

Frithjof Frockney Ohre (1910–2003) was a farmer and amateur ichthyologist. He took part in a number of Scripps Institute of Oceanography expeditions (1967–1973), on which he made a photographic record of the voyages and specimens taken. He was honoured as 'friend, willing, eager, and industrious volunteer' who helped the authors collect specimens.

Furumitsu

Naru Eagle Ray *Aetobatus narutobiei* WT White, Furumitsu & Yamaguchi, 2013

Keisuke Furumitsu is an ichthyologist at the Faculty of Fisheries, Nagasaki University, Nagasaki, Japan. He co-wrote *Age, Growth and Age at Sexual Maturity of Fan Ray* Platyrhina sinensis *(Batoidea: Platyrhinidae) in Ariake Bay, Japan* (2008).

Fylla

Roundray *Rajella fyllae* Lütken, 1887

This species is named after the vessel 'Fylla' which was used for the expeditions (1884 and 1886) to Greenland where the holotype was collected.

G

Gadig

> Brazilian Large-eyed Stingray *Dasyatis marianae*
> Gomes, Rosa & Gadig, 2000
> Brazilian Soft Skate *Malacoraja obscura* Carvalho,
> Gomes & Gadig, 2005

Dr Otto Bismarck Fazzano Gadig is an ichthyologist, biologist and taxonomist. His bachelor's degree was awarded by the Universidade Católica de Santos (1991), his master's by the Universidade Federal da Paraíba (1994) and his doctorate by the Universidade Estadual Paulista, São Paulo, Brazil where he is now a professor of vertebrate zoology and comparative anatomy and is currently based at the university's Campus Experimental do Litoral Paulista, São Vicente. He co-wrote *Abnormal Embryos of Sharpnose Sharks*, Rhizoprionodon porosus *and* Rhizoprionodon lalandii (Elasmobranchii:Carcharhinidae), *from Brazilian Coast, Western South Atlantic* (2014).

Gage

> Southern Brook Lamprey *Ichthyomyzon gagei* CL
> Hubbs & Trautman, 1937

Simon Henry Gage (1851–1944) was a histologist and embryologist at Cornell University, which awarded his first degree (1877). He became Assistant Professor of Physiology and Lecturer in Microscopical Technology (1881–1889), then Associate Professor (1889–1893). He was one of the most remarkable, influential and important figures in the history of American microscopy. The author described him as 'one of the foremost students of the lampreys', who brought this 'interesting and distinct species' to his attention. He wrote what is now considered a classic textbook, *The Microscope* (1923).

Gaimard

> Blacktip Reef Shark *Carcharhinus melanopterus* Quoy
> & Gaimard, 1824
> Pygmy Shark *Euprotomicrus bispinatus* Quoy &
> Gaimard, 1824
> Indian Speckled Carpetshark *Hemiscyllium freycineti*
> Quoy & Gaimard, 1824
> Cookiecutter Shark *Isistius brasiliensis* Quoy &
> Gaimard, 1824

Joseph (or Jean, according to some sources) Paul Gaimard (1796–1858) was a French naval surgeon, explorer and naturalist. He made a voyage to Australia and the Pacific (1817–1819) aboard the 'Uranie', during which time he kept a journal, published as *Journal du Voyage de Circumnavigation, tenu par Mr Gaimard, Chirurgien à Bord de la Corvette l'Uranie*. Though he continued with his journal, further entries were lost when the ship was wrecked off the Falklands and he had to continue his journey on board the 'Physicienne', the ship that had rescued the expedition and then been purchased as a replacement. He was aboard the 'Astrolabe', under the command of Dumont d'Urville, when it visited New Zealand (1826). He led an expedition (1838–1840) aboard the 'Récherche' to northern Europe, visiting Iceland, the Faeroe Islands, northern Norway, Archangel and Spitsbergen. His contemporary, the zoologist Henrik Krøyer, who went with Gaimard to Spitsbergen (1838), described him thus: 'He was of medium build, with curly black hair and a rather unattractive face, but with a charming and agreeable manner.' He was something of a dandy and, when visiting Iceland, handed out sketches of himself. Several fish species, two birds, a mammal and a reptile are named after him.

Galbraith

> Galbraith's Catshark *Apristurus sp. X*

Dr John K. Galbraith is a biologist who is chief scientist of the National Marine Fisheries Service, Woods Hole, USA. He took part in the expedition (2004) undertaken to enhance understanding of the occurrence, distribution and ecology of animals and animal communities along the Mid-Atlantic Ridge between Iceland and the Azores. He co-wrote *An Annotated List of Deepwater Fishes from off the New England Region, with New Area Records*. This paper mentions a species of western Atlantic *Apristurus* that might be a new species; we assume this is handily referred to as Galbraith's.

Gale

> Walking Shark *Hemiscyllium galei* Allen & Erdmann,
> 2008

Jeffrey Gale of Las Vegas, Nevada, USA is an avid underwater photographer, shark enthusiast and benefactor of the marine realm. Mr Gale successfully bid to support the conservation of this species at the Blue Auction in Monaco (2007) and has given generously to support Conservation International's Bird's Head Seascape marine conservation initiative.

Gambang

Circle-blotch Pygmy Swellshark *Cephaloscyllium circulopullum* Yano, Ahmad & Gambang, 2005
Sarawak Pygmy Swellshark *Cephaloscyllium sarawakensis* Yano, Ahmad & Gambang, 2005

Albert Chuan Gambang (b.1955) is head of Sarawak Fisheries Research Institute, Malaysia. Aberdeen University awarded his master's degree in marine and fisheries sciences. He co-wrote *Field Guide to Sharks of Malaysia and Neighbouring Countries* (2007).

García

Electric Ray sp. *Discopyge castelloi* Menni, Rincón & García, 2008

Dr Mirta Lidia García is an Argentine biologist who works at the Faculty of Natural Sciences and Museum of the Universidad Nacional de la Plata, which awarded her doctorate (1987). She co-wrote *Poromitra Crassiceps (Teleostei, Melamphaidae) Associated with the 500 Fathoms Fauna off Argentina* (2002).

Garman

Hagfish sp. *Myxine garmani* Jordan & Snyder, 1901
Natal Electric Ray *Heteronarce garmani* Regan, 1921
Brownspotted Catshark *Scyliorhinus garmani* Fowler, 1934
Freckled Skate *Leucoraja garmani* Whitley, 1939

River Stingrays *Potamotrygonidae* Garman, 1887
Ploughnose Chimaeras Family *Callorhinchidae* Garman, 1901
Long-nosed Chimaeras Family *Rhinochimaeridae* Garman, 1901
Wedgefishes *Rhynchobatidae* Garman, 1913

Freshwater Stingray genus *Potamotrygon* Garman, 1877
Guitarfish genus *Platyrhinoidis* Garman, 1881
Frilled Shark genus *Chlamydoselachus* Garman, 1884
Chimaera genus *Rhinochimaera* Garman, 1901
Catshark genus *Parmaturus* Garman, 1906
Eagle Ray genus *Aetomylaeus* Garman, 1908
Catshark genus *Apristurus* Garman, 1913
Catshark genus *Atelomycterus* Garman, 1913
Velvet Dogfish genus *Centroselachus* Garman, 1913
Catshark genus *Haploblepharus* Garman, 1913
Bull Ray genus *Pteromylaeus* Garman, 1913
Roundray genus *Urobatis* Garman, 1913
Panray genus *Zanobatus* Garman, 1913

Bignose Fanskate *Sympterygia acuta* Garman, 1877

Swellshark *Cephaloscyllium ventriosum* Garman, 1880
Brown Stingray *Dasyatis lata* Garman, 1880
Longtail Stingray *Dasyatis longa* Garman, 1880
Atlantic Guitarfish *Rhinobatos lentiginosus* Garman, 1880
Pacific Guitarfish *Rhinobatos planiceps* Garman, 1880
Pluto Skate *Fenestraja plutonia* Garman, 1881
Ackley's Ocellate Ray *Raja ackleyi* Garman, 1881 [Alt. Ocellate Ray]
Chain Catshark *Scyliorhinus rotifer* Garman, 1881
Frilled Shark *Chlamydoselachus anguineus* Garman, 1884
Smooth Skate *Malacoraja senta* Garman, 1885
Southern Eagle Ray *Myliobatis goodie* Garman, 1885
Broad Skate *Amblyraja badia* Garman, 1899
Combtooth Dogfish *Centroscyllium nigrum* Garman, 1899
Whiteface Hagfish *Myxine circifrons* Garman, 1899
Hagfish sp. *Notomyxine tridentiger* Garman, 1899
Atlantic Weasel Shark *Paragaleus pectoralis* Garman, 1906
Salamander Shark *Parmaturus pilosus* Garman, 1906
Needle Dogfish *Centrophorus acus* Garman, 1906
Mosaic Gulper Shark *Centrophorus tessellatus* Garman, 1906
Roughskin Dogfish *Centroscymnus owstoni* Garman, 1906
Rough Longnose Dogfish *Deania hystricosa* Garman, 1906
Dwarf Sawfish *Pristis clavata* Garman, 1906
Ratfish sp. *Hydrolagus barbouri* Garman, 1908
Arabian Catshark *Bythaelurus alcockii* Garman, 1913
Dwarf Gulper Shark *Centrophorus atromarginatus* Garman, 1913
Striped Smooth-Hound *Mustelus fasciatus* Garman, 1913
Freshwater Stingray sp *Potamotrygon humerosa* Garman, 1913
Raspy River Stingray *Potamotrygon scobina* Garman, 1913
Parnaiba River Stingray *Potamotrygon signata* Garman, 1913
Zipper Sand Skate *Psammobatis extent* Garman, 1913
Peruvian Eagle Ray *Myliobatis peruvianus* Garman, 1913
Spotted Roundray *Urobatis maculatus* Garman, 1913

Dr Samuel Trevor (Walton) Garman (1843–1927) was an American naturalist and zoologist most noted as an ichthyologist and herpetologist. He graduated in Illinois (1870), became a schoolteacher, and was Professor of Natural Science at a seminary in Illinois (1871–1872). He became Louis Agassiz's (q.v.) special student (1872) and (1873) became Assistant Director of the Herpetology and Ichthyology Section, Museum of Comparative Zoology, Harvard. He was in South America with Alexander Agassiz (1874) and surveyed Lake Titicaca. Harvard awarded him two honorary degrees, BS (1898) and AM (1899). He wrote a number of books as well as many papers including *The Plagiostomia (Sharks, Skates and Rays)* (1913). A bird, an amphibian and four reptiles are named after him. He was honoured in the skate name having originally described this species using a preoccupied fossil species name (*Raja ornata*).

Garrick

San Blas Skate *Dipturus garricki* Bigelow & Schroeder, 1958
Longnose Houndshark *Iago garricki* Fourmanoir & Rivaton, 1979
Azores Dogfish *Scymnodalatias garricki* Kukuev & Konovalenko, 1988
Northern River Shark *Glyphis garricki* Compagno, WT White & Last, 2008
Garrick's Catshark *Apristurus garricki* Sato, Stewart & Nakaya, 2013

White-nose Shark genus *Nasolamia* Compagno & Garrick, 1983
Squaliform Sharks genus *Scymnodalatias* Garrick, 1986

Slender Smooth-hound *Gollum attenuates* Garrick, 1954
Blackbelly Lanternshark *Etmopterus abernethyi* Garrick, 1957 NCR
[Junior Syn *Etmopterus lucifer*]
New Zealand Lantern Shark *Etmopterus baxteri* Garrick, 1957
Richardson's Ray *Bathyraja richardsoni* Garrick, 1961
South China Cookiecutter Shark *Isistius plutodus* Garrick & Springer, 1964
Smooth Deep-sea Skate *Brochiraja asperula* Garrick & Paul, 1974
Prickly Deep-sea Skate *Brochiraja spinifera* Garrick & Paul, 1974
New Zealand Smooth Skate *Dipturus innominatus* Garrick & Paul, 1974
Smooth Tooth Blacktip Shark *Carcharhinus leiodon* Garrick, 1985

Dr J. A. F. (Jack) Garrick (1928–1999) was a New Zealand ichthyologist at Victoria University College, Wellington. He specialised in elasmobranchs. He worked at Victoria University, Wellington (1950–1990) and became Professor of Zoology there (1971). He had a primary interest in the taxonomy of sharks and rays, but carried out the first exploratory deep-sea sampling using specially adapted cone nets, baited traps and longlines, regularly to depths greater than 2,000 m. Many new and rare species were obtained by the use of these innovative techniques. He was responsible for the notable discovery of the first New Zealand specimens of orange roughy (1957), which subsequently formed the basis of a multi-million dollar fishery. He collected some 721 specimens in 988 lots and deposited them at the Museum of New Zealand Te Papa Tongarewa. In the etymology of the northern river shark White says it was Garrick '…who discovered this species in the form of two newborn males from Papua New Guinea and supplied radiographs, morphometrics, drawings and other details of these specimens (since lost) to the senior author.'

Gaudiano

Sulu Gollumshark *Gollum suluensis* Last & Gaudiano, 2011

Joe Pres A. Gaudiano of WWF Philippines is an ichthyologist and head of the Marine Laboratory, Silliman University, Philippines.

Geijskes

Sharpsnout Stingray *Dasyatis geijskesi* Boeseman, 1948

Dr Dirk Cornelis Geijskes (1907–1985) was a biologist, ethnologist and entomologist. He graduated from University of Basel, Switzerland (1935). He worked in Suriname (1938–1965), firstly in agricultural research and (1954) as chief biologist and Director, Suriname Museum, Paramaribo. After returning to the Netherlands he worked at the Leiden Museum, becoming curator (1967). He wrote *Natuurwetenschappejik Onderzoek van Suriname: 1945–1965* (1967). He collected the holotype and supplied Boeseman with many fishes from Suriname. Many different taxa including fishes, insects and an amphibian are named after him.

Georgian

Antarctic Starry Skate *Amblyraja georgiana* Norman, 1938

The holotype was caught in the coastal waters of South Georgia and was named after that island.

Gerrard

Gerrard's Stingray *Himantura gerrardi* Gray, 1851
[Alt. Whitespotted Whipray]

Edward Gerrard (1810–1910) worked as an attendant in Gray's department at the British Museum (1841–1896). He was Gray's right-hand man, looking after the galleries and storerooms, and also helped Gray with shark and ray identification. He also preserved and registered bottled animals and compiled a catalogue of osteological specimens at the British Museum. Two reptiles are named after him.

Giddings

Galápagos Catshark *Bythaelurus giddingsi* McCosker, Long & CC Baldwin, 2012

Al Giddings is an underwater filmmaker and naturalist. He has also been active in the commercial film business as the cameraman for 'The Deep' (1977) and as co-producer of 'Titanic' (1997). Moreover, he is a friend of the senior author.

Gigas

Giant Skate *Dipturus gigas* Ishiyama, 1958

Gigas was a giant in Greek mythology, the child of Uranus and Gaea. The name is applied to taxa that are giants of their kind.

Gilbert, CH

Guitarfish genus *Zapteryx* Jordan & CH Gilbert, 1880

Diamond Stingray *Dasyatis dipterura* Jordan & CH Gilbert, 1880
Thornback Guitarfish *Platyrhinoidis triseriata* Jordan & CH Gilbert, 1880
Longnose Skate *Raja rhina* Jordan & CH Gilbert, 1880
Starry Skate *Raja stellulata* Jordan & CH Gilbert, 1880
Banded Guitarfish *Zapteryx exasperate* Jordan & CH Gilbert, 1880
California Ray *Raja inornata* Jordan & CH Gilbert, 1881
Sicklefin Smooth-Hound *Mustelus lunulatus* Jordan & CH Gilbert, 1882
Pacific Sharpnose Shark *Rhizoprionodon longurio* Jordan & CH Gilbert, 1882
Spiney-tail Roundray *Urotrygon aspidura* Jordan & CH Gilbert, 1882
Speckled Guitarfis *Rhinobatos glaucostigma* Jordan & CH Gilbert, 1883
Ocellate Electric Ray *Diplobatis ommata* Jordan & CH Gilbert, 1890
Brown Catshark *Apristurus brunneus* CH Gilbert, 1892
Roughtail Skate *Bathyraja trachura* CH Gilbert, 1892
Lollipop Catshark *Cephalurus cephalus* CH Gilbert, 1892
Filetail Catshark *Parmaturus xaniurus* CH Gilbert, 1892
Deep-sea Skate *Bathyraja abyssicola* CH Gilbert, 1896
Aleutian Skate *Bathyraja aleutica* CH Gilbert, 1896
Pacific Smalltail Shark *Carcharhinus cerdale* CH Gilbert, 1898
Whitenose Shark *Nasolamia velox* CH Gilbert, 1898
Rough Eagle Ray *Pteromylaeus asperrimus* CH Gilbert, 1898
Spongehead Catshark *Apristurus spongiceps* CH Gilbert, 1905
Hawaiian Lantern Shark *Etmopterus villosus* CH Gilbert, 1905
Purple Chimaera *Hydrolagus purpurescens* CH Gilbert, 1905

Dr Charles Henry Gilbert (1859–1928) was an ichthyologist and fishery biologist, whose main area of study was the Pacific salmon. He received his bachelor's degree (1879) from Butler University, Indiana, but moved to take his master's (1882) and doctorate (1883) at Indiana University, the first-ever doctorate awarded by that university. Baird asked David Starr Jordan (q.v.) to do a survey of the US west coast fisheries (1879), and Gilbert went as Jordan's assistant on an expedition from British Columbia to Southern California. This expedition lasted a year and was the start of a 50-year-long study of Pacific fishes by Gilbert and Jordan. Gilbert taught at Indiana University (1880–1884 and 1889), at the University of Cincinnati (1885–1888), and at the newly founded Stanford (1890–1925), retiring as Emeritus Professor. He served as naturalist-in-charge on cruises of the US Fish Commission's vessel 'Albatross' in Alaskan waters (1880s and 1890s), Hawaii (1902) and Japan (1906). The Gilbert Fisheries Society was established (1931) at the College of Fisheries, University of Washington, later becoming (1989) the Gilbert Ichthyology Society. Alone or jointly he described 117 new genera and 620 species of fish, writing 172 papers on them. Three reptiles are named after him.

Gilbert, CR

Carolina Hammerhead *Sphyrna gilberti* Quattro, Driggers, Grady, Ulrich & MA Roberts, 2013

Dr Carter Rowell Gilbert (b.1930) is an American zoologist and ichthyologist who first noticed the variation in vertebrae among what were thought to be scalloped hammerheads. Ohio State University awarded his bachelor's degree (1951) and his master's (1953), and the University of Michigan his doctorate (1960). He worked for Florida State Museum and the University of Florida, Gainesville (1961–1998), retiring as Curator Emeritus. Like many naturalists he is a keen philatelist, a hobby he took up at the age of six. He wrote *A Revision of the Hammerhead Sharks (Family Sphyrnidae)*, 1967. (See also **Carter**)

Gilchrist

Backwater Butterfly Ray *Gymnura natalensis* Gilchrist
  & Thompson, 1911
White-spotted Izak *Holohalaelurus punctatus*
  Gilchrist, 1914
Smalleye Catshark *Apristurus microps* Gilchrist, 1922
Izak Catshark *Holohalaelurus regain* Gilchrist, 1922
African Chimaera *Hydrolagus africanus* Gilchrist, 1922

Dr John Dow Fisher Gilchrist (1866–1926) was a
South African ichthyologist and naturalist. He
was the first marine biologist of the Cape Colony
Department of Agriculture (1895). He co-wrote *The
Freshwater Fishes of South Africa* (1913).

Gill

Basking Sharks *Cetorhinidae* Gill, 1862
Squaliform Sharks *Echinorhinidae* Gill, 1862
Nurse Sharks *Ginglymostomatidae* Gill, 1862
Bamboo Sharks *Hemiscylliidae* Gill, 1862
Numbfishes *Narcinidae* Gill, 1862
Collared Carpet Sharks *Parascylliidae* Gill, 1862
Zebra Sharks *Stegostomatidae* Gill, 1862
Rough Sharks *Oxynotidae* Gill, 1872
Ground Sharks *Pseudotriakidae* Gill, 1893
Wobbegong Sharks *Orectolobidae* Gill, 1896

Catshark genus *Cephaloscyllium* Gill, 1862
Weasel Shark genus *Chaenogaleus* Gill, 1862
Winghead Shark genus *Eusphyra* Gill, 1862
Catshark genus *Halaelurus* Gill, 1862
Requiem Shark genus *Isogomphodon* Gill, 1862
Broadfin Shark genus *Lamiopsis* Gill, 1862
Carpet Shark genus *Parascyllium* Gill, 1862
Round Stingray genus *Urotrygon* Gill, 1863
Pygmy Shark genus *Euprotomicrus* Gill, 1865
Dogfish Shark genus *Isistius* Gill, 1865

Brown Smooth-Hound *Mustelus henlei* Gill, 1863
Munda Roundray *Urotrygon munda* Gill, 1863
Grey Smooth-Hound *Mustelus californicus* Gill, 1864
Sharptooth Smooth-Hound *Mustelus dorsalis* Gill,
  1864
Bat Eagle Ray *Myliobatis californica* Gill, 1865
Sandpaper Skate *Bathyraja interrupta* Gill &
  Townsend, 1897
Blunt Skate *Rhinoraja obtuse* Gill & Townsend, 1897

Professor Theodore Nicholas Gill (1837–1914) was
an American ichthyologist, malacologist, mammolo-
gist and librarian, a zoologist at George Washington
University and associated with the Smithsonian
Institution for more than half a century. His father

wanted him to enter the Church, but this seemed an
unattractive calling to Gill, who decided to qualify as
a lawyer instead. He was fortunate enough to come
to the attention of Spencer Baird (c.1857) whom he
met in Washington while en route to the West Indies,
where he made an important collection, especially of
freshwater fishes from Trinidad. He went to Newfoundland (1859), these two trips being the only
extensive fieldwork Gill ever carried out. He was
put in charge of the Smithsonian's Library (1862),
which was transferred (1866) to the Library of Congress, where Gill served as Senior Assistant Librarian
(1866–1874). During his career he produced over 500
papers of which 388 were on ichthyology. He also
for a time edited the ornithological magazine *The
Osprey*. A marine mammal is named after him.

Girard

Spotted Spiny Dogfish *Squalus suckleyi* Girard, 1854
Inshore Hagfish *Eptatretus burgeri* Girard, 1855
Fourteen-gill Hagfish *Eptatretus polytrema* Girard, 1855
Horn Shark *Heterodontus francisci* Girard, 1855
Big Skate *Raja binoculata* Girard, 1855
Leopard Shark *Triakis semifasciata* Girard, 1855
Chestnut Lamprey *Ichthyomyzon castaneus* Girard,
  1858
Hagfish sp. *Myxine limosa* Girard, 1859

Dr Charles Frédéric Girard (1822–1895) was a French
herpetologist and ichthyologist who was Louis
Agassiz's pupil and assistant at Neuchâtel and
moved with Agassiz to the USA. He was in Cambridge, Massachusetts (1847–1850), and worked with
Baird (1850–1857), establishing the Smithsonian. He
became an American citizen (1854) and while continuing his work at the Smithsonian, studied medicine
and graduated MD from Georgetown College (1856).
He briefly visited Europe (1860). During the American Civil War he sided with the Confederacy and
supplied the Confederate army with medical and
surgical supplies. He left the USA (1864), returned
to France and practised medicine there, serving as
a physician during the siege of Paris (1870). Two
mammals and three reptiles are named after him.

Gledhill

Western Spotted Catshark *Asymbolus occiduus* Last,
  Gomon & Gledhill, 1999
Pale Spotted Catshark *Asymbolus pallidus* Last,
  Gomon & Gledhill, 1999
Orange Spotted Catshark *Asymbolus rubiginosus* Last,
  Gomon & Gledhill, 1999

Maugean Skate *Zearaja maugeana* Last & Gledhill, 2007
Pygmy Thornback Skate *Dipturus flindersi* Last &
Gledhill, 2008
Northern Sawtail Catshark *Figaro striatus* Gledhill, Last
& WT White, 2008
Western Round Skate *Irolita westraliensis* Last &
Gledhill, 2008
Arafura Skate *Okamejei arafurensis* Last & Gledhill, 2008
Thintail Skate *Okamejei leptoura* Last & Gledhill, 2008

Dr Daniel C. Gledhill is an Australian zoologist and ichthyologist. He is based in Hobart and works for CSIRO with their Fish Biogeography and Taxonomy team. He co-wrote *A Revision of the Australian Handfishes* (Lophiiformes: Brachionichthyidae), *with Descriptions of Three New Genera and Nine New Species* (2009).

Gmelin

Slender Bamboo Shark *Chiloscyllium indicum* Gmelin,
1789
Reticulate Whipray *Himantura uarnak* Gmelin, 1789
Onefin Electric Ray *Narke capensis* Gmelin, 1789
Striped Catshark *Poroderma africanum* Gmelin, 1789

Professor Johann Friedrich Gmelin (1748–1804) was a German naturalist with particular interests in botany, entomology, malacology and herpetology. He graduated MD (1768) from the University of Tübingen, becoming an adjunct professor of medicine there (1769). He became adjunct (1773) then full (1778) professor of medicine, Georg-August-Universität, Göttingen and also a professor of chemistry, botany and mineralogy. Other members of his family were with Pallas on one of his expeditions. He was editor of the thirteenth edition of Linnaeus' *Systema Naturae* (1788–1796). Three mammals, three birds and a reptile are named after him.

Gneri

Hagfish genus *Notomyxine* Nani & Gneri, 1951

Professor Francisco S. Gneri was a botanist, zoologist and ichthyologist at the Museo Argentino de Ciencias Naturales, Buenos Aires. He took part in the Argentine Antarctic Expedition (1942).

Goldsborough

Black Hagfish *Eptatretus deani* Evermann &
Goldsborough, 1907

Edmund Lee Goldsborough (1868–1953) was an American ichthyologist at the US Bureau of Fisher-ies. He wrote or co-wrote a number of papers, often descriptions of new species with Evermann or W. C. Kendall such as *Fishes of West Virginia* (1908) and *The Fishes of Alaska* (1907).

Gollum

Smoothhound Shark genus *Gollum* Compagno, 1973

This genus is named after the fictional character, Gollum, in J. R. R. Tolkien's *Lord of the Rings*. An amphibian is also named Gollum after the same character.

Gomes

Gomes' Roundray *Heliotrygon gomesi* Carvalho &
Lovejoy, 2011
[Alt. Lucifer Shark]

Brazilean Large-eyed Stingray *Dasyatis marianae*
Gomes, Rosa & Gadig, 2000
South Brazilian Skate *Dipturus mennii* Gomes &
Paragó, 2001
Porcupine Ray *Dasyatis colarensis* Santos, Gomes &
Charvet-Almeida, 2004
Brazilian Soft Skate *Malacoraja obscura* Carvalho,
Gomes & Gadig, 2005

Dr Ulisses Leite Gomes (b.1955) is a Brazilian ichthyologist. The Universidade Santa Úrsula awarded his bachelor's degree (1979) and the Universidade Federal do Rio de Janeiro his master's (1989) and doctorate (2002). He is an associate professor at the Universidade do Estado do Rio de Janeiro, where he has been working on elasmobranch taxonomy since 1988. He was honoured in the name of the roundray as a 'pioneer in the study of elasmobranch morphology and systematics in Brazil, and an esteemed colleague and collaborator of the first author'.

Gomon

Hagfish sp. *Eptatretus gomoni* Mincarone & Fernholm,
2010

Striped Stingaree *Trygonoptera ovalis* Last & Gomon,
1987
Masked Stingaree *Trygonoptera personata* Last &
Gomon, 1987
Patchwork Stingaree *Urolophus flavomosaicus* Last &
Gomon, 1987
Mitotic Stingaree *Urolophus mitosis* Last & Gomon, 1987
Coastal Stingaree *Urolophus orarius* Last & Gomon, 1987
Brown Stingaree *Urolophus westraliensis* Last &
Gomon, 1987

Western Spotted Catshark *Asymbolus occiduus* Last,
  Gomon & Gledhill, 1999
Pale Spotted Catshark *Asymbolus pallidus* Last,
  Gomon & Gledhill, 1999
Orange Spotted Catshark *Asymbolus rubiginosus* Last,
  Gomon & Gledhill, 1999
Eastern Shovelnose Stingaree *Trygonoptera imitate*
  Last & Gomon, 2008

Dr Martin Fellows Gomon (b.1945) is a US/Australian ichthyologist who first went to Australia to the Museum Victoria, Melbourne (1979) where he is Senior Curator, Ichthyology. Florida State University, Tallahassee awarded his bachelor's degree (1967) and the University of Miami his master's (1971) and doctorate (1979). He co-wrote two field guides *Fishes of Australia's South Coast* (1994) and *Fishes of Australia's Southern Coast* (2008). He has been instrumental in obtaining funding for the website 'Fishes of Australia', designed to give comprehensive information on Australia's 4,000+ species of fish. He was honoured in the hagfish name for his 'distinguished contributions to ichthyology'.

Gonzáles-Acosta

White-margin Fin Houndshark *Mustelus albipinnis*
  Castro-Aguirre, Antuna-Mendiola, González-
  Acosta & De La Cruz-Agüero, 2005

Dr Adrián Felipe González-Acosta (b.1967) is a Mexican ichthyologist at the Marine Sciences Research Centre, La Paz, Baja California, Mexico. He co-wrote *Occurrence of Hydrolagus Macrophtalmus (Chondrichthyes: Holocephali: Chimaeridae) in the Northeastern Pacific* (2010).

Goode

Southern Eagle Ray *Myliobatis goodei* Garman, 1885

Chimaera Genus *Harriotta* Goode & Bean, 1895

Pacific Longnose Chimaera *Harriotta raleighana*
  Goode & Bean, 1895
Deep-water Catshark *Apristurus profundorum* Goode
  & TH Bean, 1896
Boa Catshark *Scyliorhinus boa* Goode & TH Bean, 1896

Dr George Brown Goode (1851–1896) was an ichthyologist heavily immersed in administration at the Smithsonian, where he became Assistant Secretary. He graduated from the Wesleyan University, Middletown, Connecticut (1870) and then studied under Agassiz at Harvard (1870–1871). He returned to Wesleyan as Curator of their new natural history museum (1872), dividing his time between it and the Smithsonian in winters and working in the field with the US Fish Commission. He joined the Smithsonian full-time (1877) and became Assistant Secretary (1887). Despite having little time for research, he produced more than 100 papers, such as *Catalog of the Fishes of the Bermudas* (1876) and *Oceanic Ichthyology, A Treatise on the Deep-Sea and Pelagic Fishes of the World, Based Chiefly upon the Collections Made by the Steamers Blake, Albatross, and Fish Hawk in the Northwestern Atlantic* with Tarleton Hoffman Bean. His work on Bermuda fishes later formed the basis for his PhD, awarded by Indiana University, Bloomington (1886). He also produced the seven-volume *The Fisheries and Fishery Industries of the United States* (1884–1887). He died of pneumonia. At least five other taxa are named after him.

Goodrich

Dogfish Sharks *Squaliformes* Goodrich, 1909

Edwin Stephen Goodrich (1868–1946) was an English zoologist whose specialisms were comparative anatomy, palaeontology, evolution and embryology. He studied at the Slade School of Art but became interested in zoology and attended Merton College, Oxford, graduating in 1895. He married (1913) a fellow zoologist and they collaborated thereafter. His zoological drawings were so artistic that students would photograph the blackboard before they were erased and he exhibited his watercolours in London. He was Assistant to the Chair of comparative anatomy (1892) until becoming Professor of Zoology at Oxford (1921–1946) and Editor of the *Journal of Microscopical Science* (1920–1946). He travelled extensively to Europe, the US, North Africa, India, Sri Lanka, Malaya and Java. He wrote widely on zoological anatomy and among his books was *Living Organisms: An Account of their Origin and Evolution* (1924).

Gordo

Ghost Shark sp. *Hydrolagus lusitanicus* Moura, Figueiredo,
  Bordalo-Machado, Almeida & Gordo, 2005

Dr Leonel Paulo Sul de Serrano Gordo (b.1957) is a Portuguese marine biologist who is an assistant professor at the Department of Animal Biology of the Faculty of Sciences, University of Lisbon. His bachelor's degree was awarded (1980) and his doctorate (1992). He started his research activity into pelagic species when he joined the Portuguese Institute for

Fisheries Research (1980). Among his published work is the co-written *Embryonic Development and Maternal-embryo Relationships of the Portuguese Dogfish Centroscymnus Coelolepis* (2011).

Goto

Ginger Carpetshark *Parascyllium sparsimaculatum* Goto & Last, 2002

Dr Tomoaki Goto (b.1967) is an ichthyologist who was at the Graduate School of Fisheries Sciences, Hokkaido University, Japan where he was awarded his PhD (1997). He has also worked at the Iwate Fisheries Technology Centre. He has written many papers including *Revision of the Wobbegong Genus Orectolobus from Japan, with a Redescription of Orectolobus japonicus (Elasmobranchii: Orectolobiformes)* (2008).

Grady

Carolina Hammerhead *Sphyrna gilberti* Quattro, Driggers, Grady, Ulrich & MA Roberts, 2013

Dr James M. Grady is Professor of Biological Sciences, University of New Orleans. He is interested in evolutionary genetics, systematics and ecology of both freshwater and marine fishes. Southern University, Illinois awarded his doctorate. He co-wrote *Using Character Concordance to Define Taxonomic and Conservation Units* (2001).

Graham, A

Grahams' Skate *Dipturus grahami* Last, 2008

Alastair Graham (b.1964) is an Australian marine biologist whose bachelor's degree was awarded by Macquarie University. He worked in the Fish Department of the Australian Museum, Sydney (1988–1990) and since 1990 has been the Fish Collection Manager at the Australian National Fish Collection, CSIRO Marine and Atmospheric Research, Hobart, Tasmania. He is jointly honoured in the name of the skate with Ken Graham (below) for their 'very important, but very different contributions to the knowledge of Australian sharks and rays'. He is also honoured with the binomial *alastairi* for a species of hagfish (*Eptatretus*).

Graham, KJ

Eastern Longnose Spurdog *Squalus grahami* Last, WT White & Stevens, 2007
Grahams' Skate *Dipturus grahami* Last, 2008

Kenneth 'Ken' John Graham (b.1947), originally from New Zealand, spent his career as an ichthyologist, biologist and fisheries research scientist in Australia. Canterbury University, Christchurch awarded his bachelor's degree (1968). He worked for New South Wales State Fisheries, Sydney (1972–2008). Since 2005 he has been a research associate of the Australian Museum in Sydney. During his career as a sea-going scientist he collected tens of thousands of specimens of fishes and invertebrates, most of which now reside in the Australian Museum's mollusc, fish and marine invertebrate collections, forming the most complete record of the distribution of trawl-caught animals on the continental shelf and slope off southeast Australia. Much of his research concentrated on deep-water commercial fishery trawl surveys and by-catch assessments on board the research vessel 'Kapala'. A number of other fish and marine species are named after him, some with the binomial *kengrahami* including a new genus of shellfish named after the vessel, *Kapala kengrahami*. He was honoured in the spurdog's name as he had 'collected the type and has contributed greatly to the knowledge of southeast Australian elasmobranchs.' He is jointly honoured in the name of the skate with Alastair Graham (above) for their 'very important, but very different' contributions to the knowledge of Australian sharks and rays. He wrote *Distribution, Population Structure and Biological Aspects of* Squalus spp.(Chondrichthyes: Squaliformes) *from New South Wales and Adjacent Australian Waters* (2005) and was senior author of *Changes in Relative Abundance of Sharks and Rays on Australian South East Fishery Trawl Grounds after Twenty Years of Fishing* (2001).

Gray

Kitefin Sharks *Dalatiidae* Gray, 1851
Bullhead Sharks *Heterodontidae* Gray, 1851
Cow Sharks *Hexanchidae* Gray, 1851
Barbeled Houndsharks *Leptochariidae* Gray, 1851
Houndsharks *Triakidae* Gray, 1851

Sleeper Ray genus *Temera* Gray, 1831

Zebra Bullhead Shark *Heterodontus zebra* Gray, 1831
Finless Sleeper Ray *Temera hardwickii* Gray, 1831
Mottled Eagle Ray *Aetomylaeus maculatus* Gray, 1834
Pouched Lamprey *Geotria australis* Gray, 1851
Whitespotted Whipray *Himantura gerrardi* Gray, 1851
Chilean Lamprey *Mordacia lapicida* Gray, 1851

John Edward Gray (1800–1875) was a zoologist and entomologist. He started work at the British

Museum (1824) with a temporary appointment at 15 shillings (£0.75) per day, but became Keeper of Zoology. Gray was regarded as the leading authority on many reptiles. He was also an ardent philatelist and claimed to be the world's first stamp collector: he wrote *A Hand Catalogue of Postage Stamps for the use of the Collector* (1862). He worked at the museum with his brother George Robert Gray (1808–1872). J. E. Gray suffered a severe stroke that paralyzed his right side, including his writing hand (1869). Nevertheless, he continued to publish until the end of his life by dictating to his wife, Maria Emma, who had always worked with him as an artist and occasional co-author and after whom he named a lizard. He wrote or co-wrote over 500 scientific papers, including many describing new species such as *Description of Twelve New Genera of Fish, Discovered by Gen. Hardwicke, in India, the Greater Part in the British Museum* (1831). They have a large number of taxa named after them.

Greeley

Mountain Brook Lamprey *Ichthyomyzon greeleyi* CL Hubbs & Trautman, 1937

Dr John R. Greeley was Assistant Director of the Institute for Fisheries Research. Cornell awarded his PhD. Hubbs was a colleague of his at the University of Michigan. He collected the holotype. He co-wrote *Fishes of the Western North Atlantic* (1963).

Griffin

Griffin's Spiny Dogfish *Squalus griffini* Phillipps, 1931
[Alt. Northern Spiny Dogfish, Brown Dogfish, Greeneye, Grey Spiny Dogfish Syn. *Flakeus griffini*]

Phillipps gives no etymology but the most likely person is Louis T. Griffin (d.1935), a British-born New Zealand ichthyologist and collector. He arrived in New Zealand (1908) and joined the staff of the Auckland Museum (1908) as a preparator and was Assistant Director at the time of his death. There is a possibility it was Lawrence Edmonds Griffin (1874–1949), a US zoologist who was Custodian of Herpetology at the Carnegie Museum of Natural History (1915–1920). He wrote *A Guide for the Dissection of the Dogfish, Squalus Acanthias* (1922).

Griffith

Bancroft's Numbfish *Narcine bancroftii* Griffith & CH Smith, 1834
Scalloped Hammerhead *Sphyrna lewini* Griffith & CH Smith, 1834

[Alt. Brazilian/Lesser Electric Ray, Syn. *Narcine umbrosa*]

Edward Griffith (1790–1858) was a zoologist and one of the original members of the Zoological Society of London. His day job was as a solicitor and a Master in the Court of Common Pleas. He wrote *General and Particular Descriptions of the Vertebrated Animals* (1821) and translated (and added to) Cuvier's *Règne Animal*.

Grouser

Hagfish sp. *Eptatretus grouseri* CB McMillan, 1999

(See **McMillan, D**)

Gubanov

Arabian Carpetshark *Chiloscyllium arabicum* Gubanov, 1980

Dr Evhen Pavlovich Gubanov is an Ukrainian zoologist at the Southern Research Institute of Marine Fisheries and Oceanography, Kiev, Ukraine. He wrote *On the Biology of the Thresher Shark [Alopias vulpinus (Bonnaterre)] in the Northwest Indian Ocean* (1972).

Gudger

Greenback Skate *Dipturus gudgeri* Whitley, 1940

Dr Eugene Willis Gudger (1866–1956) was an associate in ichthyology at the AMNH, New York, which he joined (1919) and later became Curator of Fishes. His doctorate was awarded by Johns Hopkins University, Baltimore (1905). He was a world authority on whale sharks. He was honoured 'in appreciation of his work on fishes and their bibliography'.

Guggenheim

Angular Angelshark *Squatina guggenheim* Marini, 1936

Tomas Leandro Marini was appointed to the John Simon Guggenheim Memorial Foundation for studies in the fields of marine biology, oceanography and pisciculture in the United States for eight months (July 1932).

Guichenot

Redspotted Catshark *Schroederichthys chilensis* Guichenot, 1848
Yellownose Skate *Zearaja chilensis* Guichenot, 1848

Antoine Alphone Guichenot (1809–1876) was a

French zoologist, primarily in herpetology and ichthyology. He taught and researched for the Muséum National d'Histoire Naturelle, Paris and undertook collecting trips for them, most extensively in Algeria. He scaled back to just being an assistant naturalist (1856). Three reptiles are named after him.

Guitart-Manday

Longfin Mako *Isurus paucus* Guitart-Manday, 1966

Dr Darío José Guitart-Manday (1923–2000) was a Cuban ichthyologist and footballer. He was famous for his football prowess at the Secondary School of Havana but still got his bachelor's degree there (1942). The University of Havana awarded his doctorate (1951). He studied in Russia and was the first Cuban to gain a doctorate in biology in the USSR. He retired from teaching (1990).

Gunnerus

Basking Shark *Cetorhinus maximus* Gunnerus, 1765

Johan Ernst Gunnerus (1718–1773) was a Norwegian amateur botanist. He was Bishop of Nidaros (1758–1773) as well as being a professor of theology at the University of Copenhagen. Very interested in natural history, he accumulated a large collection of specimens from visits to central and northern Norway. He also encouraged others to send him specimens. He was co-founder of the Trondheim Society (1760), which became the Royal Norwegian Society of Science (1767) and was Vice-President and Director Perpetuus of the Society (1767–1773). He published a description (1765) of a basking shark and gave it its scientific name. He wrote widely on natural history, most notably *Flora Norvegica* (1766–1776).

Günther

Ray genus *Psammobatis* Günther, 1870

Short-tail Nurse Shark *Pseudoginglymostoma brevicaudatum* Günther, 1867
Whitesnout Guitarfish *Rhinobatos leucorhynchus* Günther, 1867

Copper Shark *Carcharhinus brachyurus* Günther, 1870
Daisy Stingray *Dasyatis margarita* Günther, 1870
Smalltooth Stingray *Dasyatis rudis* Günther, 1870
Crested Bullhead Shark *Heterodontus galeatus* Günther, 1870
Lamprey sp. *Lampetra ayresii* Günther, 1870
Gummy Shark *Mustelus antarcticus* Günther, 1870
Patagonian Hagfish *Myxine affinis* Günther, 1870
Japanese Sawshark *Pristiophorus japonicus* Günther, 1870
Shortnose Sawshark *Pristiophorus nudipinnis* Günther, 1870
Smallthorn Sand Skate *Psammobatis rudis* Günther, 1870
Spiny Guitarfish *Rhinobatos spinosus* Günther, 1870
Chilean Roundray *Urotrygon chilensis* Günther, 1872
Eaton's Skate *Bathyraja eatonii* Günther, 1876
Raspback Skate *Bathyraja isotrachys* Günther, 1877
Velvet Dogfish *Zameus squamulosus* Günther, 1877
Dusky Catshark *Bythaelurus canescens* Günther, 1878
La Plata Skate *Atlantoraja platana* Günther, 1880
Murray's Skate *Bathyraja murrayi* Günther, 1880
Southern Lantern Shark *Etmopterus granulosus* Günther, 1880
Peacock Skate *Pavoraja nitida* Günther, 1880
Short-tailed River Stingray *Potamotrygon brachyuran* Günther, 1880
Kai Stingaree *Urolophus kaianus* Günther, 1880
Granular Dogfish *Centroscyllium granulatum* Günther, 1887

Dr Albert Carl Ludwig Gotthilf Günther (1830–1914) was one of the giants of zoology. He was educated as a physician in Germany, then joined the British Museum (1856). He was appointed Keeper, Zoological Department (1857). He became a naturalised British subject (1862) and changed his second and third Christian names to Charles Lewis. He became President of the Biological Section, British Association for the Advancement of Science (1880), and was President of the Linnean Society (1881–1901). He wrote a vast number of papers and books among which is the eight-volume *Catalogue of the Fishes in the British Museum* (1859–1870). Three mammals, two birds, 26 amphibians and 67 reptiles are named after him.

H

Haacke

Western Shovelnose Ray *Aptychotrema vincentiana* Haacke, 1885

Dr Johann Wilhelm Haacke (1855–1912) was a German zoologist. The University of Jena awarded his doctorate (1878), after which he was an assistant there and at Kiel University. He moved to New Zealand (1881) and went from there to Australia where he was Director of the Natural History Museum in Adelaide (1882–1884). He returned to Europe and was Director of Frankfurt Zoo (1888–1893) and then taught at the Darmstadt University of Technology.

Haas

Taiwan Torpedo *Torpedo formosa* Haas & Ebert, 2006

Diane Lee Haas (b.1977) graduated with a master's degree from Moss Landing Marine Laboratories, California (2011). She currently works as an environmental scientist for the California Department for Fish and Wildlife. She co-wrote *First Record of Hermaphroditism in the Bering Skate*, Bathyraja interrupta (2008) and wrote *Age, Growth, and Reproduction of the Aleutian Skate*, Bathyraja aleutica *(Gilbert, 1896), from Alaskan Waters* (2011). She is a creative cook, well known for her fish cupcakes! She tells us that decoration of her cakes have included roughtail skate *Bathyraja trachura*, red abalone *Haliotis rufescens* and chinook salmon *Oncorhynchus tshawytscha*.

Habarer

Graceful Catshark *Proscyllium habereri* Hilgendorf, 1904

Dr Karl Albert Haberer (1864–1941) was a German naturalist, student of politics, and collector in Japan and China. The Boxer Rebellion (1899) forced him to stay in Peking (Beijing) for longer than he planned, but he bought some 'Dragon Bones' (fossilised bones used in traditional Chinese medicine) in a market. Among them was a human-like molar that eventually led to the discovery of 'Peking Man' (*Homo erectus pekinensis*). He wrote *Schädel und Skeletteile aus Peking* (1902). He collected the type of the catshark. A bird is named after him.

Haeckel

Freckled Catshark *Scyliorhinus haeckelii* Miranda-Ribeiro, 1907
Smallspine Spookfish *Harriotta haeckeli* Karrer, 1972

Dr Ernst Haeckel (1834–1919) was an evolutionary biologist, zoologist, philosopher and artist. He qualified as a physician in Berlin (1857). He studied zoology at Friedrich-Schiller-Universität Jena (1859–1862) and was Professor of Comparative Anatomy (1862–1909). He travelled extensively in the Canary Islands (1866–1867) and in Dalmatia, Egypt, Turkey and Greece (1869–1873). He met Thomas Huxley and Charles Darwin, whose theories he embraced and promoted. He wrote *The Riddle of the Universe* (1901). An asteroid, 12323 Häckel, is named after him, as are two mountains, one in the USA and the other in New Zealand, and a reptile. A research vessel has been named after him and some fishes are named after the vessel or both it and the man.

Hagiwara

Izu Catshark *Scyliorhinus tokubee* Shirai, Hagiwara & Nakaya, 1992

Soichi Hagiwara is a zookeeper and ichthyologist at the Shimoda Floating Aquarium, Japan. He co-wrote *Scyliorhinus Tokubee sp. nov. from Izu Peninsula, Southern Japan (Scyliorhinidae, Elasmobranchii)* (1992).

Hale

Hale's Wobbegong *Orectolobus halei* Whitley, 1940
[Alt. Gulf Wobbegong, Banded Carpet Shark]

Herbert Mathew Hale (1895–1963) worked at the South Australian Museum, Adelaide (1914–1960). He became the Director's Assistant (1917), Assistant in zoology (1922), Zoologist (1925) and Director (1928).

Haller

Haller's Roundray *Urobatis halleri* Cooper, 1863

George Morris Haller (1851–1889) was the young son of an American army officer, Granville Owen Haller. Haller Junior was injured in the foot while wading along the shore in San Diego Bay (1862). He was treated by Dr Cooper who suspected that the boy had been hit by a stingray and set out to examine the local round stingrays. Rightly believing it to be new to science he described the ray and named it after his patient. George Haller fought at Gettysburg (alongside his father) at the age of 12 (1863). In later life he became a lawyer. He was drowned when the canoe in which he was travelling capsized.

Hallstrom

Papuan Epaulette Carpetshark *Hemiscyllium hallstromi* Whitley, 1967

Sir Edward John Lees Hallstrom (1886–1970) was born in Coonamble, New South Wales, Australia. He was a pioneer of refrigeration, a philanthropist and leading aviculturist. He began work in a furniture factory aged 13, but later opened his own factory to make ice-chests and then wooden cabinets for refrigerators. He eventually designed and manufactured the first popular domestic Australian refrigerator. Hallstrom made generous donations to medical research, children's hospitals and the Taronga Zoo in Sydney, becoming an honorary life director there. He commissioned (1940) the artist Cayley to paint all the Australian parrots. Twenty-nine large watercolours were produced and presented to the Royal Zoological Society of New South Wales. He visited New Guinea (1950). There is a research collection of 1,600 rare books on Asia and the Pacific at the University of New South Wales Library known as the Hallstrom Pacific Collection, purchased with funds Hallstrom gave to the Commonwealth government (1948) for the purpose of establishing a library of Pacific affairs and colonial administration. When he was Trustee and Chairman of Taronga Zoological Park the shark holotype and paratype were kept there alive in captivity. A mammal and three birds are named after him.

Hamilton, A

Blind Electric Ray *Typhlonarke aysoni* A Hamilton, 1902

Augustus Hamilton (1853–1913) was a New Zealand ethnologist and biologist. He became a museum director.

Hamilton, F

Ganges Stingray *Himantura fluviatilis* F Hamilton, 1822

Dr Francis Hamilton-Buchanan (1762–1829) was an ichthyologist and physician who qualified at Glasgow (1783) and was to have been a ship's surgeon, but his health was bad and remained poor until 1794, when he became well enough to work. He joined the East India Company's Bengal service as an assistant surgeon. He collected botanical specimens as he travelled. His botanical drawings were so admired that a number were presented to Joseph Banks, to whom he regularly sent specimens. He studied the fishes of the Ganges and was often employed on survey work on all sorts of subjects, including fisheries. He became Superintendent of the Calcutta Botanical Gardens (1814). His family name at birth was Buchanan, but he dropped it and took the name Hamilton, his mother's maiden name (1815). He signed his name as 'Francis Hamilton' or 'Francis Hamilton (formerly Buchanan).' He wrote *Account of the Fishes of the Ganges* (1822). A bird and a reptile are named after him.

Hamlyn-Harris

Purple Eagle Ray *Myliobatis hamlyni* Ogilby, 1911

Dr Ronald Hamlyn-Harris (1874–1953) was an English-born entomologist. He studied in Naples, Italy (1901) and was awarded a doctorate by the Eberhard Karl University, Tübingen, Germany (1902). He went to Australia (1903) and became a schoolmaster at Toowoomba Grammar School, Queensland. He revitalised science teaching, raised funds for a new laboratory and gave popular lectures. He was Director of the Queensland Museum, Brisbane (1910–1917), which came as great relief to Ogilby who had been trying to combine being an administrator with conducting research. Ogilby and Hamlyn-Harris became life-long friends and would go fishing together. Hamlyn-Harris then ran a fruit farm (1917–1922). He was put in charge of charge of the Australian Hookworm Campaign's central laboratory in Brisbane (1922–1924). He was Brisbane's city entomologist (1926–1934) and later a full-time lecturer at the University of Queensland in Brisbane (1936–1942).

Hanner

Argentine Skate *Dipturus argentinensis* Díaz de Astarloa, Mabragaña, Hanner & Figueroa, 2008

Dr Robert H. Hanner became the Associate Director for the Canadian Barcode of Life Network at the Biodiversity Institute of Ontario, University of Guelph (2005), where he is an associate professor. Eastern Michigan University awarded his bachelor's degree and the University of Oregon his doctorate. Earlier in his career he was a curatorial associate at the American Museum of Natural History, New York. He co-wrote *Molecular Approach to the Identification of Fish in the South China Sea* (2012).

Hardwicke

Finless Sleeper Ray *Temera hardwickii* Gray, 1831

Major-General Thomas Hardwicke (1756–1835) served in the Bengal army of the East India Company. He was an amateur naturalist and collector who was the first to make the red panda *Ailurus fulgens* widely known, through a paper that he wrote (1821), *Description of a New Genus . . . from the Himalaya Chain of Hills between Nepaul and the Snowy Mountains*. Cuvier stole a march on Hardwicke in formally naming the red panda because Hardwicke's return to England was delayed. He collected reptiles in India and published on them (1827) with Gray. He also collected the type of the ray. Five reptiles, a mammal and a bird are named after him.

Hardy

McMillan's Catshark *Parmaturus macmillani* Hardy, 1985

Ratfish sp. *Hydrolagus pallidus* Hardy & Stehmann, 1990

Dr Graham S. Hardy (b.1947) is a herpetologist who worked at the National Museum of New Zealand Te Papa Tongarewa, Wellington, where he was Curator of reptiles and fishes (1976–1989). He completed his doctorate and lectured in vertebrate zoology at Victoria University, Wellington (1975). He wrote *Fish Types in the National Museum of New Zealand* (1990). A reptile is named after him. One informant advises us that he last heard that Hardy was living in a fishing village in Japan.

Harriott

Chimaera Genus *Harriotta* Goode & Bean, 1895

Thomas Harriott (c.1560–1621) was an English astronomer, mathematician, ethnographer and translator, who published the first English work on American natural history (1588). He is sometimes credited with the introduction into Britain of the potato. When in Carolina he learned the Algonquian language. Sir Walter Raleigh, whom Bean & Goode also honoured in the name of a fish, hired him as a mathematics tutor.

Harrisson

Dumb Gulper Shark *Centrophorus harrissoni* McCulloch, 1915

Charles Turnbull Harrisson (1869–1914) was a zoologist and biologist from Tasmania who took part in the Mawson Antarctic Expedition (1911–1914) and was responsible for collecting many of the most interesting fish species. After returning from the Antarctic he became Government Biologist to the Fisheries Bureau and was on board the research vessel 'Endeavour' when it sank near Macquarie Island. The original description was published in a volume that starts with the following words: '*In memoriam* H. C. Dannevig, director; G. W. C. Pim, master; C. T. Harrisson, biologist; and eighteen others, comprising the crew of the F. I. S. 'Endeavour' who were lost at sea in December, 1914.' No distress signal was sent and no trace ever found.

Hasselt

Hasselt's Bamboo Shark *Chiloscyllium hasseltii* Bleeker, 1852

Dr Johan Coenraad van Hasselt (1797–1823) was a Dutch biologist. He graduated in medicine at the University of Groningen (1820). However, he was more interested in natural history, like his fellow student and close friend Heinrich Kuhl. They undertook collecting trips together in Europe and visited a number of natural history museums, where they met famous zoologists of the time. They subsequently published several papers. They were sent to Java to study its natural history (1820). They started work en route studying pelagic fauna, as well as that of Madeira, the Cape of Good Hope and the Cocos Islands before arriving in Java. When Kuhl died after less than a year on the island, van Hasselt continued collecting, until he himself died two years later. A mammal, an amphibian and a bird are named after him.

Heald

Heald's Skate *Dipturus healdi* Last, WT White & Pogonoski, 2008
[Alt. Leyland's Skate)

David I. Heald is a zoologist, ichthyologist and botanist who has had a most varied career. The University of Western Australia awarded his bachelor's degree (1969). He worked as a diving marine scientist at Watermans Marine Research Laboratory, Department of Fisheries and Wildlife, Western Australia (1970–1987). He made a complete career change and worked for a variety of groups as a financial planner and investment consultant (1988–2008). Since then he has worked at a call centre, as a gas meter reader, offering garden services and since 2011 as a team member for a petrol service station. He wrote *The Commercial Shark Fishery in Temperate Waters of*

Western Australia (1987). It was he who discovered the species off Western Australia (early 1980s).

Heckel

Apron Ray genus *Discopyge* Heckel, 1846

Apron Ray *Discopyge tschudii* Heckel, 1846

Johann Jakob Heckel (1790–1857) was an Austrian zoologist and taxidermist with a particular interest in ichthyology. He became Director of the Fish Collection at the Vienna Natural History Museum. He wrote more than 60 works including *The Freshwater Fishes of the Austrian Danubian Monarchy.*

Hector

New Zealand Eagle Ray *Myliobatis tenuicaudatus* Hector, 1877

Dr Sir James Hector (1834–1907) was a Scottish-born Canadian geologist who took his medical degree at Edinburgh and, as both geologist and surgeon, was part of the Palliser expedition to western North America (1857–1860). He discovered and named many landmarks in the Rockies, including Kicking Horse Pass, the route later taken by the Canadian Pacific Railway. He returned to Scotland via the Pacific Coast, the California goldfields and Mexico. He became Director of the Geological Survey of New Zealand (1865) and eventually Curator of the Colonial Museum in Wellington (Museum of New Zealand Te Papa Tongarewa). He wrote *Outlines of New Zealand Geology* (1886). Two mammals and a bird are named after him.

Heemstra

East African Skate *Okamejei heemstrai* McEachran & Fechhelm, 1982

Stingray family *Hexatrygonidae* Heemstra & MM Smith, 1980

Stingray genus *Hexatrygon* Heemstra & MM Smith, 1980

Hoary Catshark *Apristurus canutus* S Springer & Heemstra, 1979

Smallfin Catshark *Apristurus parvipinnis* S Springer & Heemstra, 1979

Sixgill Stingray *Hexatrygon bickelli* Heemstra & MM Smith, 1980

Dragon Stingray *Himantura draco* Compagno & Heemstra, 1984

Dwarf Smooth-Hound *Mustelus minicanis* Heemstra, 1997

Gulf Smooth-Hound *Mustelus sinusmexicanus* Heemstra, 1997

Ornate Sleeper Ray *Electrolux addisoni* Compagno & Heemstra, 2007

Dr Phillip Clarence Heemstra (b.1941) is an American marine ichthyologist and scientific diver who is currently Curator Emeritus with the South African Institute for Aquatic Biodiversity at Grahamstown. His doctorate was awarded by the University of Miami (1973). He spent time on board the 'Fridtjof Nansen' research vessel in the Indian Ocean. With his wife, Elaine, he has carried out a number of underwater fish surveys. They co-wrote *Coastal Fishes of Southern Africa* (2004). He is a member of the editorial board of *Copeia*. He is reported to be one of the few people on the planet to have tasted coelacanth though we don't know if he had chips with it. He was honoured in the skate name for having made available to the authors specimens of the new species and for being 'extremely cooperative' in generally supplying them with South African elasmobranch material.

Heller

Galapagos Shark *Carcharhinus galapagensis* Snodgrass & Heller, 1905

Edmund Heller (1875–1939) was an American zoologist. Stanford University awarded his BA in zoology (1901); during his time there he collected in the Mojave and Colorado Desserts. He led several expeditions to Africa (1909–1912), which he wrote about in collaboration with Theodore Roosevelt as *Life-histories of African Game Animals* (1914) and *China* (1916). He took part in a number of other expeditions including to British Columbia (1914) and Peru (1915). He became Curator of Mammals at the Field Museum, Chicago (1926–1928) during which time he took part in other collecting trips. Then he was Director of Washington Park Zoo (1928–1935) and the Fleishhacker Zoo, San Francisco (1935–1929). Five birds, three mammals and three reptiles are named after him.

Hemprich

Arabian Smooth-Hound *Mustelus mosis* Hemprich & Ehrenberg, 1899

Wilhelm Friedrich Hemprich (1796–1825) was a physician, traveller and collector. He co-wrote *Natural*

Historical Journeys in Egypt and Arabia (1828) with Ehrenberg, whom he had met whilst studying medicine in Berlin. They were invited to serve (1820) as naturalists on an expedition to Egypt and they continued to journey and collect in the region, including the Lebanon and the Sinai Peninsula before returning to Egypt and going on to Ethiopia. Hemprich died of fever in the Eritrean port of Massawa. A mammal, three birds and two reptiles are named after him.

Henle

Brown Smooth-Hound *Mustelus henlei* Gill, 1863
Bigtooth River Stingray *Potamotrygon henlei* Castelnau, 1855

Guitarfishes *Rhinobatidae* Müller & Henle, 1837
Guitarfishes *Rhynchobatus* Müller & Henle, 1837
Mackerel Sharks family *Lamnidae* Müller & Henle, 1838
Whale Sharks *Rhincodontidae* Müller & Henle, 1839
Rays *Urolophidae* Müller & Henle, 1841

Electric Ray genus *Narcine* Henle, 1834
Gulper Shark genus *Centrophorus* Müller & Henle, 1837
Bamboo Shark genus *Chiloscyllium* Müller & Henle, 1837
Requiem Shark genus *Galeocerdo* Müller & Henle, 1837
Nurse Shark genus *Ginglymostoma* Müller & Henle, 1837
Bamboo Shark genus *Hemiscyllium* Müller & Henle, 1837
Stingray genus *Himantura* Müller & Henle, 1837
Sawshark genus *Pristiophorus* Müller & Henle, 1837
Carpet Shark genus *Stegostoma* Müller & Henle, 1837
Fanskate genus *Sympterygia* Müller & Henle, 1837
Stingray genus *Taeniura* Müller & Henle, 1837
Reef Shark genus *Triaenodon* Müller & Henle, 1837
Stingray genus *Urolophus* Müller & Henle, 1837
Stingray genus *Urogymnus* Müller & Henle, 1837
Sliteye Shark genus *Loxodon* Müller & Henle, 1838
Fanray genus *Platyrhina* Müller & Henle, 1838
Spadenose Shark genus *Scoliodon* Müller & Henle, 1838
Fiddler Ray genus *Trygonorrhina* Müller & Henle, 1838
Sand Shark genus *Odontaspididae* Müller & Henle, 1839
Houndshark genus *Triakis* Müller & Henle, 1839
Deepwater Dogfish genus *Centroscyllium* Müller & Henle, 1841
Guitarfish genus *Rhinidae* Müller & Henle, 1841
Rio Skate genus *Rioraja* Müller & Henle, 1841
Stingaree genus *Trygonoptera* Müller & Henle, 1841

Porcupine River Stingray *Potamotrygon hystrix* Müller & Henle, 1834
Grey Bamboo Shark *Chiloscyllium griseum* Müller & Henle, 1838

Brownbanded Bamboo Shark *Chiloscyllium punctatum* Müller & Henle, 1838
Blackspotted Catshark *Halaelurus buergeri* Müller and Henle, 1838
Dark Shyshark *Haploblepharus pictus* Müller & Henle, 1838
Shortfin Devil Ray *Mobula kuhlii* Müller & Henle, 1838
Leopard Catshark *Poroderma pantherinum* Müller & Henle, 1838
Narrowmouthed Catshark *Schroederichthys bivius* Müller & Henle, 1838
Spadenose Shark *Scoliodon laticaudus* Müller & Henle, 1838
Yellow-spotted Catshark *Scyliorhinus capensis* Müller & Henle, 1838
Spinetail Mobula *Mobula japonica* Müller & Henle, 1838
Pigeye Shark *Carcharhinus amboinensis* Müller & Henle, 1839
Spinner Shark *Carcharhinus brevipinna* Müller & Henle, 1839
Whitecheek Shark *Carcharhinus dussumieri* Müller & Henle, 1839
Silky Shark *Carcharhinus falciformis* Müller & Henle, 1839
Pondicherry Shark *Carcharhinus hemiodon* Müller & Henle, 1839
Finetooth Shark *Carcharhinus isodon* Müller & Henle, 1839
Bull Shark *Carcharhinus leucas* Müller & Henle, 1839
Blacktip Shark *Carcharhinus limbatus* Müller & Henle, 1839
Hardnose Shark *Carcharhinus macloti* Müller & Henle, 1839
Spot-tail Shark *Carcharhinus sorrah* Müller & Henle, 1839
Ganges Shark *Glyphis gangeticus* Müller & Henle, 1839
Speartooth Shark *Glyphis glyphis* Müller & Henle, 1839
Japanese Topeshark *Hemitriakis japonica* Müller & Henle, 1839
Daggernose Shark *Isogomphodon oxyrhynchus* Müller & Henle, 1839
Broadfin Shark *Lamiopsis temminckii* Müller & Henle, 1839
Barbeled Houndshark *Leptocharias smithii* Müller & Henle, 1839
Sliteye Shark *Loxodon macrorhinus* Müller & Henle, 1839
Brazillian Sharpnose Shark *Rhizoprionodon lalandii* Müller & Henle, 1839
Banded Houndshark *Triakis scyllium* Müller & Henle, 1839
Eagle Ray *Aetomylaeus milvus* Müller & Henle, 1841
Short-snouted Shovelnose Ray *Aptychotrema bougainvillii* Müller & Henle, 1841

African Softnose Skate *Bathyraja smithii* Müller & Henle, 1841

Whip Stingray *Dasyatis akajei* Müller & Henle, 1841

Bennett's Stingray *Dasyatis bennetti* Müller & Henle, 1841

Pale-edged Stingray *Dasyatis zugei* Müller & Henle, 1841

Tentacled Butterfly Ray *Gymnura tentaculata* Müller & Henle, 1841

Dwarf Whipray *Himantura walga* Müller & Henle, 1841

Cuckoo Skate *Leucoraja naevus* Müller & Henle, 1841

Blue-spotted Stingray *Neotrygon kuhlii* Müller & Henle, 1841

Ocellate Spot Skate *Okamejei kenojei* Müller & Henle, 1841

Ocellate River Stingray *Potamotrygon motoro* Müller & Henle, 1841

Large-tooth Sawfish *Pristis perotteti* Müller & Henle, 1841

Lesser Sandshark *Rhinobatos annulatus* Müller & Henle, 1841

Bluntnose Guitarfish *Rhinobatos blochii* Müller & Henle, 1841

Brazilian Guitarfish *Rhinobatos horkelii* Müller & Henle, 1841

Widenose Guitarfish *Rhinobatos obtusus* Müller & Henle, 1841

Brown Guitarfish *Rhinobatos schlegelii* Müller & Henle, 1841

Rough Cownose Ray *Rhinoptera adspersa* Müller & Henle, 1841

Flapnose Ray *Rhinoptera javanica* Müller & Henle, 1841

Rio Skate *Rioraja agassizii* Müller & Henle, 1841

Smallnose Fanskate *Sympterygia bonapartii* Müller & Henle, 1841

Blotched Fantail Ray *Taeniura meyeni* Müller & Henle, 1841

Southern Fiddler Ray *Trygonorrhina fasciata* Müller & Henle, 1841

Common Stingaree *Trygonoptera testacea* Müller & Henle, 1841

New Ireland Stingaree *Urolophus armatus* Müller & Henle, 1841

Sepia Stingaree *Urolophus aurantiacus* Müller & Henle, 1841

Striped Panray *Zanobatus schoenleinii* Müller & Henle, 1841

Lesser Guitarfish *Zapteryx brevirostris* Müller & Henle, 1841

New Zealand Rough Skate *Zearaja nasuta* Müller & Henle, 1841

Dr Friedrich Gustav Jakob Henle (1809–1885) was a German physician, anatomist and zoologist whose main subjects were ichthyology and human biology. He was Professor of Anatomy in Zurich (1840–1844), at Heidelberg (1844–1852) and at Göttingen (1852–1885). Neither description of the eponymous fish states that they are named after him but it is clear from the material discussed in the pages around the descriptions that he and his work were the cause of the discussions. Along with Johann Müller he produced the first authoritative work on sharks (1839–1841); together they were giants among contemporary ichthyologists.

Henry

Walking Shark *Hemiscyllium henryi* Allen & Erdmann, 2008

Wolcott Henry is a professional under water photographer who is based in Washington DC. He is a certified Smithsonian Institution science diver and has been contracted to the National Geographic Society since 1997. He co-wrote *Wild Ocean* (1999). He was honoured for his generous support of Conservation International's marine initiatives, including the taxonomy of New Guinea fishes.

Hensley

Hagfish sp. *Eptatretus mendozai* Hensley, 1985

Hagfish sp. *Myxine mcmillanae* Hensley, 1991

Dannie Alan Hensley (1944–2008) was an American ichthyologist. San Bernardino Valley College awarded his first degree (1966) and the University of South Florida his PhD (1978). He published his first paper before graduating (1965) and was still publishing four decades later, for example, *Revision of the Genus Asterorhombus (Pleuronectiformes: Bothidae)* (2005). He was an ichthyologist at the Marine Research Laboratory Florida (1974–1978) and then became a post-doctoral research associate at Florida Atlantic University. From there he moved to the University of Puerto Rico for 28 years, becoming Assistant Professor (1980–1984) then Associate Professor (1984–1991) and finally full Professor (1991).

Herklots

Longfin Catshark *Apristurus herklotsi* Fowler, 1934

Dr Geoffrey Alton Craig Herklots (1902–1986) was a British biologist, botanist and ornithologist at the University of Hong Kong (1928–1941) who was interned under the Japanese occupation (1942–

1945). He returned to England (1945) and joined the Colonial Service and was Principal of the Imperial College of Tropical Agriculture, Trinidad (1953–1960). Fowler honoured him adding 'with many fond memories of the China Sea and Java'.

Hermann

Zebra Shark *Stegostoma fasciatum* Hermann, 1783

Johannis Herrmann (sometimes Hermann) (1738–1800) was Professor of Medicine and Natural History, Université de Strasbourg. He wrote *Observationes Zoologicae Quibus Novae Complures Aliaeque Animalium Species, Describuntur et Illustrantur Edidit Fridericus Ludovicus Hammer* (1804). Two reptiles are named after him.

Herre

Houndshark genus *Hemitriakis* Herre, 1923

Whitefin Japanese Topeshark *Hemitriakis leucoperiptera* Herre, 1923
Weasle Shark sp. *Hemigaleus machlani* Herre, 1929

Dr Albert William Christian Theodore Herre (1868–1962) was an ichthyologist, ecologist, botanist and lichenologist. He gained his bachelor's degree in botany and later took his master's (1905) and doctorate (1908), both in ichthyology, at Stanford. He became acting head, Biology Department, University of Nevada (1909–1910), was vice-principal of a high school, Oakland (1910–1912), taught at a school in Washington State (1912–1915), and was then head of the Science Department, Western Washington College of Education (1915–1919). He then went to the Philippines, where he was Chief of Fisheries, Bureau of Science, Manila (1919–1928). He was Curator of Zoology, Natural History Museum, Stanford (1928–1946). After retiring he returned to the Philippines (1947) as a member of the Fishery Program, US Fish and Wildlife Bureau, and then worked in the School of Fisheries, University of Washington (1948–1957). After his second retirement he researched and collected lichens (1957–1962). A reptile is named after him.

Herwig

Cape Verde Skate *Raja herwigi* G Krefft, 1965

This species is named after the research vessel 'Walther Herwig', from which the holotype was caught.

Higman

Smalleye Smooth-Hound *Mustelus higmani*
S Springer & RH Lowe, 1963

James B. Higman (1922–2009) was an ichthyologist. Western Maryland College awarded his bachelor's degree and the University of Miami his master's. He was a captain in the Chemical Corp, US Army (1942–1946), fighting from Utah Beach, Normandy to the River Elbe. He taught at the Rosenstiel School of Marine and Atmospheric Science, University of Miami for 30 years and was a professor there. He co-wrote *A Review of Shrimp Capture and Culture Fisheries of the United States* (1993). He collected the holotype while serving as an observer on an exploratory fishing expedition for the Bureau of Commercial Fisheries, US Department of the Interior. He was honoured for 'initial interest in the species and his care in the preparation of excellent notes on its natural history.'

Hildebrand

Southern Stingray *Dasyatis americana* Hildebrand & Schroeder, 1928
Peruvian Butterfly Ray *Gymnura afuerae* Hildebrand, 1946
Stingray sp. *Urotrygon caudispinosus* Hildebrand, 1946
Peruvian Stingray *Urotrygon peruanus* Hildebrand, 1946
Stingray sp. *Urotrygon serrula* Hildebrand, 1946

Dr Samuel Frederick Hildebrand (1883–1949) was born of immigrant parents who never learned to speak English and was brought up as a farm boy in Indiana. He worked for the US Bureau of Fisheries, Washington DC (1910–1949), as a scientific assistant (1910–1914), Director of the US Fisheries Biological Stations, Beaufort, North Carolina (1914–1918 and 1925–1931) and Key West, Florida (1918–1919), ichthyologist in Washington (1919–1925) and senior ichthyologist (1931–1939). He took part in a number of expeditions including two with Meek to Panama (1910–1911 and 1912), two on his own (1935 and 1937) and one to Central America with Foster (1924).

Hilgendorf

Catshark genus *Proscyllium* Hilgendorf, 1904
Graceful Catshark *Proscyllium habereri* Hilgendorf, 1904

Franz Martin Hilgendorf (1839–1904) was a German zoologist and palaeontologist. He entered the University of Berlin (1859) to study philology and moved to Tübingen University (1861) where he was awarded a doctorate (1863) for a thesis on a geological subject. He worked at, and studied in, the Zoological Museum in Berlin (1863–1868). He was Director of the Hamburg Zoological Gardens (1868–1870). He was a private lecturer at the Polytechnic Institute in Dresden (1871–1872) and then became a lecturer at the Imperial Medical Academy in Tokyo (1873–1876). On his return to Germany he became an assistant to Peters and worked in various departments of the Berlin Museum until ill health forced him to retire (1876–1903). A mammal is named after him.

Hill

Caribbean Lantern Shark *Etmopterus hillianus* Poey, 1861

Richard Hill (1795–1872) was a Jamaican anti-slavery activist, judge and naturalist who corresponded with Poey. He was honoured for his *Contributions to the Natural History of the Shark* (1850) and other writings on fishes. A bird is named after him.

Hiyama

Leadhued Skate *Notoraja tobitukai* Hiyama, 1940

Dr Yoshio Hiyama (1909–1988) was a Japanese ichthyologist who practised the art of gyotaku (illustrating through fish rubbing), which he used as scientific illustrations of Japanese fish species. He worked at the Department of Fisheries, Faculty of Agriculture, University of Tokyo and was (by 1971) Professor Emeritus. He wrote *Systematic List of Fishes of the Ryukyu Islands* (1951).

Ho

Taiwanese Wedgefish *Rhynchobatus immaculatus* Last, Ho & R-R Chen, 2013

Dr Hsuan-Ching 'Hans' Ho (b.1978) is a marine biologist and ichthyologist. The National Taiwan Ocean University, Keelung awarded his bachelor's (2000), master's (2002) and doctorate (2010). He was a post-doctoral fellow of the Biodiversity Research Center, Academia Sinica (2010), where he has spent over ten years working on fish taxonomy. He is now an assistant researcher, Department of Exhibition, National Museum of Marine Biology & Aquarium and assistant professor, Institute of Marine Biodiversity and Evolutionary Biology, National Dong Hwa University. With more than 30 new species to his name, he authored well over 100 publications, including being co-author of *The Complete Mitochon-drial Genome of the Shortfin Mako, Isurus oxyrinchus (Chondrichthyes, Lamnidae)* (2013). The new species of worm eel, *Pylorobranchus hoi*, was named after him in 2012.

Hoff

Butterfly Skate *Bathyraja mariposa* Stevenson, Orr, Hoff & McEachran, 2004
Leopard Skate *Bathyraja panthera* Orr, Stevenson, Hoff, Spies & McEachran, 2011

Dr Gerald Raymond 'Jerry' Hoff (b.1962) is an ichthyologist and research fisheries biologist with the National Oceanic and Atmospheric Administration, Seattle, USA, where he has spent his career working on life history and taxonomic issues of deepwater fish species. His current interest is in reproduction in skates and their nursery habitat and its conservation. The University of West Florida, Pensacola, awarded his bachelor's degree (1988), the University of Texas at Austin his master's (1993) and the University of Washington, Seattle his doctorate (2007). He wrote *A Nursery Site of the Alaska Skate* (Bathyraja parmifera) *in the Eastern Bering Sea* (2008) and co-wrote *Results of the 2010 Eastern Bering Sea Upper Continental Slope Survey of Groundfish and Invertebrate Resources* (2011).

Holland

Yellow-spotted Skate *Okamejei hollandi* Jordan & RE Richardson, 1909

Dr William Jacob Holland (1848–1932) was a Jamaican-born American Presbyterian minister, entomologist and palaeontologist and Director of the Carnegie Museum (1898–1922). He was honoured for giving support to the authors' study of Taiwanese fishes. A bird is named after him.

Holt

Deep-water Ray *Rajella bathyphila* Holt & Byrne, 1908
Straightnose Rabbitfish *Rhinochimaera atlantica* Holt & Byrne, 1909

Ernest William Lyons Holt (1864–1922) was a British marine naturalist, ichthyologist and soldier who served in the Nile Campaign (1884–1885) and the Third Burmese War (1886–1887) during which he was invalided home. He studied zoology at St Andrew's University (1888). He worked for the Marine Biological Association in Grimsby (1892–1894) and at the Plymouth Marine Laboratory (1895–1898). He worked in Ireland (1899–1914) for the Department of Agriculture and Technical Instruction for Ireland as scientific adviser and fisheries inspector (1908) and Chief Inspector (1914).

Horkel

Brazilian Guitarfish *Rhinobatos horkelii* Müller & Henle, 1841

Dr Johann Horkel (1769–1846) was a German physician and botanist who had an example of this species in alcohol that he provided to the authors. He studied medicine at the University of Halle (1787), where he stayed as a lecturer and associate professor of medicine (1804–1810). He was Professor of Plant Physiology, University of Berlin (1810–1846).

Hortle

Hortle's Whipray *Himantura hortlei* Last, Manjaji-Matsumoto & Kailola, 2006

Kent Gregory Hortle is an Australian fisheries and environmental consultant whose bachelor's degree in zoology was awarded by Monash University, Melbourne (1979). From 1980 onwards he has worked as a biologist or environmental scientist in Asia and Australia. He provided the first photographs and fresh specimens of the whipray, captured during sampling while he was Environmental Monitoring Superintendent at the Freeport mine in Papua, Indonesia (1996–2001). Since 2001 he has worked primarily on fisheries of the Mekong River basin in Laos, Cambodia, Thailand and Vietnam as an advisor to the Mekong River Commission or as a consultant. He has published numerous papers and reports covering environment, fish and fisheries, particularly in the Mekong region.

Houttuyn

Electric Ray genus *Torpedo* Houttuyn, 1764

Maarten (Martinus) Houttuyn (Houttujin) (1720–1798) was a Dutch naturalist who wrote widely on natural history topics. He is most well known as a botanist.

Howell-Rivero

Chimaera sp. *Chimaera cubana* Howell-Rivero, 1936
Dwarf Catshark *Scyliorhinus torrei* Howell-Rivero, 1936
Cuban Dogfish *Squalus cubensis* Howell-Rivero, 1936

Dr Luis Hugo Howell-Rivero (1899–1986) was a Cuban biologist and anthropologist at the University of Havana. The Instituto de Segunda Enseñanza, Havana awarded his bachelor's degree (1925) and

the University of Havana his doctorate (1930). He researched West Indian fishes at Harvard's Museum of Comparative Zoology (1934–1935). He wrote *Some New, Rare, and Little Known Fishes from Cuba* (1936). (See **Rivero**)

Hu

South China Catshark *Apristurus sinensis* Chu & Hu, 1981
Narrow Legskate *Anacanthobatis stenosoma* Li & Hu, 1982

Hu Ai-Sun is a Chinese zoologist.

Huang

Hagfish sp. *Eptatretus nelsoni* Kuo, Huang & Mok, 1994
Hagfish sp. *Eptatretus sheni* Kuo, Huang & Mok, 1994
Hagfish sp. *Paramyxine fernholmi* Kuo, Huang & Mok, 1994
Hagfish sp. *Paramyxine wisneri* Kuo, Huang & Mok, 1994
Hagfish sp. *Eptatretus rubicundus* Kuo, Lee & Mok, 2010

Kao-Fong Huang is a Taiwanese ichthyologist at the National Sun Yat-Sen University, Kaohsiung, Taiwan. He has written widely, including the co-written *Hagfishes of Taiwan* (1993).

Hubbs, CL

Kern Brook Lamprey *Entosphenus hubbsi* Vladykov & Kott, 1976
Giant Hagfish *Eptatretus carlhubbsi* CB McMillan & Wisner, 1984
Hagfish sp. *Myxine hubbsi* Wisner & CB McMillan, 1995

Brook Lamprey sub-genus *Lethenteron* Creaser & CL Hubbs, 1922
Cow Stingray *Dasyatis ushiei* Jordan & CL Hubbs, 1925
Ratfish sp. *Hydrolagus eidolon* Jordan & Hubbs, 1925
Southern Brook Lamprey *Ichthyomyzon gagei* CL Hubbs & Trautman, 1937
Mountain Brook Lamprey *Ichthyomyzon greeleyi* CL Hubbs & Trautman, 1937
Silver Lamprey *Ichthyomyzon unicuspis* CL Hubbs & Trautman, 1937
Salmon Shark *Lamna ditropis* CL Hubbs & Follett, 1947
Pit-Klamath Brook Lamprey *Entosphenus lethophagus* CL Hubbs, 1971
Hagfish sp. *Eptatretus minor* Fernholm & CL Hubbs, 1981
Hagfish sp. *Eptatretus multidens* Fernholm & CL Hubbs, 1981

Professor Carl Levitt (Leavitt) Hubbs (1894–1979) was Professor of Biology at the Scripps Institution of Oceanography, California. Many of the Hubbs family members were ichthyologists, so it is no wonder that many aquatic species carry the scientific name *hubbsi*. *Octopus hubbsorum* was named for Carl as well as his wife, Laura Cornelia (Clark) Hubbs (1893–1988), and their son Clark Hubbs. Professor and Mrs Hubbs had three children, all of whom became ichthyologists. Their daughter Frances married yet another ichthyologist, Robert Rush Miller. He was honoured in the name of the giant hagfish because he was a 'giant in ichthyology'. The beaked whale is named after him in both the vernacular and the scientific form.

Hubbs, LC

Hagfish sp. *Eptatretus laurahubbsae* CB McMillan & Wisner, 1984

Laura Cornelia Hubbs née Clark (1893–1988) was an ichthyologist and the wife of Carl Levitt Hubbs (above) as well as a friend and co-worker of the authors and contributed to the life and work of her husband. Both her bachelor's degree (1915) and her master's (1916) were awarded by Stanford University. She worked part-time at the University of Michigan Museum of Zoology (1929–1944). (See **Laura Hubbs**)

Huidobro-Campos

Gulf Angelshark *Squatina heteroptera* Castro-Aguirre, Espinoza-Pérez & Huidobro-Campos, 2007
Mexican Angelshark *Squatina mexicana* Castro-Aguirre, Espinoza-Pérez & Huidobro-Campos, 2007

Dr Leticia Huidobro-Campos is a marine biologist. She was a professor at the Institute of Biology, Department of Zoology, Universidad Nacional Autónoma de México (2006). She is now at the Instituto Nacional de la Pesca, Mexico City. She co-wrote *24 Peces Invasores en el Centro de México* (2014).

Hulley

Roughnose Legskate *Cruriraja hulleyi* Aschliman, Ebert & Compagno, 2010

Taillight Shark genus *Euprotomicroides* Hulley & Penrith, 1966
Bottlenose Skate genus *Rostroraja* Hulley, 1972

Taillight Shark *Euprotomicroides zantedeschia* Hulley & Penrith, 1966

Roberts Bigmouth Skate *Amblyraja robertsi* Hulley, 1970
Yellow-spotted Skate *Leucoraja wallacei* Hulley, 1970
Ghost Skate *Rajella dissimilis* Hulley, 1970
Smoothback Skate *Rajella ravidula* Hulley, 1970
African Pygmy Skate *Neoraja stehmanni* Hulley, 1972

Dr Percy Alexander 'Butch' Hulley (b.1941) is a South African zoologist and ichthyologist. He is Curator of Fishes and Deputy Director of the Iziko South African Museum. He has described many species new to science. He was honoured for his 'pioneering work on southern African skate', among which the legskate occurs.

Human

Human's Whaler Shark *Carcharhinus humani* WT White & Weigmann, 2014

Natal Shyshark *Haploblepharus kistnasamyi* Human & Compagno, 2006
Honeycomb Izak *Holohalaelurus favus* Human, 2006
Grinning Izak *Holohalaelurus grennian* Human, 2006

Dr Brett A. Human (d.2011) was an Australian marine biologist and a science diver at the Department of Aquatic Zoology, Western Australia Museum, Welshpool. He co-wrote *Is the Megamouth Shark Susceptible to Mega-distortion? Investigation of the Effects of Twenty-two years of Fixation and Preservation on a Large Specimen of Megachasma Pelagios (Chondrichthyes: Megachasmidae)* (2012). Before working in Western Australia he had been a researcher at the South African Museum. The etymology of the Whaler Shark reads: 'Named after the late Dr Brett Human, for important contributions to shark taxonomy in South Africa and Oman in the western Indian Ocean region, and who is sorely missed by his colleagues.'

Hureau

Kerguelen Sandpaper Skate *Bathyraja irrasa* Hureau & Ozouf-Costaz, 1980

Jean-Claude Hureau (b.1935) is a French ichthyologist who retired from his professorship at the Muséum National d'Histoire Naturelle, Paris (2001). He carried out considerable research in the Kerguelen Islands as well as on the fishes of Terre Adélie, Antarctica, where he spent over a year during 1960–1962 at the Dumont d'Urville Research Station.

Hutchins

Western Wobbegong *Orectolobus hutchinsi* Last, Chidlow & Compagno, 2006

Dr J. Barry Hutchins (b.1946) was Curator of Fishes at the Western Australian Museum (1998–2007), which he joined as a technical officer (1972) and where he worked until he retired. He is now a research associate. The University of New South Wales, Sydney awarded his bachelor's degree (1968). After service in Vietnam as a corporal in the infantry, Australian Army (1969–1970), he worked on prawn and scallops trawlers in Queensland (1970–1972) before joining the Western Australian Museum. Murdoch University awarded his first class honours degree in ichthyology (1979) and a doctorate (1988). He has written many papers on Australian fishes, including three field guides: his first book was *The Fishes of Rottnest Island* (1979) and he was senior author of the *The Marine and Estuarine Fishes of South-western Australia* (1983) and *Sea Fishes of Southern Australia* (1986), which was based on his seven-month survey of the near-shore waters of southern Queensland, New South Wales, Victoria, Tasmania and South Australia. Dr Hutchins was the first (1983) person to recognise that this wobbegong was a new species. Several other fishes are named after him.

Hutton

New Zealand Torpedo *Torpedo fairchildi* Hutton, 1872
Short-tail Stingray *Dasyatis brevicaudata* Hutton, 1875

Frederick Wollaston Hutton (1836–1905) was an English geologist and zoologist who settled in New Zealand. He served in the Indian Mercantile Marine, and then in the army (1855–1865). He saw service in the Crimean War and the Indian Mutiny. He wrote *Catalogue of the Birds of New Zealand* (1871). The Royal Society of New Zealand established (1909) the Hutton Memorial Fund in his memory. It awards the Hutton Medal and provides grants for the encouragement of research into the zoology, botany and geology of New Zealand. He described two *Elasmobranchii*. Four birds are named after him.

I

Iago

Houndshark genus *Iago* Compagno & S Springer, 1971

The name of this genus derives from Iago, the villain in Shakespeare's play *Othello*.

Ichihara

Japanese Velvet Dogfish *Zameus ichiharai* Yano & Tanaka, 1984

Dr Tadayoshi Ichihara was a scientist at the Whales Research Institute, Tokyo, Japan (1960s–1980s). Controversially he declared the Indian Ocean population of blue whales to be a separate (smaller) species at a time when whaling of blue whales was banned (1960s). This was seen by many as an attempt to get around the ban. However, he was correct. He wrote a number of papers including with Yano and Tanaka including *Notes on a Pacific Sleeper Shark* (1982). He was honoured because he had suggested to the authors that they study this shark.

Iglésias

Black Roughscale Catshark *Apristurus melanoasper* Iglésias, Nakaya & Stehmann, 2004

Chimaera sp. *Chimaera opalescens* Luchetti, Iglésias & Sellos, 2011

Dogfish Shark sp. *Squalus formosus* WT White & Iglésias, 2011

Milk-eye Catshark *Apristurus nakayai* Iglésias, 2012

Domino Skate *Bathyraja leucomelanos* Iglésias & Lévy-Hartmann, 2012

Samuel Paco Iglésias is a French ichthyologist at the Station de Biologie Marine et Marinarium de Concarneau, belonging to the Muséum National d'Histoire Naturelle. Among his published papers is the co-written *Molecular Phylogeny and Node Time Estimation of Bioluminescent Lantern Sharks (Elasmobranchii: Etmopteridae)* (2010).

Indrambarya

Hagfish sp. *Eptatretus indrambaryai* Wongratana, 1983

Professor Dr Boon Indrambarya (1907–1994) was a leading Thai marine biologist. He graduated from Cornell University in fish biology and worked as an assistant with the US Fish Commission (1929) and later at the Department of Fisheries, Ministry of Agriculture, Bangkok. He was the founder of remote sensing in Thailand (1970s). He was also Director of the Environmental and Ecological Research Institute, the Applied Scientific Research Corporation of Thailand. He was honoured in the name of the hagfish as being 'one of the senior-most pioneer fisheries biologists of Thailand'. A building is named after him at the Faculty of Fisheries, Kasetsart University, Bangkok, where he was the Dean (1944–1962) and from where he retired as Professor Emeritus (1974). The National College of Education, Evanston, Illinois awarded him an honorary doctorate (1963).

Indroyono

Indonesian Houndshark *Hemitriakis indroyonoi* WT White, Compagno & Dharmadi, 2009

Dr Dwisuryo Indroyono Soesilo (b.1955) is an engineer and geologist. He is Secretary and Deputy Senior Minister of the Coordinating Ministry for People's Welfare of the Republic of Indonesia and was Chairman, Marine and Fisheries Research Agency, Jakarta (2009). Institut Teknologi Bandung, Bandung, Indonesia awarded his bachelor's degree (1979), the University of Michigan his master's (1981) and the University of Iowa his doctorate (1987). The shark was 'named for Dr. Indroyono Soesilo, who has provided a great deal of support for shark research in Indonesia and was a strong advocate for the production of the field guide to sharks and rays of Indonesia.'

Ios

White-headed Hagfish *Myxine ios* Fernholm, 1981

IOS stands for the Institute of Oceanographic Sciences, England, which supplied the type specimen.

Iredale

Wedgenose Skate *Dipturus whitleyi* Iredale, 1938

Tom Iredale (1880–1972) was an English-born Australian artist, ornithologist and malacologist who was completely self-taught. He grew up in England where he was apprenticed to a pharmacist (1899–1901). He was an avid birdwatcher and egg collector. It is thought he had tuberculosis: advised to take a sea voyage for his health, he went to New Zealand. He took a post as a clerk in Christchurch (1902–1907) during which time he married. He joined an expedition to the Kermadec Islands (1908) and lived there for nearly a year living on the birds he was studying and molluscs collected on the shoreline, where he developed an interest in malacology. He went to Queensland (1909) and collected thousands of specimens there. He returned to England and became a clerk, and later secretary to Gregory MacAlister Mathews, the Australian ornithologist, after working with him for a number of years at the British Museum (1909–1923). While Matthews is credited with writing *Birds of Australia*, Iredale is said to have written much of the text. He returned to Australia (1923), taking a position as a conchologist at the Australian Museum, Sydney (1924–1944). He published widely on birds, ecology and shells. He also wrote 'popular' science articles for newspapers under various pseudonyms, including *Garrio*, which is Latin for 'I chatter'. Among many taxa named for him are several gastropods and fish and ten birds.

Irolita

Softnose Skate genus *Irolita* Whitley, 1931

Irolita is a character in a fairy story written by Marie-Catherine Le Jumel de Barneville, Baroness d'Aulnoy (1650–1705).

Irvine

Spineback Guitarfish *Rhinobatos irvinei* Norman, 1931

Dr Frederick Robert Irvine (1898–1962) was a botanist and general naturalist who made a collection of fishes near Accra, Gold Coast (Ghana) that included the holotype of the guitarfish. He wrote *The Fishes and Fisheries of the Gold Coast* (1947).

Ishihara

Abyssal Skate *Bathyraja ishiharai* Stehmann, 2005

Purple-black Skate *Bathyraja caeluronigricans* Ishiyama & Ishihara, 1977
Commander Skate *Bathyraja lindbergi* Ishiyama & Ishihara, 1977
White-blotched Skate *Bathyraja maculate* Ishiyama & Ishihara, 1977
Smallthorn Skate *Bathyraja minispinosa* Ishiyama & Ishihara, 1977
Bottom Skate *Bathyraja pseudoisotrachys* Ishiyama & Ishihara, 1985
Borneo Sand Skate *Okamejei cairae* Last, Fahmi & Ishihara, 2010

Dr Hajime Ishihara (b.1950) is an ichthyologist and environment specialist. He was connected with the

Iraq Sea Line Project under the Japan International Corporation Agency and is now with W & I Associates Corporation, Fujisawa, Japan. Tokyo University of Fisheries awarded his bachelor's degree (1975) and his master's (1977) and the University of Tokyo his doctorate (1990). He is currently working on the reconstruction of the Iraq Crude Oil Pipeline and has been based in Amman, Jordan since 2009. He has written *First Record of a Skate*, Anacanthobatis borneensis *from the East China Sea* (1984) and co-wrote, with Matthias Stehmann, *A Second Record of the Deep-water Skate* Notoraja subtilispinosa *from the Flores Sea, Indonesia* (1990). In honouring him in the name of the abyssal skate, Stehmann described him as his 'skatology colleague and friend of more than 25 years, who devoted his life's research to chondrichthyan fishes, producing important revisions of North Pacific *Bathyraja*'.

Ishikawa

Japanese Spurdog *Squalus japonicus* Ishikawa, 1908

Professor Dr Chiyomatsu Ishikawa (1861–1935) was a Japanese zoologist, evolutionary theorist and ichthyologist. After graduating from Tokyo University he spent time studying in Germany under August Weismann. He was at the Naples Zoological Station (1887). He was a zoologist at the College of Agriculture, Imperial University, Tokyo and Curator of the Imperial Museum. He was largely responsible for bringing Darwinism to Japan. At some stage he was also principal of the Dokkyo Middle School, Tokyo. An amphibian and at least one fish are named after him.

Ishiyama

Plain Pygmy Skate *Fenestraja ishiyamai* Bigelow & Schroeder, 1962

Jointnose Skate genus *Rhinoraja* Ishiyama, 1952
Skate genus *Bathyraja* Ishiyama, 1958
Skate genus *Notoraja* Ishiyama, 1958
Skate genus *Okamejei* Ishiyama, 1958

Dusky-pink Skate *Bathyraja diplotaenia* Ishiyama, 1952
Dusky-purple Skate *Bathyraja matsubarai* Ishiyama, 1952
Skate sp. *Breviraja abasiriensis* Ishiyama, 1952
White-bellied Softnose Skate *Rhinoraja longicauda* Ishiyama, 1952
Bigtail Skate *Dipturus macrocauda* Ishiyama, 1955
Eremo Skate *Bathyraja trachouros* Ishiyama, 1958

Giant Skate *Dipturus gigas* Ishiyama, 1958

Sharpspine Skate *Okamejei acutispina* Ishiyama, 1958

Browneye Skate *Okamejei schmidti* Ishiyama, 1958

Oda's Skate *Rhinoraja odai* Ishiyama, 1958

Hokkaido Skate *Bathyraja simoterus* Ishiyama, 1967

Purple-black Skate *Bathyraja caeluronigricans*
Ishiyama & Ishihara, 1977

Commander Skate *Bathyraja lindbergi* Ishiyama &
Ishihara, 1977

White-blotched Skate *Bathyraja maculate* Ishiyama &
Ishihara, 1977

Smallthorn Skate *Bathyraja minispinosa* Ishiyama &
Ishihara, 1977

Notoro Skate *Bathyraja notoroensis* Ishiyama &
Ishihara, 1977

Bottom Skate *Bathyraja pseudoisotrachys* Ishiyama &
Ishihara, 1985

Boeseman's Skate *Okamejei boesemani* Ishiyama,
1987

Dr Reizo Ishiyama (1912–2008) was an ichthyologist who graduated from Tokyo Fisheries College (Tokyo University of Marine Science and Technology). After graduating he developed tuberculosis and worked as an engineer at small fish-farming factory in Tokorozawa, Saitama prefecture in order to be able to live a quiet life and recover from his illness. After he had recovered he was an assistant at the Faculty of Agriculture, Kyoto University (1947–1951). He worked at Shimonoseki College of Fisheries as Assistant Professor (1951–1953) and as Professor (1953–1967). He then was Professor at Tokyo University of Fisheries (1967–1975). Over his career he wrote many arti- cles and papers including *Studies on the Rajid Fishes (Rajidae) found in the Waters around Japan* (1958) and, jointly with Ishihara, *Five New Species of Skates in the Genus* Bathyraja *from the Western North Pacific, with Reference to their Interspecific Relationships* (1977). He was honoured in the plain pygmy skate name for his work on Japanese batoids.

Iwama

Long-tailed River Stingray *Plesiotrygon iwamae* Rosa,
Castello & Thorson, 1987

Satoko Iwama (d.>1987) was a Brazilian zoologist who was a student and teacher at the Instituto de Botânica, São Paulo. He wrote *The Pollen Spectrum of the Honey of Tetragonisca angustula angustula Latreille (Apidae, Meliponinae)* (1975) as part of his master's thesis.

Iwatsuki

Hyuga Fanray *Platyrhina hyugaensis* Iwatsuki,
Miyamoto & Nakaya, 2011

Yellow-spotted Fanray *Platyrhina tangi* Iwatsuki,
Zhang & Nakaya, 2011

Dr Yukio Iwatsuki is a professor at Miyazaki University, Japan, from which he graduated with a bachelor's degree (1983) and a master's (1985). The University of Tokyo awarded his doctorate (1988). He is editor of Ichthyological Research, Japan. He co-wrote *Timor snapper,* Lutjanus timorensis *(Quoy et Gaimard), Collected from Japanese Waters* (1994).

J

Jacobsen

Eastern Banded Catshark *Atelomycterus marnkalha*
Jacobsen & MB Bennett, 2007

Dr Ian P. Jacobsen is an ichthyologist in the Fish Department, Queensland Museum. The University of Queensland, School of Biomedical Sciences awarded his doctorate. He co-wrote *A Taxonomic Review of the Australian Butterfly Ray* Gymnura australis *(Ramsay & Ogilby, 1886) and Other Members of the Family Gymnuridae (Order Rajiformes) from the Indo-West Pacific* (2009).

James

Eastern Pacific Black Ghostshark *Hydrolagus melanophasma* James, Ebert, Long & Didier, 2009

Kelsey C. James is an American ichthyologist. She graduated from the Pacific Shark Research Center (2011) and took her master's there (2012). She is presently a PhD student at the University of Rhode Island. Her publications include, as lead author, the description of the ghostshark, *A New Species of Chimaera*, Hydrolagus melanophasma *sp. nov. (Chondrichthyes: Chimaeriformes: Chimaeridae) from the Eastern North Pacific* (2009).

Jayakar

Oman Cownose Ray *Rhinoptera jayakari* Boulenger, 1895

Colonel Dr Atmaram Sadashiv G. Jayakar (1844–1911) was an Indian surgeon and entomologist. The Indian Medical Service sent him to Muscat (1878), and during his 21 years in the Oman area (1879–1900) he studied the local wildlife and collected specimens, which he donated to Natural History Museum, London (1885–1899), including the ray holotype. He spent so long in Muscat that he acquired the nickname 'Muscati'. He compiled a book of local proverbs published later as *Omani Proverbs* (1987). He has a mammal and three reptiles named after him.

Jenkins, JT

Jenkins' Whipray *Himantura jenkinsii* Annandale, 1909

James Travis Jenkins (1876–1959) was an ichthyolo-gist who became fishery advisor to the Government of Bengal. He was a close collaborator of Annandale and helped him collect the holotype. He wrote *The Sea Fisheries* (1920).

Jenkins, OP

Pacifric Cownose Ray *Rhinoptera steindachneri* Evermann & OP Jenkins, 1891

Dr Oliver Peebles Jenkins (1850–1935) was a physiologist and ichthyologist. His doctorate was awarded by Indiana University. He was Professor of Physiology at Leland Stanford Junior University (1891–1895) and Acting Professor of Physiology, Cooper Medical College (1895–1912), holding this position until the College became a department of Stanford University. He was appointed Professor of Physiology and eventually retired as Professor Emeritus of Physiology and Histology at Stanford University.

Jensen, AS

Shorttail Skate *Amblyraja jenseni* Bigelow & Schroeder, 1950

Spinytail Skate *Bathyraja spinicauda* AS Jensen, 1914

Professor Adolf Severin Jensen (1866–1953) was a zoologist, ichthyologist and malacologist. He did much work on the fauna of Greenland and made several expeditions there including the Tjalfe Expedition (1908–1909), which he led. He became malacological Curator, Zoological Museum, Københavns Universitet (1892) and was Professor of Zoology there (1917–1936). He wrote *The Fishes of East-Greenland* (1904).

Jensen, K

Sulu Sea Skate *Okamejei jensenae* Last & Lim, 2010

Dr Kirsten Jensen is a cestode parasitologist who is Associate Professor of Organismal Biology at the University of Kansas. The University of Connecticut awarded her doctorate (2001). The etymology for this species states that, during an extensive field survey of the fish markets of Borneo, conducted over the past decade, Dr Jensen captured digital images of all chondrichthyan specimens, and sampled and provided illustrations of most species in a field guide to the sharks and rays of Borneo (Last *et al.*). Along with a close colleague, Dr Janine Caira, she has gained a broad knowledge of the taxonomy of the chondrichthyan fauna, as well as their invertebrate parasites. She has published several papers alone or with others such as *A Monograph on the Lecanicephalidea (Platyhelminthes: Cestoda)* (2005).

Jenyns

Southern Hagfish *Myxine australis* Jenyns, 1842

Reverend Leonard Jenyns (1800–1893) was a clergyman and amateur naturalist. He became vicar of Swaffham Priory, Cambridgeshire (1828). His published natural history papers (1830s) were critically acclaimed. This led to an invitation (1836) from Charles Darwin to document the fish collection made by Darwin on the 'Beagle'. He wrote the four-part *Fish*, which Darwin edited (1840–1842). He moved (1849) to Swainswick near Bath where he founded the Bath Natural History Society. He published an autobiography (1887).

Jeong

Korean Skate genus *Hongeo* Jeong & Nakabo, 2009

Korea Skate *Raja koreana* Jeong & Nakabo, 1997
Skate sp. *Okamejei mengae* Jeong, Nakabo & Wu, 2007
Wu's Skate *Dipturus wuhanlingi* Jeong & Nakabo, 2008

Choong-Hoon Jeong is a Korean ichthyologist at the Department of Oceanography, Inha University, Korea (1995). He co-wrote *Raja Koreana, a New Species of Skate (Elasmobranchii, Rajoidei) from Korea* (1997).

Jespersen

Jespersen's Hagfish *Myxine jespersenae* Møller, Feld, Poulsen, Thomsen & Thormar, 2005

Dr Åse Jespersen (b.1955) is a Danish biologist, Associate Professor Emeritus in the Department of Biology, Marine Biological Section, Helsingør, part of the University of Copenhagen. She started her career as a research assistant at the State Serum Institute (1983) and held posts in pharmacology and biology. Her specialism is the reproductive systems of marine invertebrates on which she has published widely, such as the co-written *Sex, Seminal Receptacles, and Sperm Ultrastructure in the Commensal Bivalve* Montacuta phascolionis (2000). She was honoured for her contributions to the reproductive biology of hagfishes.

Johannis Davis

Travancore Skate *Dipturus johannisdavisi* Alcock, 1899

John Davis (1550–1605) is better remembered for having a strait rather than a skate named after him. The Davis Strait is between Greenland and Labrador. This 'celebrated Elizabethan navigator and explorer' is too famous a figure to need a biography here but this is what Alcock says in his original description: 'John Davis who – though best known for his Arctic voyages – piloted three expeditions to the East Indies and lost his life in Indian seas.'

Jones

Strickrott's Hagfish *Eptatretus strickrotti* Møller & Jones, 2007

Dr W. Joe Jones is a marine biologist who became a lab director at the University of South Carolina (2007), which awarded all his degrees, bachelor's (1995), master's (1997) and doctorate (2001). He was visiting research scientist at the Netherlands Institute for Sea Research (NIOZ) (1995) then a teaching assistant at the University of South Carolina (1995–2001), where he has remained for much of his career. He was a post-doctoral researcher at Universität Konstanz, Germany (2002) then Research Associate, Monterey Bay Aquarium Research Institute, Moss Landing, California (2003–2007), before returning to South Carolina as Director of the University's Environmental Genomics Core Facility. He has published papers throughout his career, often collaborating with co-authors, such as *Absence of Cospeciation between Deep-sea Mytilids and their Thiotrophic Endosymbionts* (2008). He said of the hagfish, which was taken by an Alvin submersible at a thermal vent in the East Pacific: 'We saw this little thing swimming like a worm and I told Bruce [Bruce Strickrott, the Alvin pilot], "There is no way you are going to catch it!"'

Jordan

Jordan's Chimaera *Chimaera jordani* Tanaka (I), 1905

Goblin Shark family *Mitsukurinidae* Jordan, 1898
Shark genus *Deania* Jordan & Snyder, 1902
Thornback Ray family *Platyrhinidae* Jordan, 1923

Guitarfish genus *Zapteryx* Jordan & CH Gilbert, 1880
Whiptail Stingray genus *Dasyatidae* Jordan, 1888
Sleeper Shark genus *Somniosidae* Jordan, 1888
Requiem Sharks genus *Carcharhinidae* Jordan & Evermann, 1896
Goblin Shark genus *Mitsukurina* Jordan, 1898
Dogfish Shark genus *Zameus* Jordan & HW Fowler, 1903
Guitarfish genus *Tarsistes* Jordan, 1919

Diamond Stingray *Dasyatis dipterura* Jordan & CH
    Gilbert, 1880
Thornback Guitarfish *Platyrhinoidis triseriata* Jordan &
    CH Gilbert, 1880
Longnose Skate *Raja rhina* Jordan & CH Gilbert, 1880
Starry Skate *Raja stellulata* Jordan & CH Gilbert, 1880
Banded Guitarfish *Zapteryx exasperate* Jordan &
    Gilbert, 1880
California Ray *Raja inornata* Jordan & CH Gilbert, 1881
Sicklefin Smooth-Hound *Mustelus lunulatus* Jordan &
    CH Gilbert, 1882
Pacific Sharpnose Shark *Rhizoprionodon longurio*
    Jordan & CH Gilbert, 1882
Spiney-tail Roundray *Urotrygon aspidura* Jordan & CH
    Gilbert, 1882
Speckled Guitarfish *Rhinobatos glaucostigma* Jordan
    & CH Gilbert, 1883
Ohio Lamprey *Ichthyomyzon bdellium* Jordan, 1885
Ocellate Electric Ray *Diplobatis ommata* Jordan & CH
    Gilbert, 1890
Equatorial Ray *Raja equatorialis* Jordan & Bollman, 1890
Spade Sand Skate *Psammobatis rutrum* Jordan, 1891
Giant Electric Ray *Narcine entemedor* Jordan & Starks,
    1895
Roger's Roundray *Urotrygon rogersi* Jordan & Starks,
    1895
Southern Banded Guitarfish *Zapteryx xyster* Jordan &
    Evermann, 1896
Goblin Shark *Mitsukurina owstoni* Jordan, 1898
Silver Chimaera *Chimaera phantasma* Jordan &
    Snyder, 1900
Hagfish sp. *Myxine garmani* Jordan & Snyder, 1901
Blackbelly Lantern Shark *Etmopterus lucifer* Jordan &
    Snyder, 1902
Whitefin Dogfish *Centroscyllium ritteri* Jordan & HW
    Fowler, 1903
[Alt. Blotchy Swellshark]
Acutenose Skate *Dipturus tengu* Jordan & HW Fowler,
    1903
Shortspine Spurdog *Squalus mitsukurii* Jordan &
    Snyder, 1903
Gecko Catshark *Galeus eastmani* Jordan & Snyder,
    1904
Spookfish *Hydrolagus mitsukurii* Jordan & Snyder, 1904
Pink Whipray *Himantura fai* Jordan & Seale, 1906
Blacktip Sawtail Catshark *Galeus sauteri* Jordan & RE
    Richardson, 1909
Yellow-spotted Skate *Okamejei hollandi* Jordan & RE
    Richardson, 1909
Guitarfish sp. *Tarsistes philippii* Jordan, 1919
Cow Stingray *Dasyatis ushiei* Jordan & CL Hubbs, 1925
Ratfish sp. *Hydrolagus eidolon* Jordan & CL Hubbs,
    1925

Dr David Starr Jordan (1851–1931) was a leading American ichthyologist, physician, educator, peace activist, believer in eugenics and, not least, founding President of Stanford University. Oddly he was educated at a local girls' high school and took his botany degree at Cornell. He studied further at Butler University and the Indiana University School of Medicine. His early career was spent teaching at several small colleges before joining the natural history faculty at Indiana Bloomington University (1879). He then became the nation's youngest ever University President at Indiana University (1885) and subsequently (1891–1913) became founding President at Stanford and later Chancellor before retiring (1916). His stance on peace (he was President of the World Peace Foundation (1910–1914), opposing US involvement in the First World War) stemmed from his belief that war killed off the strongest people in the gene pool. In his latter years he advocated compulsory sterilisation under the auspices of the Human Betterment Foundation. A cloud also hangs over him concerning the death of Jane Stanford, President of the Stanford University board of trustees. She died of strychnine poisoning while on holiday in Hawaii. Jordan rushed there and hired a physician to investigate the cause of death, which was quickly declared to be heart failure. Mrs Stanford was reportedly planning to have Jordan removed from his position at the University. He refused to learn his students' names on the grounds that he would forget the name of a fish for every student's name he learned. Apart from many scientific papers he wrote many longer works including, against war, *War and Waste* (1913), an autobiography, *Days of a Man* (1922) and much on zoology including the four-volume *Fishes of North and Middle America* (1896–1900). He described or co-described 38 species of *Elasmobranchii*. Twenty-nine species of fish and two amphibians are named after him.

Joung

Shortfin Smooth Lanternshark *Etmopterus joungi*
    Knuckey, Ebert & Burgess, 2011

Dr Shoou-Jeng Joung (b.1958) is an ichthyologist at the National Taiwan Ocean University, Keelung, where he is an associate professor and Chairman, Department of Environmental Biology and Fisheries Science. He has made a particular study of genetic markers in whalesharks. He has published more than 30 papers, including recently co-writing *Estimation of Life History Parameters of the Sharpspine Skates, Okamejei acutispina, in the Northeastern Waters of*

Taiwan (2011) and *Fisheries, Management, and Conservation for the Whale Shark,* Rhincodon typus *in Taiwan* (2012). He was honoured for his contributions to chondrichthyan research in Taiwan and for his assistance and support during field surveys conducted by the second and third authors in Taiwanese fish markets.

Jowett

Hagfish sp. *Neomyxine biniplicata* LR Richardson & Jowett, 1951

Joy P. Jowett was a zoologist at the Department of Zoology, Victoria University College, Wellington, New Zealand. She co-wrote the description in *A New Species of Myxine (cyclostomata) from Cook Strait* (1951).

K

Kabeya

Viper Dogfish *Trigonognathus kabeyai* Mochizuki & Ohe, 1990

Hiromichi Kabeya was captain of the trawler 'Seiryo-Maru' from which the holotype was caught (1986). He was still master of the vessel in 2006.

Kailola

Hortle's Whipray *Himantura hortlei* Last, Manjaji-Matsumoto & Kailola, 2006

Dr Patricia 'Tricia' J. Kailola is an Australian biologist, fish taxonomist and fisheries scientist consultant and honorary fellow at the University of the South Pacific, Suva, Fiji. She was at the Department of Zoology, University of Adelaide (1989) and at the Department of Primary Industries and Energy, Bureau of Resource Sciences, Fisheries Research & Development Corporation, Australia (1993). She is a research associate at the Australian Museum, Sydney. She has written several books on tropical fishes and *Australian Fisheries Resources* (1993). She has made enormous contributions to the systematics of *Ariidae* and a sea catfish is named after her.

Kamohara

Kamohara's Sand-shark *Pseudocarcharias kamoharai* Matsubara, 1936
Bareskin Dogfish *Centroscyllium kamoharai* Abe, 1966

Saddle Carpetshark *Cirrhoscyllium japonicum* Kamohara, 1943

Dr Toshiji Kamohara (1901–1972) was an ichthyologist who graduated from Tokyo Imperial University (1926). He was a professor at a high school in Kochi (1928). He was an artilleryman in the Japanese army (1938–1939). He was a professor in the Zoology Department, Kochi University (1949–1965), retiring as Kochi University Professor Emeritus. He described 52 new species of which 45 are still valid. He secured the type specimen of the sand-shark from a fish market and presented it to Matsubara. He was honoured in the dogfish name for 'his generosity to all ichthyologists'.

Kampa

Longnose Catshark *Apristurus kampae* LR Taylor, 1972

Dr Elizabeth Maitland Boden née Kampa (1922–1986) was an oceanographer who was chief scientist on board the research vessel 'Argos' (1970) and deposited the holotype at the Scripps Institution of Oceanography, where she was on the staff (1944–1977). The University of California Los Angeles awarded her doctorate (1950). She joined the University of Hawaii (1977).

Kan

Miller Lake Lamprey *Entosphenus minimus* Bond & Kan, 1973

Dr Ting Tien Kan is a Hong Kong Chinese biologist and ichthyologist. The University of Oregon awarded his PhD (1975); the thesis was on the systematics, variation, distribution and biology of lampreys in that state. He worked as a research assistant at the AMNH (1967) and later worked at the fisheries section, Department of Biology, University of Papua New Guinea. He co-wrote *The Fishery Resources of Papua New Guinea* (1989).

Kanak

New Caledonian Catshark *Aulohalaelurus kanakorum* Séret, 1990

The Kanaky are an indigenous people in New Caledonia, where this shark was first taken. As Séret puts it, the shark name is 'dedicated to the Melanesian people of New Caledonia'.

Kapala

Kapala Stingaree *Urolophus kapalensis* Yearsley & Last, 2006

The F.R.V. 'Kapala' (formerly of the NSW Fisheries Research Institute, Australia), which collected type, was honoured for the 'extremely valuable fish collections made by the vessel over almost three decades'.

Karaman, MS

Drin Brook Lamprey *Eudontomyzon stankokaramani* MS Karaman, 1974

Dr Mladen Stanko Karaman (1937–1991) was a Yugoslav (Serbian) zoologist who was an expert on isopods. Originally from Skopje (now in the Former Yugoslav Republic of Macedonia), he lived in Split (now in Croatia) (1940–1950) and in Dubrovnik (now in Croatia) (1950–1953), returning to Skopje (1953) where he graduated from the University (1962). He moved to Pristina (now in Kosovo) (1963) as an assistant professor at the Faculty of Invertebrate Zoology. His doctorate was awarded by the University of Ljubljana (now in Slovenia) (1964) and he became Associate Professor (1969–1976). He became full Professor of Comparative Morphology and Invertebrate Taxonomy at what was known at the time as the Kragujevac Faculty of Sciences, Belgrade (then in Yugoslavia and now in Serbia) and was Dean of the Faculty (1980–1982). He was one of many zoologists and biologists produced by this family over four generations. See the next entry for details of his father.

Karaman, SL

Drin Brook Lamprey *Eudontomyzon stankokaramani*
    MS Karaman, 1974

Stanko Luka Karaman (1889–1959) was the describer's father and also a biologist and zoologist. He founded and managed both the Museum of Natural Science and the zoo in Skopje, Yugoslavia (1926). He became the first director of the Institute of Biology, Dubrovnik (1950) but returned to Skopje as Curator of the Museum of Natural Science (1953). His wife, Zora (1907–1974), became a professor in the Faculty of Forestry at the University. His son described him as 'the greatest explorer of freshwater fish fauna in Yugoslavia'. See the entry above for his son with details of places that were in the former Yugoslavia.

Karnasuta

Mekong Stingray *Dasyatis laosensis* TR Roberts &
    Karnasuta, 1987

Dr Jaranthada Karnasuta (b.1949) is a Thai ichthyologist and zoologist. He had degrees in zoology from Kasetsart University and in fisheries from the University of Michigan; the University of Alberta, Canada awarded his doctorate in zoology. He was Permanent Secretary, Department of Fisheries, Ministry of Agriculture and Cooperatives, Bangkok, Thailand (2003) and was Director General, Department of Fisheries of Thailand (2005–2007). He wrote *Systematic Revision of Southeastern Asiatic Cyprinid Fish Genus Osteochilus with Description of Two New Species and a Subspecies* (1993).

Karrer

Smallspine Spookfish *Harriotta haeckeli* Karrer, 1972

Dr Christine Karrer is a German ichthyologist at the Zoologisches Museum der Universität Hamburg, having previously worked at the Zoologisches Museum an der Humboldt Universität zu Berlin. Other fishes are named after her, such as a conger eel. Among her very many papers she wrote *Anguilliformes du Canal de Mozambique* (1983).

Kato

Sharpfin Houndshark *Triakis acutipinna* Kato, 1968

Dr Susumu Kato (b.1933) is an American marine biologist who is Hawaiian-born to Japanese parents. He was employed by the National Marine Fisheries Service (1985), having earlier (1960s and 1970s) been involved in schemes like the farming of sea urchins. He co-wrote *Field Guide to Eastern Pacific and Hawaiian Sharks* (1967).

Kaup

Numbfish genus *Narke* Kaup, 1826

Johann Jakob von Kaup (1803–1873) was a German zoologist, ornithologist and palaeontologist who became the director of the Grand Duke's natural history 'cabinet' in Darmstadt. He was a proponent of 'natural philosophy'; he believed in an innate mathematical order in nature and he attempted biological classifications based on the Quinarian system. He wrote *Classification der Säugethiere und Vögel* (1844). An amphibian and four birds are named after him.

Kemper

Bahamas Ghostshark *Chimaera bahamaensis* Kemper,
    Ebert, Didier & Compagno, 2010
Cape Chimaera *Chimaera notafricana* Kemper, Ebert,
    Compagno & Didier, 2010

Jenny M. Kemper (b.1985) is a biologist who received her bachelor's degree in biological sciences from Florida State University (2007) and her master's degree in marine science from the Pacific Shark Research Center, part of Moss Landing Marine Laboratory, California State University (2012). Her thesis was *Food Habits and Trophic Ecology of Two Common Skate Species*, Bathyraja interrupta *and* Raja rhina, *in Prince William Sound, Alaska*. She has also been involved in several projects pertaining to the systematics, taxonomy and biology of chimaeroid fishes. She is now a PhD student in the Marine Biomedicine and Environmental Sciences program at the Medical University of South Carolina. Her current work involves using molecular approaches to study the phylogenetics and developmental biology of cartilaginous fishes.

Kendall

Aguja Skate *Bathyraja aguja* Kendall & Radcliffe, 1912

Dr William Converse Kendall (1861–1939) qualified as a physician at Georgetown University, Washington DC (1885). Bowdoin University awarded him an honorary doctorate in science (1935) to celebrate the 50 years since he qualified. He worked for the US Bureau of Fisheries (1890–1930), retiring as Director. He was a world authority on salmon and trout.

Kenoje

Ocellate Spot Skate *Okamejei kenojei* Müller & Henle, 1841

The original description has no etymology. The holotype was caught of the southwest coast of Japan and discovered in the fish market in Nagasaki. Müller and Henle adapted the Japanese name of a species, which is *keno-ei*. (See **Zuge**)

Kessler

Caspian Lamprey *Caspiomyzon wagneri* Kessler, 1870

Karl Fedorovich Kessler (1815–1881) was a Russian-German zoologist and collector, who was one of the founders of the St Petersburg Society of Naturalists (1868) and its President (1868–1879). He took part in the (1874) Fedtschensko expedition to Turkestan and the Aralo-Caspian Expedition (1877). He wrote about fishes and other vertebrates in European Russia in his reports on the two expeditions. Several freshwater fishes, such as Kessler's loach *Nemacheilus kessleri*, are also named after him. He was Dean of St Petersburg University (1880). He was a colleague of Nicolai Petrovitch Wagner (q.v.). Two birds are named after him.

Kistnasamy

Natal Shyshark *Haploblepharus kistnasamyi* Human & Compagno, 2006

Lined Catshark *Halaelurus lineatus* Bass, D'Aubrey & Kistnasamy, 1975
[Alt. Banded Catshark]
Thorny Lantern Shark *Etmopterus sentosus* Bass, D'Aubrey & Kistnasamy, 1976

Nadaraj 'Nat' Kistnasamy (b.1938) is a retired South African shark researcher who worked for 45 years at the Oceanographic Research Institute, Durban and was the Natal shyshark's original discoverer. He was honoured 'for outstanding efforts and pioneering work in the systematics and taxonomy of the chondrichthyan fauna of southern Africa'. He turned to shark research when he failed to find a job as an electrician, for which he was originally trained.

Kittipong

Kittipong's Stingray *Himantura kittipongi* Vidthayanon & TR Roberts, 2005
[Alt. Roughback Whipray]

Khun Jarutanin Kittipong is a Thai aquarium fish dealer in Bangkok and prominent fish expert. He provided (2004) the original five specimens that formed the basis for the description.

Klunzinger

Snaggletooth Shark *Hemipristis elongatus* Klunzinger, 1871

Carl Benjamin Klunzinger (1834–1914) was a German physician and zoologist. He studied medicine at Tübingen and Würzburg. He journeyed to Cairo (1862) where he learnt Arabic and worked as a physician at Kosseir, a Red Sea port (1864) where he amassed a collection of fishes and other marine organisms (1864–1869). He compared his collections with those in Stuttgart, Frankfurt and Berlin (1869), distinguishing many new species. He again collected by the Red Sea (1872–1875) before returning to Stuttgart, where he became Professor of Zoology at the University (1884). Among other works he wrote *Synopsis der Fische des Rothen Meeres* (1870).

Knapp

Hagfish sp. *Myxine knappi* Wisner & CB McMillan, 1995

Dr Leslie William Knapp (b.1929) is an ichthyologist at the Department of Vertebrate Zoology, National Museum of Natural History, Smithsonian. Cornell awarded his bachelor's degree (1952) and PhD (1964) and the University of Missouri his master's (1958). He was honoured in the name of the hagfish for supplying the authors with study material. He has published widely over 50 years including a contribution to *A Checklist of the Fishes of the South China Sea* (2000).

Kner

Spotted Houndshark *Triakis maculate* Kner &
Steindachner, 1867

Dr Rudolf Kner (1810–1869) was an Austrian zoologist specialising in ichthyology. He studied medicine in Linz and Vienna, receiving his degrees in medicine and surgery (1835) from the University of Vienna. He then worked with Heckel (q.v.) and others at the National History Museum, Vienna (1836–1841). He was Professor of Natural Science at the University of Lemburg, Austria (now Lviv, Ukraine) (1841–1849). He returned to the University of Vienna as the first Professor of Zoology in Austria (1849).

Knuckey

Shortfin Smooth Lanternshark *Etmopterus joungi*
Knuckey, Ebert & Burgess, 2011

James D. S. Knuckey (b.1986) is an American icthyologist. San Jose State University, San Jose, California awarded his first degree and he is currently (2014) an ichthyology student working towards his master's degree on the systematics of soft-nosed skates from the eastern North Pacific at the Pacific Shark Research Center, Moss Landing Marine Laboratories, California. He has worked for San Jose State University and the Marine Science Institute, Redwood City, California. His publications include IUCN Red List Assessments for little-known sharks and skates and several chapters and the introduction for a forthcoming book entitled *Fishes of the Western Indian Ocean*.

Koch

Lantern Shark sp. *Etmopterus compagnoi* Fricke &
Koch, 1990

Isabel Koch is an ichthyologist and biologist who was awarded her diploma in biology at the University of Stuttgart and worked at the Staatliches Museum für Naturkunde, Stuttgart (1989–1996). Since 1996 she has been the Curator of Reptiles, Amphibians, Fishes and Invertebrates at Wilhelma Zoological and Botanical Gardens, Stuttgart. This institution is noted for its breeding programmes, results being shared through annual presentations for the European Union of Aquarium Curators. She co-wrote the article describing this species, *A New Species of the Lantern Shark Genus Etmopterus from Southern Africa (Elasmobranchii: Squalidae)* (1990) and since then has written a number of popular books and articles for visitors to the zoo and for aquarists and divers.

Kondyurin

Spiny Dogfish ssp. *Squalus acanthias ponticus*
Myagkov & Kondyurin, 1986

V. V. Kondyurin was a Russian ichthyologist. He has dropped from sight and our informant says that it is many years since he worked but he does not know that he is dead.

Konovalenko

Azores Dogfish *Scymnodalatias garricki* Kukuev &
Konovalenko, 1988
Sparsetooth Dogfish *Scymnodalatias oligodon*
Kukuev & Konovalenko, 1988

Ivan Ivanovitch Konovalenko (b.1947) is a fisheries biologist who graduated from the Kaliningrad Technical Institute of Fishing Industry and Agriculture. He worked at Atlantic Scientific Research Institute of Marine Fisheries & Oceanography AtlantNIRO, Kaliningrad, Russia (1967–1991) during which period he was a member of several scientific expeditions in the Atlantic and Pacific oceans and specialised in the study of pelagic fish fauna of the southeast Pacific. Since 1992 he has been an administrator. Among his publications is the co-written *First Discovery of Nesiarchus nasutus Johnson, 1862* (Gempylidae) *in the Southeastern Pacific* (1985) and *Two New Species of Sharks of the Genus* Scymnodalatias (Dalatiidae) *from the North Pacific and Southeastern Pacific Oceans* (1988).

Konstantinou

Springer's Sawtail Catshark *Galeus springeri*
Konstantinou & Cozzi, 1998

Hera Konstantinou (b.1952) is a member of the faculty at the Department of Wildlife & Fisheries Sciences, Texas A & M University. She is an ichthyologist and a crypto-zoologist and is open about her interest in mermaids. In private life she is Mrs Cozzi, wife of the junior author. Her mother is known as the 'shark lady' (see **Clark E**). See **Cozzi** for details of their interest in training dogs.

Kotlyar

Smalldisk Torpedo *Torpedo microdiscus* Parin &
Kotlyar, 1985
Semipelagic Torpedo *Torpedo semipelagica* Parin &
Kotlyar, 1985
Smalleye Lantern Shark *Etmopterus litvinovi* Parin &
Kotlyar, 1990

Dense-scale Lantern Shark *Etmopterus pycnolepis* Kotlyar, 1990

Dr Alexander Nikolaevich Kotlyar (b.1950) is a biologist and ichthyologist who was an associate curator of ichthyology at the Zoological Museum of Moscow University in the 1990s and is now Chief Scientist, Laboratory of Oceanic Ichthyofauna, P.P. Shirshov Institute of Oceanology, Russian Academy of Science. He has more than 500 publications to his name including *Beryciformes Fishes of the World Ocean* (1996) and *Dictionary of Animal Names in Five Languages: Fishes. Latin, Russian, English, German and French* (1989), covering approximately 11,700 names.

Kott

Modoc Brook Lamprey *Entosphenus folletti* Vladykov & Kott, 1976

Kern Brook Lamprey *Entosphenus hubbsi* Vladykov & Kott, 1976

Alaskan Brook Lamprey *Lethenteron alaskense* Vladykov & Kott, 1978

Lamprey sp. *Lethenteron matsubarai* Vladykov & Kott, 1978

Klamath River Lamprey *Entosphenus similis* Vladykov & Kott, 1979

Macedonia Brook Lamprey *Eudontomyzon hellenicus* Vladykov, Renaud, Kott & Economidis, 1982

Dr Edward Kott was a Canadian biologist who was Professor of Biology at the Department of Biology, Wilfrid Laurier University (1967–2003), after which he joined the board of governors. The University of Toronto awarded his PhD. His interest in lampreys is reflected in his published works, which includes *Liver and Muscle Composition of Mature Lampreys* (1971).

Krefft, G

Krefft's Skate *Malacoraja kreffti* Stehmann, 1978

Hagfish sp. *Nemamyxine kreffti* CB McMillan & Wisner, 1982

Brazilian Blind Electric Ray *Benthobatis kreffti* Rincón, Stehmann & Vooren, 2001

Cape Verde Skate *Raja herwigi* G Krefft, 1965

Thickbody Skate *Amblyraja frerichsi* G Krefft, 1968

Whitemouth Skate *Bathyraja schroederi* G Krefft, 1968

Broadbanded Lantern Shark *Etmopterus gracilispinis* G Krefft, 1968

Brazillian Skate *Rajella sadowskii* G Krefft & Stehmann, 1974

Thintail Skate *Dipturus leptocaudus* G Krefft & Stehmann, 1975

Roughskin Skate *Dipturus trachydermus* G Krefft & Stehmann, 1975

Dr Gerhard Krefft (1912–1993) was a German ichthyologist and herpetologist whose great uncle was Johann Ludwig Gerard Krefft (below). His doctorate was awarded by Hamburg University (1938). He went to the Canary Islands (1939) but was recalled to Germany after six months and spent the Second World War in the German army. He founded the fish collection at the Institut für Seefischerei, Hamburg, now held by the Zoological Museum in Hamburg. He was leader of the expeditions of the research vessel 'Walther Herwig' until his retirement (1977). He collected the Brazilian blind electric ray (1968) and was honoured for his contributions to elasmobranch systematics, particularly in respect of elasmobranchs from the southwest Atlantic. He was honoured in the skate name for his 'numerous publications on zoogeography and taxonomy of cartilaginous fishes (particularly the *Rajidae*) and Atlantic meso- and bathypelagic bony fishes; for his leadership of the Ichthyology Group of the Institute of Sea Fisheries and its extensive scientific fish reference collection; and for his knowledge and encouragement during more than ten years of collaboration with Stehmann'.

Krefft, JLG

Reef Manta Ray *Manta alfredi* JLG Krefft, 1868

Johann Ludwig (Louis) Gerard Krefft (1830–1881) was a German-born Australian adventurer, artist, zoologist, and palaeontologist. He emigrated from Germany to the USA (1851) and worked as an artist in New York, then sailed for Australia (1852) to join the gold rush. He was a miner (1852–1857), then joined the National Museum in Melbourne as a collector and artist. He seems to have had a temper: he feuded with the Museum trustees and was dismissed (1874). He refused to accept the dismissal and barricaded himself in his office. He was later carried out of the building, still sitting on his chair, deposited in the street with the door locked behind him. He felt he had been hard done by and so set up a rival 'Office of the Curator of the Australian Museum' and successfully sued the trustees for a substantial sum of money. That was the end of his career and he never worked seriously again, but wrote natural history articles for the Sydney press. He wrote *The Snakes of Australia* (1869) and *The Mammals of Australia* (1871).

Three reptiles, three amphibians, two birds and a mammal are named after him.

Krempf

Marbled Freshwater Whipray *Himantura krempfi* Chabanaud, 1923

Dr Armand Krempf (b.1879) was a French marine biologist, who first went to Vietnam (1903) as part of a scientific expedition to Hanoi. He was the founding Director of the Oceanographic and Fisheries Service, Nha Trang Institute of Oceanography, Indochina (Vietnam) (1922–1931). He wrote *Carcass on Coast of Annam, 1883* (1925) about a large, enigmatic 'armour-plated' creature reportedly washed up on the Vietnamese coast. The identity of this oddity remains a mystery. A bird is named after him.

Krupp

Slender Weasel Shark *Paragaleus randalli* Compagno, Krupp & KE Carpenter, 1996

Dr Friedhelm Krupp (b.1954) is a marine biologist who was Curator of Ichthyology at the Senckenberg Research Institute and Natural History Museum, Frankfurt (1987–1991 and 2002–2011). The University of Mainz, Germany awarded all his degrees – bachelor's and master's (both 1982) and doctorate (1985). He also studied at Stirling, Scotland and Amman, Jordan. He has spent much time in the Middle East and was instrumental in the creation of nature reserves among the oil fields and production platforms in the Arabian Gulf off the coast of Saudi Arabia. Since 2011 he has been Director, Qatar Museum of Natural History, Doha. He co-wrote *Protecting the Gulf's Marine Ecosystems from Pollution* (2007), one of over 120 scientific publications.

Kuhl

Shortfin Devil Ray *Mobula kuhlii* Müller & Henle, 1838
Kuhl's Stingray *Neotrygon kuhlii* Müller & Henle, 1841

Ocellated Eagle Ray *Aetobatus ocellatus* Kuhl, 1823

Dr Heinrich Kuhl (1797–1821) was a German naturalist and zoologist. He became an assistant to Coenraad Temminck at the Rijksmuseum van Natuurlijke Historie, Leiden. He travelled to Java (1820) with his friend Johan Coenraad van Hasselt (q.v.) to study the fauna of the Dutch East Indies. After less than a year in Java, Kuhl died in Buitenzorg (Bogor) of a liver infection brought on by the tropical climate and overexertion. The description of the ocellated eagle ray is contained in a letter from van Hasselt to Temminck which only arrived in Leiden after Kuhl's death. His collections are housed at Leiden and were studied by the authors of the stingray. Seven birds, six mammals, three reptiles and an amphibian are named after him as well as other fishes.

Kukuev

Mid-Atlantic Skate *Rajella kukujevi* Dolganov, 1985

Azores Dogfish *Scymnodalatias garricki* Kukuev & Konovalenko, 1988
Sparsetooth Dogfish *Scymnodalatias oligodon* Kukuev & Konovalenko, 1988

Dr Yefim Izrailevich Kukuev (b.1947) is a Russian ichthyologist. He enrolled at the Kaliningrad Technical Institute, Faculty of Ichthyology and Fishery (1966) and studied the collections of Atlantic fishes at the Atlantic Scientific Research Institute of Marine Fisheries & Oceanography (AtlantNIRO). His doctorate was awarded by the P.P. Shirshov Institute of Oceanology of the Russian Academy of Sciences, Moscow (1980). He is presently the head of the sector of fauna taxonomy in FSUE 'AtlantNIRO'. He has written or co-written more than 160 scientific papers including *Status of Fanfishes of the Genus Pteraclis from the South-Eastern Pacific Ocean (Perciformes: Bramidae)* (2009) and has described 14 new species from 5 fish families.

Kuo

Hagfish sp. *Myxine kuoi* Mok, 2002

Hagfish sp. *Eptatretus chinensis* Kuo & Mok, 1994
Hagfish sp. *Eptatretus nelsoni* Kuo, Huang & Mok, 1994
Hagfish sp. *Eptatretus sheni* Kuo, Huang & Mok, 1994
Hagfish sp. *Paramyxine fernholmi* Kuo, Huang & Mok, 1994
Hagfish sp. *Paramyxine wisneri* Kuo, Huang & Mok, 1994
Hagfish sp. *Myxine formosana* Mok & Kuo, 2001
Hagfish sp. *Eptatretus rubicundus* Kuo, Lee & Mok, 2010

Dr Chien-Hsien Kuo is a Taiwanese molecular biologist. He was a researcher at the Department of Biology, National Taiwan Normal University, Taipei and is now Assistant Professor, Department of Aquatic Bioscience, National Chiayi University, Taiwan. Among his many published papers is the co-written *Phylogeny of Hagfish Based on the Mitochondrial 16S rRNA gene* (2003). He was honoured in the hagfish name for his contributions to hagfish taxonomy.

Kurup

Smoothhound sp. *Mustelus mangalorensis* Cubelio,
Remya & Kurup, 2011

Dr B. Madhusoodana Kurup is Vice-Chancellor, Kerala University of Fisheries and Ocean Sciences. He is a former director of the School of Industrial Fisheries, Cochin, India. As of 23 June 2012 he was under investigation for 'financial misappropriations', allegations which he denied, and the Cochin University of Science and Technology had started the process to terminate his employment, according to a report in 'The Times of India'.

Kux

Turkish Brook Lamprey *Lampetra lanceolata* Kux &
Steiner, 1972

Zdeněk Kux (1923–1990) was a Czech zoologist and Curator of Zoology at the Moravian Museum, Brno, Moravia, Czechoslovakia (Czech Republic) (1950–1988) He wrote *Ichthyofauna of the Western Part of the Carpathian Arc and Adjacent Lowlands: Bionomics Carpathian Lamprey (Lampetra Danford Regan)* (1966).

L

Labrador

Freshwater Stingray sp. *Potamotrygon labradori*
Castex, Maciel & Achenbach, 1963
[Junior Syn. *Potamotrygon motoro*]

José Sánchez Labrador (1718–1798) was a Spanish naturalist and Jesuit monk in Argentina. He was expelled by the authorities to Italy having been accused of espionage (1767). He wrote *Historia de las Regiones del Rio de la Plata*.

Lacépède

Broadnose Sevengill Shark *Notorynchus cepedianus*
Péron, 1807

Chimaera Genus *Callorhinchus* Lacépède, 1798

Clubnose Guitarfish *Rhinobatos thouin* Anonymous,
referred to Lacépède, 1798
Undulate Ray *Raja undulata* Lacépède, 1802
Bottlenose Skate *Rostroraja alba* Lacépède, 1803
[Alt. Spearnose Skate, White Skate Syn. *Raja alba,
Raja bicolor, Raja bramante, Raja marginat, Raja
rostellata*]
Crossback Stingaree *Urolophus cruciatus* Lacépède, 1804

Bernard Germaine Etienne de la Ville, Comte de Lacépède (1756–1825), was a French naturalist. He came to the attention of Buffon, whose work on the classification of animals he was encouraged to continue. Buffon also got him a job at the Jardin du Roi (later Jardin des Plantes) (1785). Lacépède was active in politics and during the 'Terror' lived in Normandy to avoid the guillotine. After his return to Paris he gave up scientific work for a political career and held several offices of state. He was a good musician and composer. He wrote poetry, political treatises and even extended Buffon's work. His own zoological output included *Histoire Naturelle des Quadrupèdes Ovipares, Serpents, Poissons et Cétacées* (1825) and *Histoire Naturelle des Poissons* (1798). Two mammals and two reptiles are named after him. (See **Cépède**)

Lafont

Blonde Ray *Raja brachyura* Lafont, 1873
[Junior Syn. *Raja asterias* Gunther, 1870, Syn. *Raia

brachyura* Lafont, 1871, *Raja blanda, Betaraia
blanda, Raja oculata*]

Alexandre Lafont was a French naturalist and member of the Linnaean Society of Bordeaux who published on cephalopods (1869–1871). He was co-founder and a financial backer of the Société Scientifique d'Arcachon (1866). He described the ray in *Description d'une Nouvelle Espèce de Raie* (1873). He is also commemorated in the name of a gastropod.

Lahille

Argentine Torpedo *Torpedo puelcha* Lahille, 1926

Dr Fernando Lahille (1861–1940) was a French physician and ichthyologist who graduated in medicine at the University of Paris (1893) and emigrated from France to Argentina to work as a hydro-biologist at the Museum in La Plata (1893–1904). He taught in Mariano Acost Normal School (1904–1910) and became Professor of General Zoology, Faculty of Agronomy and Veterinary Medicine, University of Buenos Aires (1910). Lahille Island in the Antarctic is named after him.

Lakeside

Hagfish sp. *Eptatretus lakeside* Mincarone & McCosker
2004

This species is named after the Lakeside Foundation of California for supporting the senior author's work.

Lal Mohan

Stripenose Guitarfish *Rhinobatos variegatus* Nair & Lal
Mohan, 1973

Dr Richard Samuel Lal Mohan (b.1937) was principal scientist at the Central Marine Fisheries Institute, Cochin, India, formerly principal scientist of the Indian Council of Agriculture Research. His bachelor's degree was awarded by Scott Christian College, Nagercoil (1960); Annamalai University awarded his master's and Madurai University his doctorate. He did post-doctoral work at Duke University, North Carolina. He currently works in private management consultancy and is Chairman of the Conservation of Nature Trust, Nagercoil area. He wrote *Whales and Dolphins of India* (1999).

Laland

Brazillian Sharpnose Shark *Rhizoprionodon lalandii*
Müller & Henle, 1839

(See **Delalande**)

Lalanne

Seychelles Spurdog *Squalus lalannei* Baranes, 2003

Maurice Jean Leonard Loustau-Lalanne (b.1955) has held a number of posts in the government of the Seychelles and was the Seychelles principal secretary for the environment at the time of the discovery of this species. He became (2010), and still is, Ambassador and Principal Secretary, Ministry of Foreign Affairs. He was interim Chairman, Air Seychelles (2011) and is Chairman of both the Seychelles Islands Foundation and the Botanical Gardens Foundation. He was honoured for his 'help in organising the expedition that collected the type, his kindness, and his friendship'.

Lamna

Mackerel Skark Genus *Lamna* Cuvier, 1816

The genus name is derived from the Greek lamia, a large and voracious shark. However, originally this comes from Lamia in Greek mythology, daughter of King Belos, who avenged the murder of her own children by killing the children of others, and who behaved so cruelly that her face turned into a nightmarish mask.

Lamotte

Brook Lamprey *Lampetra lamottei* Lesueur, 1827
[Syn. *Petromyzon lamottenii, Lethenteron appendix* (DeKay 1842)]

Lamotte was one of the party, led by the French explorer Philip Francis Renault, that discovered (1720) the location of what became the Mine Lamotte, at one time an important source of lead. The holotype was acquired by Lesueur in a cave near the mine, as he described in his *American Ichthyology, or Natural History of the Fishes of North America, with Coloured Figures from Drawings Executed from Nature* (1827).

Lana

Lana's Sawshark *Pristiophorus lanae* Ebert & Wilms, 2013

Lana Ebert is a marine biologist and the senior author's niece. He tells us that he named this shark after her in recognition of her graduating from the University of San Francisco. J. E. Norton collected the shark holotype off Baltazar Island, Philippines (1966).

Last

Last's Numbfish *Narcine lasti* Carvalho & Séret, 2002

Skate genus *Insentiraja* Yearsley & Last, 1992
Deepsea Skate genus *Brochiraja* Last & McEachran, 2006

Plain Maskray *Neotrygon annotate* Last, 1987
Painted Maskray *Neotrygon leylandi* Last, 1987
Striped Stingaree *Trygonoptera ovalis* Last & Gomon, 1987
Masked Stingaree *Trygonoptera personata* Last & Gomon, 1987
Patchwork Stingaree *Urolophus flavomosaicus* Last & Gomon, 1987
Mitotic Stingaree *Urolophus mitosis* Last & Gomon, 1987
Coastal Stingaree *Urolophus orarius* Last & Gomon, 1987
Brown Stingaree *Urolophus westraliensis* Last & Gomon, 1987
Eastern Looseskin Skate *Insentiraja laxipella* Yearsley & Last, 1992
Pale Skate *Notoraja ochroderma* McEachran & Last, 1994
Blotched Catshark *Asymbolus funebris* Compagno, Stevens & Last, 1999
Western Spotted Catshark *Asymbolus occiduus* Last, Gomon & Gledhill, 1999
Pale Spotted Catshark *Asymbolus pallidus* Last, Gomon & Gledhill, 1999
Dwarf Catshark *Asymbolus parvus* Compagno, Stevens & Last, 1999
Orange Spotted Catshark *Asymbolus rubiginosus* Last, Gomon & Gledhill, 1999
Variagated Catshark *Asymbolus submaculatus* Compagno, Stevens & Last, 1999
Tailspot Lantern Shark *Etmopterus caudistigmus* Last, Burgess & Séret, 2002
Pink Lantern Shark *Etmopterus dianthus* Last, Burgess & Séret, 2002
Lined Lantern Shark *Etmopterus dislineatus* Last, Burgess & Séret, 2002
Blackmouth Lantern Shark *Etmopterus evansi* Last, Burgess & Séret, 2002
Pygmy Lantern Shark *Etmopterus fusus* Last, Burgess & Séret, 2002
False Lantern Shark *Etmopterus pseudosqualiolus* Last, Burgess & Séret, 2002
Ginger Carpetshark *Parascyllium sparsimaculatum* Goto & Last, 2002
Chesterfield Island Stingaree *Urolophus deforgesi* Séret & Last, 2003
New Caledonian Stingaree *Urolophus neocaledoniensis* Séret & Last, 2003
Butterfly Stingaree *Urolophus papilio* Séret & Last, 2003

Coral Sea Stingaree *Urolophus piperatus* Séret & Last, 2003

Spotted Shovelnose Ray *Aptychotrema timorensis* Last, 2004

Magnificent Catshark *Proscyllium magnificum* Last & Vongpanich, 2004

Bareback Shovelnose Ray *Rhinobatos nudidorsalis* Last, Compagno & Nakaya, 2004

Goldeneye Shovelnose Ray *Rhinobatos sainsburyi* Last, 2004

Bali Catshark *Atelomycterus baliensis* WT White, Last & Dharmadi, 2005

Australian Weasel Shark *Hemigaleus australiensis* WT White, Last & Compagno, 2005

Roughnose Stingray *Pastinachus solocirostris* Last, Manjaji & Yearsley, 2005

Enigma Skate *Brochiraja aenigma* Last & McEachran, 2006

White-lipped Skate *Brochiraja albilabiata* Last & McEachran, 2006

Smooth Blue Skate *Brochiraja leviveneta* Last & McEachran, 2006

Small Prickly Skate *Brochiraja microspinifera* Last & McEachran, 2006

Tubemouth Whipray *Himantura lobistoma* Manjaji-Matsumoto & Last, 2006

Australian Grey Smooth-Hound *Mustelus ravidus* WT White & Last, 2006

White-fin Smooth-Hound *Mustelus widodoi* WT White & Last, 2006

Ghost Skate *Notoraja hirticauda* Last & McEachran, 2006

Western Wobbegong *Orectolobus hutchinsi* Last, Chidlow & Compagno, 2006

Jimbaran Shovelnose Ray *Rhinobatos jimbaranensis* Last, WT White & Fahmi, 2006

Indonesian Shovelnose Ray *Rhinobatos penggali* Last, WT White & Fahmi, 2006

Kapala Stingaree *Urolophus kapalensis* Yearsley & Last, 2006

Southern Mandarin Dogfish *Cirrhigaleus australis* WT White, Last & Stevens, 2007

Indonesian Speckled Catshark *Halaelurus maculosus* WT White, Last & Stevens, 2007

Rusty Catshark *Halaelurus sellus* WT White, Last & Stevens, 2007

White-tip Catshark *Parmaturus albimarginatus* Séret & Last, 2007

White-clasper Catshark *Parmaturus albipenis* Séret & Last, 2007

Beige Catshark *Parmaturus bigus* Séret & Last, 2007

Velvet Catshark *Parmaturus lanatus* Séret & Last, 2007

Eastern Highfin Spurdog *Squalus albifrons* Last, WT White & Stevens, 2007

Western Highfin Spurdog *Squalus altipinnis* Last, WT White & Stevens, 2007

Bighead Spurdog *Squalus bucephalus* Last, Séret & Pogonoski, 2007

Fatspine Spurdog *Squalus crassispinus* Last, Edmunds & Yearsley, 2007

Greeneye Spurdog *Squalus chloroculus* Last, WT White & Motomura, 2007

Edmund's Spurdog *Squalus edmundsi* Last, WT White & Stevens, 2007

Eastern Longnose Spurdog *Squalus grahami* Last, WT White & Stevens, 2007

Indonesian Shortsnout Spurdog *Squalus hemipinnis* WT White, Last & Yearsley, 2007

Western Longnose Spurdog *Squalus nasutus* Last, Marshall & WT White, 2007

Bartail Spurdog *Squalus notocaudatus* Last, WT White & Stevens, 2007

Kermadec Spiny Dogfish *Squalus raoulensis* Duffy & Last, 2007

Maugean Skate *Zearaja maugeana* Last & Gledhill, 2007

Bighead Catshark *Apristurus bucephalus* WT White, Last & Pogonoski, 2008

Starry Catshark *Asymbolus galacticus* Séret & Last, 2008

Sombre Catshark *Bythaelurus incanus* Last & Stevens, 2008

Whitefin Swellshark *Cephaloscyllium albipinnum* Last, Motomura & WT White, 2008

Cook's Swellshark *Cephaloscyllium cooki* Last, Séret & WT White, 2008

Painted Swellshark *Cephaloscyllium pictum* Last, Séret & WT White, 2008

Flagtail Swellshark *Cephaloscyllium signourum* Last, Séret & WT White, 2008

Speckled Swellshark *Cephaloscyllium speccum* Last, Séret & WT White, 2008

Saddled Swellshark *Cephaloscyllium variegatum* Last & WT White, 2008

Narrowbar Swellshark *Cephaloscyllium zebrum* Last & WT White, 2008

Dwarf Black Stingray *Dasyatis parvonigra* Last & WT White, 2008

Deepwater Skate *Dipturus acrobelus* Last, WT White & Pogonoski, 2008

Pale Tropical Skate *Dipturus apricus* Last, WT White & Pogonoski, 2008

Grey Skate *Dipturus canutus* Last, 2008

Longnose Skate *Dipturus confuses* Last, 2008

Endeavour Skate *Dipturus endeavouri* Last, 2008

False Argus Skate *Dipturus falloargus* Last, 2008

Pygmy Thornback Skate *Dipturus flindersi* Last & Gledhill, 2008

Grahams' Skate *Dipturus grahami* Last, 2008

Heald's Skate *Dipturus healdi* Last, WT White & Pogonoski, 2008

Blacktip Skate *Dipturus melanospilus* Last, WT White & Pogonoski, 2008

Ocellate Skate *Dipturus oculus* Last, 2008

Queensland Deepwater Skate *Dipturus queenslandicus* Last, WT White & Pogonoski, 2008

Weng's Skate *Dipturus wengi* Séret & Last, 2008

Northern Sawtail Catshark *Figaro striatus* Gledhill, Last & WT White, 2008

Phallic Catshark *Galeus priapus* Séret & Last, 2008

Northern River Shark *Glyphis garricki* Compagno, WT White & Last, 2008

Black-spotted Whipray *Himantura astra* Last, Manjaji-Matsumoto & Pogonoski, 2008

Freshwater Whipray *Himantura dalyensis* Last & Manjaji-Matsumoto, 2008

Leopard Whipray *Himantura leopard* Manjaji-Matsumoto & Last, 2008

Western Round skate *Irolita westraliensis* Last & Gledhill, 2008

Sawback Skate *Leucoraja pristispina* Last, Stehmann & Séret, 2008

White Spotted Gummy Shark *Mustelus stevensi* WT White & Last, 2008

Eastern Spotted Gummy Shark *Mustelus walkeri* WT White & Last, 2008

Peppered Maskray *Neotrygon picta* Last & WT White, 2008

Blue Skate *Notoraja azurea* McEachran & Last, 2008

Broken Ridge Skate *Notoraja lira* McEachran & Last, 2008

Blotched Skate *Notoraja sticta* McEachran & Last, 2008

Thintail Skate *Okamejei leptoura* Last & Gledhill, 2008

Arafura Skate *Okamejei arafurensis* Last & Gledhill, 2008

Thintail Skate *Okamejei leptoura* Last & Gledhill, 2008

Floral Banded Wobbegong *Orectolobus floridus* Last & Chidlow, 2008

Dwarf Spotted Wobbegong *Orectolobus parvimaculatus* Last & Chidlow, 2008

Network Wobbegong *Orectolobus reticulatus* Last, Pogonoski & WT White, 2008

Elongate Carpetshark *Parascyllium elongatum* Last & Stevens, 2008

Sandy Skate *Pavoraja arenaria* Last, Mallick & Yearsley, 2008

Mosaic Skate *Pavoraja mosaic* Last, Mallick & Yearsley, 2008

False Peacock Skate *Pavoraja pseudonitida* Last, Mallick & Yearsley, 2008

Dusky Skate *Pavoraja umbrosa* Last, Mallick & Yearsley, 2008

Tropical Sawshark *Pristiophorus delicatus* Yearsley, Last & WT White, 2008

Challenger Skate *Rajella challengeri* Last & Stehmann, 2008

Eyebrow Wedgefish *Rhynchobatus palpebratus* Compagno & Last, 2008

Eastern Angelshark *Squatina albipunctata* Last & WT White, 2008

Indonesian Angelshark *Squatina legnota* Last & WT White, 2008

Western Angelshark *Squatina pseudocellata* Last & WT White, 2008

Yellow Shovelnose Stingaree *Trygonoptera galba* Last & Yearsley, 2008

Eastern Australian Sawshark *Pristiophorus peroniensis* Yearsley, Last & WT White, 2008

Whitefin Chimaera *Chimaera argiloba* Last, WT White & Pogonoski, 2008

Southern Chimaera *Chimaera fulva* Didier, Last & WT White, 2008

Longspine Chimaera *Chimaera macrospina* Didier, Last & WT White, 2008

Shortspine Chimaera *Chimaera obscura* Didier, Last & WT White, 2008

Western Legskate *Sinobatis bulbicauda* Last & Séret, 2008

Eastern Leg Skate *Sinobatis filicauda* Last & Séret, 2008

Blue Legskate *Sinobatis caerulea* Last & Séret, 2008

Eastern Shovelnose Stingaree *Trygonoptera imitate* Last & Gomon, 2008

Sapphire Skate *Notoraja sapphire* Séret & Last, 2009

Ningaloo Maskray *Neotrygon ningalooensis* Last, WT White & Puckridge, 2010

Borneo Sand Skate *Okamejei cairae* Last, Fahmi & Ishihara, 2010

Sulu Sea Skate *Okamejei jensenae* Last & Lim, 2010

Indonesian Wobbegong *Orectolobus leptolineatus* Last, Pogonoski & WT White, 2010

Narrowtail Stingray *Pastinachus gracilicaudus* Last & Manjaji-Matsumoto, 2010

Starrynose Stingray *Pastinachus stellurostris* Last, Fahmi & Naylor, 2010

Broadnose Wedgefish *Rhynchobatus springeri* Compagno & Last, 2010

Sulu Gollumshark *Gollum suluensis* Last & Gaudiano, 2011

New Zealand Eureka Skate *Brochiraja heuresa* Last & Séret, 2012

Deep-sea Skate sp. *Brochiraja vittacauda* Last & Séret, 2012

Arabian Whipray *Himantura randalli* Last, Manjaji-Matsumoto & Moore, 2012

Deepwater Skate sp. *Notoraja alisae* Séret & Last, 2012

Deepwater Skate sp. *Notoraja fijiensis* Séret & Last, 2012

Deepwater Skate sp. *Notoraja inusitata* Séret & Last, 2012
Deepwater Skate sp. *Notoraja longiventralis* Séret & Last, 2012
Merauke Stingray *Dasyatis longicauda* Last & White, 2013
Stingray sp. *Himantura javaensis* Last & White, 2013
Ridgeback Skate *Dipturus amphispinus* Last & Alava, 2013
Taiwanese Wedgefish *Rhynchobatus immaculatus* Last, Ho & R-R Chen, 2013

Dr Peter Robert Last is an eminent Australian ichthyologist who is the senior taxonomist and Senior Principal Research Scientist at CSIRO Marine and Atmospheric Research, Hobart. The University of Tasmania awarded both his bachelor's degree (1975) and his doctorate (1983). He joined the Tasmanian Fisheries Development Authority as a research scientist (1978) and then moved to the CSIRO Division of Fisheries (1984) as Curator of the Australian National Fish Collection. The Muséum National d'Histoire Naturelle, Paris made him an honorary professor (1997). He has written or co-written over 220 papers, which include descriptions of 153 species. He is senior author of *Sharks and Rays of Australia* (2009) and *Sharks and Rays of Borneo* (2010).

Latham

Knifetooth Sawfish *Anoxypristis cuspidate* Latham, 1794
Longnose Sawshark *Pristiophorus cirratus* Latham, 1794
Leichhardt's Sawfish *Pristis microdon* Latham, 1794
Smalltooth Sawfish *Pristis pectinata* Latham, 1794

Dr John Latham (1740–1837) was a British physician, naturalist and author. He played a leading role in the formation of the Linnean Society of London (1788) and was a Fellow of the Royal Society. He was a practising physician in Kent, England until his retirement (1796). He wrote the descriptions of birds in *The Voyage of Governor Phillip to Botany Bay* (1789), and wrote the *Index Ornithologicus* (1790) and the *General History of Birds* (1821–1828), which he commenced at the age of 81. He designed, sketched and coloured the illustrations himself. He knew all the important English naturalists and collectors of his day, so was able to examine practically all the specimens and drawings of Australian birds and other taxa which reached England. In his later books he provided the first published descriptions and scientific names of many iconic Australian birds. He has been called the 'grandfather' of Australian ornithology. Twelve birds are named after him.

Laura Hubbs

Hagfish sp. *Eptatretus laurahubbsae* CB McMillan & Wisner, 1984

(See **Hubbs, L**)

Láurusson

Iceland Catshark *Apristurus laurussonii* Sæmundsson, 1922

Gísli Láurusson (1865–1935) was an Icelander described as 'Hr Direktur'. He was a goldsmith, watchmaker and farmer with an interest in wildlife, particularly of that in the surrounding seas and birds and their conservation. He also had an interest in local history. He was on the board of several companies including two trawlers (Draupnir and Herjólfur). He is mentioned as supplying Sæmundsson with a fish of a different species. He was honoured 'for his long and invaluable support of the study of Icelandic fishes, and for carefully encouraging the skilful fishermen of Vestmannaeyjar Island' (which is close to the type locality).

Lay

Spotted Ratfish *Hydrolagus colliei* Lay & ET Bennett, 1839

George Tradescant Lay (1800–1845) was a British naturalist, diplomat and missionary. He was naturalist aboard 'HMS Blossom' under Beechey (q.v.) (1825–1828) during which he collected in the Pacific, California, China, eastern Russia, Mexico, Hawaii and South America. He then became a missionary in China (1836–1839) during which he studied Chinese language and culture. He returned to England but was posted back to China as Consul (1843–1845), where he died of fever. A genus of flowering plant is named after him.

Leichhardt

Leichhardt's Sawfish *Pristis microdon* Latham, 1794
[Alt. Largetooth/Freshwater Sawfish]

We think it may have acquired its vernacular name from the fact that it is found in the Leichhardt River in Queensland rather than directly from Dr Frederick Wilhelm Ludwig Leichhardt (1813–1848), who explored in Australia and gave his name to the river.

Lempriere

Thornback skate *Dentiraja lemprieri* J Richardson, 1845

Deputy Assistant Commissary-General Thomas James Lempriere (1796–1852) was interned by the French as an enemy alien (1803) but was quickly released when they calculated he was only seven. He emigrated from France to Van Diemen's Land (Tasmania) (1822), where he became a merchant and banker. He joined the Commissariat Department (1826) as a storekeeper at the penal settlements on Sarah Island and Maria Island, Macquarie Harbour. He became Deputy Assistant Commissary General (1837) and Assistant Commissary General (1844) and Coroner for Tasmania (1846). He was recalled to England (1849) for transfer as Assistant Commissary General in Hong Kong, but his health failed and he was invalided home (1851). He died on the voyage and is buried in Aden. He cut a tide gauge into a rock near Port Arthur, which shows that the sea level there has risen about 13.5 cm since the 1840s. He is honoured, according to the original description as someone for 'whose exertions the ichthyology of Van Diemen's Land is much indebted'.

Leopold

White-blotched River Stingray *Potamotrygon leopoldi* Castex & Castelo, 1970
[Alt. Xingu River Ray, Polka-Dot Stingray, Black Devil Stingray, Yellow-Spotted Black Amazon Stingray, Black Ray, Eclipse Ray]

Léopold III (1901–1983) was King of the Belgians (1934–1951) before abdicating in favour of his son. After this he followed his passion for anthropology and entomology, travelling the world. Among other places he spent time in Senegal and explored the Orinoco and Amazon with Heinrich Harrer. He spent time with the indigenous peoples of the area and wrote *La Fête Indienne, Souvenirs d'un Voyage chez les Indiens du Haut-Xingù* (1967). He was also sponsor of scientific studies at the Institut Royal des Sciences Naturelles de Belgique.

Lesson

Tawny Nurse Shark *Nebrius ferrugineus* Lesson, 1831

René Primevère Lesson (1794–1849) was a French ornithologist and naturalist of enormous influence and importance. Whilst he was best known as a zoologist, he was also a skilled botanist and pharmacist. He was employed on the 'Coquille' (1822) as botanist, and then on the 'Astrolabe' (1826–1829) as naturalist and collector. The Astrolabe's surgeon and botanist was René's brother, Pierre Adolphe Lesson. Both voyages were in the Pacific and called at many islands, including New Zealand. He published a considerable number of ornithological texts including *Manuel d'Ornithologie* (1828), *Centurie Zoologique* (1830) and *Traité d'Ornithologie* (1831). He became Deputy Chief Pharmacist (1832) and later (1839) Chief Pharmacist for the French Navy at Rochefort. He was awarded the Légion d'honneur (1847). A mammal and 16 birds are named after him.

Lesueur

Bluntnose Stingray *Dasyatis say* Lesueur, 1817
Dusky Shark *Carcharhinus obscurus* Lesueur, 1818
Sand Devil *Squatina dumeril* Lesueur, 1818
Tiger Shark *Galeocerdo cuvier* Péron & Lesueur, 1822
Atlantic Stingray *Dasyatis sabina* Lesueur, 1824
Bullnose Eagle Ray *Myliobatis freminvillii* Lesueur, 1824
Brook Lamprey *Lampetra lamottei* Lesueur, 1827
[Syn. *Petromyzon lamottenii, Lethenteron appendix* (DeKay 1842)]

Charles Alexandre Lesueur (Le Sueur) (1778–1846) was a French naturalist, artist and explorer. At 23 he set sail for Australia and Tasmania aboard 'Le Géographe' as an assistant gunner. Baudin (q.v.) appointed him as an official expedition artist when the original artists jumped ship in Mauritius. During the next four years he and fellow naturalist François Péron collected more than 100,000 zoological specimens representing 2,500 new species, and Lesueur made 1,500 drawings. From these drawings he produced a series of watercolours on vellum, which were published (1807–1816) in the expedition's official report, *Voyage de Decouvertes aux Terres Australes*. Lesueur lived in the USA (1815–1837) and undertook some local travels and collecting. He met Audubon (1824) and so admired his work that he suggested he should try again to get it published in France. He returned to France (1837) when two close friends, Thomas Say (q.v.) and Joseph Barabino, died. Lesueur was appointed Curator of the Muséum d'Histoire Naturelle du Havre (1845), which was created to house his drawings and paintings. A bird is named after him.

Lévy-Hartmann

Domino Skate *Bathyraja leucomelanos* Iglésias & Lévy-Hartmann, 2012

Dr Lauriana Lévy-Hartmann (b.1978) is a French biologist and ichthyologist in New Caledonia where she was born and educated. The Université Paris 6 Pierre et Marie Curie-Jussieu awarded her first degree and the University of New Caledonia her doctorate (2011). Her doctoral thesis was *Iden-*

tification Génétique des Populations Ichtyques Marines de Beryx Splendens de la ZEE de la Nouvelle-Calédonie (2011).

Lewin

> Scalloped Hammerhead *Sphyrna lewini* Griffith & CH Smith, 1834

John William Lewin (1770–1819) was an English naturalist and engraver. He went to Sydney, Australia (1800) and collected widely there until his death. His father (some sources say his older brother), William Lewin, was the author of a seven-volume work, *Birds of Great Britain*. Lewin accompanied James Grant on his survey expeditions to the Bass Strait and then to the Hunter River. In Sydney he earned a meagre living as a portrait artist. Governor Macquarie appointed Lewin to the position of City Coroner (1814). He also accompanied Macquarie and made drawings during the construction of the road across the Blue Mountains. Macquarie commissioned Lewin to draw plants collected by the Surveyor-General, Henry Oxley, when exploring the country beyond Bathurst, the Liverpool Plains and New England District. As well as natural history, Lewin also painted landscapes and portraits of aboriginals. He wrote *Prodromus Entomology, Natural History of Lepidopterous Insects of New South Wales* and *Birds of New Holland* (1808), the 1813 edition of which was the first illustrated book to be engraved and printed in Australia. He also wrote *A Natural History of the Birds of New South Wales* (1838). Five birds are named after him.

Castro in *The Sharks of North America* (2011) suggests the shark was named after Danish military surgeon and anatomist Ludwig Lewin Jacobson (1783–1843), but cites no supporting evidence.

Leyland

> Painted Maskray *Neotrygon leylandi* Last, 1987

> Leyland's Skate *Dipturus healdi* Last, WT White & Pogonoski, 2008
> [J Alt. of Heald's Skate]

Guy Geoffrey Leyland (b.1950) is Principal Executive Officer, Western Australia Fishing Industry Council. His BSc in zoology was awarded by the University of Western Australia. He supplied the majority of the Australian material used by Last in the revision of the genus. He co-wrote *Continental Shelf Fishes of Northern and North-western Australia: An Illustrated Guide* (1985) and contributed to *Trawled*

Fishes of Southern Indonesia and Northwestern Australia (1984) whilst he was working for CSIRO, Division of Fisheries Research. His work with the fishing industry in Western Australia has focused on shoring up the statutory basis for rights-based fisheries management including quota-based management and statutory compensation for fishers displaced from marine reserves. He is currently the fishing industry project leader for facilitating third-party environmental certification of Western Australian fisheries through the London-based Marine Stewardship Council, which provides independent validation of the performance of fisheries.

Li

> South China Legskate *Anacanthobatis nanhaiensis* Meng & Li, 1981
> Narrow Legskate *Anacanthobatis stenosoma* Li & Hu, 1982
> Dogfish Shark sp. *Squalus acutirostris* Chu, Meng & Li, 1984
> Humpback Catshark *Apristurus gibbosus* Chu, Meng & Li, 1985
> Broadmouth Catshark *Apristurus macrostomus* Meng, Chu & Li, 1985
> South China Cookiecutter Shark *Isistius labialis* Meng, Chu & Li, 1985
> Smalldorsal Catshark *Apristurus micropterygeus* Meng, Chu & Li, 1986

Dr Sheng Li is an ichthyologist at the South China Sea Fisheries Research Institute, Guangzhou.

Lim

> Sulu Sea Skate *Okamejei jensenae* Last & Lim, 2010

Annie Pek Khiok Lim is a senior laboratory assistant at the Regional Fisheries Biosecurity Centre Sarawak, Malaysia. She has co-written *Field Guide to Sharks of the Southeast Asian Region* (2012).

Lima

> Freshwater Stingray sp. *Potamotrygon limai* Fontenelle, JP da Silva & Carvalho, 2014

Dr José Lima de Figueiredo aka 'Ze Lima' is a Brazilian ichthyologist who is a former researcher and curator of fishes at the Zoological Museum of the University of São Paulo. He is a professor at the University of São Paulo, which awarded his bachelor's degree (1969) and doctorate (1981). He co-wrote *The Northernmost Record of* Bassanago albescens *and Comments on the Occurrence of* Rhynchoconger guppyi

(Teleostei: Anguilliformes: Congridae) *along the Brazilian Coast* (2011).

Limboonkeng

Amoy Fanray *Platyrhina limboonkengi* Tang, 1933
[Syn. *Platyrhina sinensis*]

Dr Lim Boon Keng (1869–1957) was a Singaporean Chinese physician who, in addition to being very interested in natural history, promoted social and educational reforms in Singapore in the early part of the twentieth century. He was expected to co-operate during the Japanese occupation of Singapore in the Second World War, and, in order to avoid this, he developed the technique of getting very drunk. An area of Singapore and a station on the MRT system are named Boon Keng after him.

Linck

Sawfish *Pristis* Linck, 1790

Smoothhound Shark genus *Mustelus* Linck, 1790
Guitarfish genus *Rhinobatos* Linck, 1790

Heinrich Friedrich Linck described these fishes. The species are all described in the article, *Versuch einer Eintheilung der Fische nach den Zähnen* in 'Magasin Neuestes aus der Physik und Naturgeschichte, Gotha' (1790). He is, in all probability, Johann Heinrich Friedrich Linck (1734–1807), a German naturalist who owned the Golden Lion pharmacy in Leipzig. He collected natural history specimens and published on fishes.

Lindberg

Commander Skate *Bathyraja lindbergi* Ishiyama & Ishihara, 1977

Giant Stumptail Stingray *Dasyatis gigantean* Lindberg, 1930

Georgii Ustinovich Lindberg (1894–1976) was a Russian ichthyologist who worked at the Zoological Institute, Russian Academy of Science. He was honoured in the name of the skate for his 'great work' on western North Pacific zoogeography.

Linnaeus/ Linné

Dogfish Shark genus *Squalus* Linnaeus, 1758
Skate genus *Raja* Linnaeus, 1758

Ploughnose Chimaera *Callorhinchus callorynchus* Linnaeus, 1758

Great White Shark *Carcharodon carcharias* Linnaeus, 1758
Rabbitfish *Chimaera monstrosa* Linnaeus, 1758
Common Stingray *Dasyatis pastinaca* Linnaeus, 1758
Blue Skate *Dipturus batis* Linnaeus, 1758
Longnosed Skate *Dipturus oxyrinchus* Linnaeus, 1758
Velvet Belly Lantern Shark *Etmopterus spinax* Linnaeus, 1758
School Shark *Galeorhinus galeus* Linnaeus, 1758
[Alt. Tope Shark, Soupfin Shark, Snapper Shark]
Spiny Butterfly Ray *Gymnura altavela* Linnaeus, 1758
River Lamprey *Lampetra fluviatilis* Linnaeus, 1758
Shagreen Skate *Leucoraja fullonica* Linnaeus, 1758
Common Smooth-Hound *Mustelus mustelus* Linnaeus, 1758
Common Eagle Ray *Myliobatis Aquila* Linnaeus, 1758
Atlantic Hagfish *Myxine glutinosa* Linnaeus, 1758
Angular Roughshark *Oxynotus centrina* Linnaeus, 1758
Sea Lamprey *Petromyzon marinus* Linnaeus, 1758
Blue Shark *Prionace glauca* Linnaeus, 1758
Common Sawfish *Pristis pristis* Linnaeus, 1758
Thornback Ray *Raja clavata* Linnaeus, 1758
Brown Ray *Raja miraletus* Linnaeus, 1758
Common Guitarfish *Rhinobatos rhinobatos* Linnaeus, 1758
Small-spotted Catshark *Scyliorhinus canicula* Linnaeus, 1758
Nursehound *Scyliorhinus stellaris* Linnaeus, 1758
Bonnethead *Sphyrna tiburo* Linnaeus, 1758
Smooth Hammerhead *Sphyrna zygaena* Linnaeus, 1758
Spiny Dogfish *Squalus acanthias* Linnaeus, 1758
Angelshark *Squatina squatina* Linnaeus, 1758
Common Torpedo *Torpedo torpedo* Linnaeus, 1758

Carl Linné (1707–1778) is now much better known by the Latinised form of his name, Carolus Linnaeus (or just Linnaeus). Late in life (1761) he was ennobled and so could call himself Carl von Linné. In the natural sciences he was undoubtedly one of the great heavyweights of all time, ranking with Darwin and Wallace. Primarily a botanist, he nevertheless invented the system, published in *Systema Naturae*, that is still in use today, albeit with modifications, for naming, ranking, and classifying all living organisms. He entered Lunds Universitet (1727) to study medicine and transferred to Uppsala Universitet (1728). At that time botany was part of medical training. His first expedition was to Lapland (1732), followed by an expedition to central Sweden (1734). He went to the Netherlands (1735) and finished his studies as a physician there before enrolling at Universiteit Leiden. He returned to Sweden (1738), lecturing and practising medicine in Stockholm. Appointed Professor at Uppsala (1742), he restored the University's botanical garden. He bought the manor estate of Hammarby outside Uppsala, where he built a small museum for his extensive personal collections (1758). This house and garden still exist and are run by Uppsala University. His son, also Carl, succeeded to his professorship at Uppsala, but was never noteworthy as a botanist. When Carl the Younger died (1783) with no heirs, his mother and sisters sold the elder Linnaeus' library, manuscripts and natural history collections to the English natural historian Sir James Edward Smith, who founded the Linnean Society of London to take care of them. Surprisingly few taxa have been named after Linnaeus, and even fewer still regarded as valid, but five birds, four reptiles, three mammals and two amphibians are among them.

Litvinov

Smalleye Lantern Shark *Etmopterus litvinovi* Parin & Kotlyar, 1990

Dr Feodor F. Litvinov (1954–2011) was a Russian ichthyologist and fisheries expert at the Atlantic Scientific Research Institute of Marine Fisheries and Oceanography, Kaliningrad, Russia, where he was senior scientist. He contributed to longer works and wrote a number of papers, such as *Ecological Characteristics of the Spiny Dogfish* Squalus mitsukurii *from Nazca and Sala-y-Gomez Submarine Ridges* (1990). He was honoured as he helped collect the type aboard the research vessel 'Professor Schtockman'.

Liu, FH

Mottled Skate *Raja pulchra* FH Liu, 1932

Liu Fah-Hsuen was a Taiwanese ichthyologist at Fisheries College, National Taiwan University, Taipei. Before the Chinese Revolution he was at the National Tsing Hua University in Peking (Beijing), China. He wrote *The Fisheries Schools of Taiwan* (1966).

Liu, JX

Blackfin Gulper Shark *Centrophorus isodon* Chu, Meng & JX Liu, 1981
Requiem Shark sp. *Carcharhinus macrops* JX Liu, 1983

Liu Ji-Xing is at Shanghai Fisheries College (now Shanghai Ocean University) and South China Sea Fisheries Research Institute, Guangzhou. His papers include the co-written *Description of a New Species of Scyliorhinidae from China* (1983).

Lloris

Slender Electric Ray *Narcine rierai* Lloris & Rucabado, 1991

Dr Domingo Lloris Samo is a Spanish ichthyologist whose bachelor's degree (1977) and doctorate in biology were awarded by the University of Barcelona (1984). He is retired from the Institute of Marine Sciences, Barcelona, where he worked (1973–2012) and where he founded the Biological Collections of the Institut de Cièncas del Mar. He co-wrote *Encyclopaedia of Living Marine Resources of the Mediterranean* (1999).

Lloyd

Lloyd's Electric Ray *Heteronarce mollis* Lloyd, 1907

Reversed Skate *Amblyraja reversa* Lloyd, 1906
Skate sp. *Okamejei philipi* Lloyd, 1906
Smoothtail Mobula *Mobula thurstoni* Lloyd, 1908
Andaman Legskate *Cruriraja andamanica* Lloyd, 1909

Captain Dr Richard E. Lloyd MB DSc was a physician in the Indian Medical Service and was surgeon naturalist of the Marine Survey of India. He was also an accomplished artist. He took part in a cruise of the research vessel 'Investigator' (1906) and wrote up the results of this and previous voyages (1909). He was (1908) Acting Professor of Biology at the Medical College of Bengal and wrote a biology textbook for students there, *An Introduction to Biology for Students in India*. He was promoted to major (1913). He also wrote *A Description of the Deep-sea Fish Caught by the R.I.M.S. Ship 'Investigator' Since the Year 1900, with Supposed Evidence of Mutation in* Malthopsis (1909).

Loboda

Freshwater Stingray sp. *Potamotrygon amandae* Loboda & Carvalho, 2013
Freshwater Stingray sp. *Potamotrygon pantanensis* Loboda & Carvalho, 2013

Thiago Silva Loboda (b.1982) is a Brazilian ichthyologist whose bachelor's degree (2005) and master's (2010) were awarded by the University of São Paulo, where he works in the Department of Zoology, Institute of Biosciences. He is now studying at the same institution for his doctorate, which he hopes to conclude in 2015. He co-wrote *Systematic Revision of the* Potamotrygon motoro *(Müller & Henle, 1841) Species Complex in the Paraná-Paraguay Basin, with Description of Two New Ocellated Species* (Chondrichthyes: Myliobatiformes: Potamotrygonidae) (2013).

Lockington

Pacific Hagfish *Eptatretus stoutii* Lockington, 1878

William Neale Lockington (1840–1902) was an English critic, architect and zoologist who became Curator of zoology for the California Academy of Science (1875–1881). He wrote a number papers including *Descriptions of New Genera and Species of Fishes from the Coast of California* (1880). He has a number of fish species named after him.

Long, DJ

Galapagos Gray Skate *Rajella eisenhardti* DJ Long & McCosker, 1999
White Spot Chimaera *Hydrolagus alphus* Quaranta, Didier, DJ Long & Ebert, 2006
Galápagos Ghostshark *Hydrolagus mccoskeri* Barnett, Didier, DJ Long & Ebert, 2006
Eastern Pacific Black Ghostshark *Hydrolagus melanophasma* James, Ebert, DJ Long & Didier, 2009
Jaguar Catshark *Bythaelurus giddingsi* McCosker, DJ Long & CC Baldwin, 2012

Dr Douglas J. Long is chief Curator of natural sciences at Oakland Museum of California, a professor of biology at St. Mary's College of California, and a research associate in ichthyology, ornithology and mammalogy at the California Academy of Sciences. University of California (Riverside) awarded his BSc in biological anthropology, UC (Berkeley) (1988) his master's in palaeontology (1990) and his PhD in integrative biology (1994). During this time he was also a curatorial assistant at UC Museum of Palaeontology (1988–1994) and a graduate lecturer. He undertook post-doctoral work at the California Academy of Sciences Department of Ichthyology (1995–1997). He was a research associate at the Marine Mammal Center (1993–1996), Collections Manager at CAS (1998–2005) and Acting Curator (2003–2005). He has written contributions to books and at least 60 papers such as *Sharks, Rays, and Chimaeras* in *The New Book of Knowledge* (2003) and *Animal: The Definitive Visual Guide to the World's Wildlife* (2001).

Long, JJ

Aleutian Dotted Skate *Rhinoraja longi* Raschi & McEachran, 1991
[Syn. *Bathyraja taranetzi, Rhinoraja tanaetzi*]

James John Long (b.1952) was a fisheries biologist (1976–1993) and lives in Shoreline, Washington State. He sent the describers the type series of this skate along with many other fishes from the Aleutian Islands. His extensive collections of Alaska marine fishes are deposited at Oregon State University in Corvallis, Oregon State.

López

Denticled Roundray *Urotrygon cimar* López & Bussing, 1998

Myrna Isabel López Sanchez (b.1937) is a biologist at the Escuela de Biología & Centro de Investigación del Mar y Limnología, Universidad de Costa Rica. She co-wrote *Peces de la Isla del Coco y Peces Arrecifales de la Costa Pacífica de América Central Meridional* (2005). In private life she is Mrs Bussing.

Lovejoy

Roundrays genus *Heliotrygon* Carvalho & Lovejoy, 2011

Gomes' Roundray *Heliotrygon gomesi* Carvalho & Lovejoy, 2011
Rosa's Roundray *Heliotrygon rosai* Carvalho & Lovejoy, 2011
Long-tailed River Stingray *Plesiotrygon nana* Carvalho & Ragno, 2011
Tiger Ray sp. *Potamotrygon tigrina* Carvalho, Sabaj Pérez & Lovejoy, 2011

Dr Nathan Richard Lovejoy is Professor of Biological Sciences, University of Toronto at Scarborough, Ontario, Canada, having previously been Assistant Professor, Department of Zoology, University of Manitoba, Canada (2001–2005). He was post-doctoral research fellow, Museum of Vertebrate Zoology, University of California, Berkeley (1999–2001). The University of Toronto awarded his bachelor's degree (1991) and his master's (1993) and Cornell University his doctorate (1999). Among his more than 30 publications he co-wrote *The Biogeography of Marine Incursions in South America* (2011).

Lowe, RH

Smalleye Smooth-Hound *Mustelus higmani* S Springer & RH Lowe, 1963

Dr Rosemary Helen Lowe-McConnell (b.1921) is a British biologist and ichthyologist. The University of Liverpool awarded her bachelor's and master's degrees and her doctorate. She wrote *Recent Research in the African Great Lakes: Fisheries, Biodiversity and Cichlid Evolution* (2003).

Lowe, RT

Madeiran Ray *Raja maderensis* RT Lowe, 1838
Smooth Lantern Shark *Etmopterus pusillus* RT Lowe, 1839
Birdbeak Dogfish *Deania calcea* RT Lowe, 1839
Bigeye Thresher *Alopias superciliosus* RT Lowe, 1841
Madeira Butterfly Ray *Gymnura hirundo* RT Lowe, 1843

Richard Thomas Lowe (1802–1874) was a British botanist, ichthyologist and malacologist as well as being a clergyman. He graduated from Christ's College, Cambridge (1825) and then took holy orders. He became a pastor in Madeira (1832), where he also studied the local wildlife and wrote a book on its plant life. He died off Sicily in a shipwreck.

Lübbert

Lubbert's Guitarfish *Rhynchobatus luebberti*
Ehrenbaum, 1915
[Alt. African Wedgefish]

Hans Julius Lübbert (1870–1951) studied politics and law at universities in Kiel and Rome but caught severe malaria and had to change course. He was in the civil service (1904–19) working as a fisheries inspector at Cuxhaven and became Director of the Hamburg State Fisheries Service (1922). In retirement he researched and taught but as he was of Jewish descent he was expelled from all societies and public bodies; in the aftermath of the Second World War he was re-instated as Fisheries Director (1945) to rebuild the local fishing industry. He and Ehrenbaum co-wrote *Handbuch der Seefischerei Nordeuropas* (1930).

Luchetti

Chimaera sp. *Chimaera opalescens* Luchetti, Iglésias & Sellos, 2011

Elena A. Luchetti is an Italian ichthyologist at the Station de Biologie Marine et Marinarium de Concarneau, part of the Muséum National d'Histoire Naturelle. She was lead author of *Chimaera Opalescens n. sp., a New Chimaeroid (Chondrichthyes: Holocephali) from the Northeastern Atlantic* (2011) that described this chimaera, and *Écologie et Biologie de Chimaera Monstrosa* (Chondrichtyens Holocéphale) (2006).

Lucifer

Blackbelly Lantern Shark *Etmopterus lucifer* Jordan & Snyder, 1902

The fallen angel Lucifer, of the Judeo-Christian tradition, is referred to in the fourteenth chapter of the Biblical book of Isaiah: 'How art thou fallen from heaven, O Lucifer, son of the morning! How art thou cut down to the ground, which didst weaken the nations!' The name Lucifer, meaning 'light-bearer', was given to the Morning Star (Venus). Presumably the shark was named to emphasise the synergy between the words 'lantern' and 'light', without reference to any 'demonic' quality. Two mammals and four birds also have 'Lucifer' in their names.

Lucifora

Shortnose Eagle Ray *Myliobatis ridens* Ruocco, Lucifora, Díaz de Astarloa, Mabragaña & Delpiani, 2012

Luis Oscar Lucifora is an ichthyologist at the Centro de Investigaciones Ecológicas Subtropicales, Consejo Nacional de Investigaciones Científicas y Técnicas, Misiones, Argentina. He has written a number of articles and contributions to larger works since the late 1990s through to the present day (2013), including the co-written *Food Habits, Selectivity, and Foraging Modes of the School Shark*, Galeorhinus galeus (2006).

Lusitania

Lowfin Gulper Shark *Centrophorus lusitanicus* Barbosa du Bocage & Brito Capello, 1864

Lusitania is the Latin name for Portugal.

Lütken

Roundray *Rajella fyllae* Lütken, 1887

Professor Christian Frederik Lütken (1827–1901) was a Danish naturalist. He was a professional soldier in the Danish army, resigning his commission (1852) to concentrate on natural history. He lectured (1856–1862) at the Zoology Department, University of Copenhagen. He then taught at the Polytechnic School (1877–1881) before returning to the University, becoming Professor of Zoology (1885). He suffered a stroke that left him paralysed (1899) and forced him to retire. He sometimes co-wrote with J.T. Reinhardt (q.v.) and, among other works including very successful books popularising natural history, he wrote *The Ichthyological Results. Danish Ingolf Expedition, II* (1898). Two amphibians are named after him.

M

Mabragaña

Joined-fins Skate *Bathyraja cousseauae* Díaz de
  Astarloa & Mabragaña, 2004
Argentine Skate *Dipturus argentinensis* Díaz de
  Astarloa, Mabragaña, Hanner & Figueroa, 2008
Shortnose Eagle Ray *Myliobatis ridens* Ruocco, Lucifora,
  Díaz de Astarloa, Mabragaña & Delpiani, 2012

Dr Ezequiel Mabragaña is an Argentine ichthyologist at the Department of Marine Sciences, Universiad Nacional de Mar del Plata. He co-wrote *Feeding Ecology of* Bathyraja macloviana (Rajiformes, Arhynchobatidae): *A Polychaete-feeding Skate from the South-west Atlantic* (2005).

Machlan

Weasle Shark sp. *Hemigaleus machlani* Herre, 1929

Perry Lester Machlan (1880–1941) was acting Collector of Customs, Sitankai, Phillipines when Herre described the species. He honoured his 'esteemed friend ... for assisting in his study of the fishes of the Sulu Archipelago'. Machlan was in the Philippines from 1903 or even earlier until his death.

Maciel

Large-spot River Stingray *Potamotrygon falkneri*
  Castex & Maciel, 1963
Freshwater Stingray sp. *Potamotrygon labradori* Castex,
  Maciel & Achenbach, 1963 [JS. *Potamotrygon
  motoro*]

Ignacio O. Maciel co-wrote *Corografia de las Islas del Rio Parana* (1959). Like Castex he was a Jesuit priest.

Mack

Whiskery Shark *Furgaleus macki* Whitley, 1943

George Mack (1899–1963) was a British-born ichthyologist and ornithologist who emigrated to Western Australia (1919) and worked at the National Museum of Victoria, Melbourne (1923–1945) and at the Queensland Museum (1946–1963). He became Director just before his death.

Mackay

Mackay's Torpedo *Torpedo mackayana* Metzelaar, 1919

Donald Jacob Baron Mackay (1839–1921), who was born a Dutch national in the Hague, became eleventh Lord Reay on the death of his father (1876) and changed his nationality to British (1877). He had a distinguished career in politics and as a colonial administrator and governor, including being Governor of Bombay (1885–1890), and served as Under-Secretary of State for India (1894–1895). He maintained very close contact with the Dutch community and the Netherlands and was one of the promoters of the expedition that collected fishes, including this one, in the Dutch West Indies (1904–1905).

Macklot

Hardnose Shark *Carcharhinus macloti* Müller & Henle,
  1839

Heinrich Christian Macklot (1799–1832) was a taxidermist who was appointed to assist members of the Dutch Natural Science Commission. He went on an expedition to New Guinea and Timor (1828–1830). Three birds, a reptile and a mammal are named after him. There is no etymology in the original description. However, the only known specimen of the speartoothed shark, which Müller and Henle described at the same time in *Systematische Beschreibung der Plagiostomen* was a stuffed individual from New Guinea waters collected by Macklot.

Maclay

Japanese Bullhead Shark *Heterodontus japonicus*
  Maclay & Macleay, 1884

Nicholas Miklouho-Maclay (1846–1888) was a Russian explorer, ethnologist, anthropologist and biologist. His education at St Petersburg University came to an abrupt end when he was expelled and debarred for 'breaking the rules'. He left Russia and completed his education in Germany studying the humanities at Heidelberg, medicine at Leipzig and zoology at Jena under Haeckel who made him assistant on his expedition to the Canary Isles (1866). There he became interested in sponges and sharks. He left for Australia (1878), where he established a marine biological station at Watsons Bay in Sydney. He travelled extensively conducting research in the Middle East, Australia, New Guinea and Oceania, living for the most part in Sydney. He was a liberal who was prominent in opposition to slavery and colonialism. He became a close friend of Sir William Macleay (q.v.). He left Australia (1887) to present his work to the Imperial Russian Geographical Society, taking his young family with him. Unfortunately he died there from a previously undiagnosed brain tumour. (His family returned to Sydney where they lived on a pension granted by the Tsar.)

Macleay

Australian Marbled Catshark *Atelomycterus macleayi* Whitley, 1939

Australian Bull Ray *Myliobatis australis* Macleay, 1881
Shortnose Spurdog *Squalus megalops* Macleay, 1881
Mangrove Whipray *Himantura granulate* Macleay, 1883
Cowtail Stingray *Pastinachus atrus* Macleay, 1883
Sydney Skate *Dipturus australis* Macleay, 1884
Japanese Bullhead Shark *Heterodontus japonicus* Maclay & Macleay, 1884
Sandyback Stingaree *Urolophus bucculentus* Macleay, 1884

Sir William John Macleay (1820–1891) was a Scottish medical student who followed his uncle Alexander to Sydney (1838), where he became an all-round naturalist. He wrote widely on entomology, ichthyology and zoology and took part in several collecting expeditions. He published a two-volume *Descriptive Catalogue of Australian Fishes* (1881). The whole of the Macleay family were avid naturalists and collectors, so prolific that the Macleay Museum, University of Sydney was built (1887) to house their vast collection. Alexander Macleay's insect collection was added to by his son, William Sharp Macleay (1792–1865), and expanded to include all aspects of natural history by William's cousin, William John Macleay. A mammal, a reptile and two birds are named after him. There is no etymology in Whitley's description of the cat shark, so which Macleay it was named after remains a mystery. However, Whitely was educated in zoology at Sydney Museum so would have known what giants among Australian zoologists the family were.

Macloviana

Patagonian Skate *Bathyraja macloviana* Norman, 1937

The original description has no etymology. However, it is not named for a person called MacLove or some such, but is derived from the Latin name for the town St Malo in Brittany, France. People from this town are called 'Malouines'. The Spanish name 'Malvinas' for the Falkland Islands is a derivative of their French name 'les Malouines'. Louis Antoine de Bougainville christened the islands (1764) in reference to the Saint Malo, from which his expedition departed. So several species from that area are also called *macloviana* after these islands.

Macneill

Shorttail Torpedo *Torpedo macneilli* Whitley, 1932

Francis 'Frank' Alexander McNeill (1896–1969) was an Australian zoologist and ichthyologist at the Australian Museum in Sydney. He served (1914–1918) with the Australian Light Horse in the First World War including fighting at Gallipoli (1915). He was in charge of crustacea (1922–1961) and was the Australian Museum's representative on the Great Barrier Reef Committee (1945). He collected the holotype and illustrated it in an earlier publication as *T. fairchildi*. Whitley wrote his obituary.

Magdalena

Magdalena River Stingray *Potamotrygon magdalenae* AHA Duméril, 1865

As the vernacular name explains, this species is not named after a person but after a river.

Mallick

Sandy (Peacock) Skate *Pavoraja arenaria* Last, Mallick & Yearsley, 2008
Mosaic Skate *Pavoraja mosaic* Last, Mallick & Yearsley, 2008
False Peacock Skate *Pavoraja pseudonitida* Last, Mallick & Yearsley, 2008
Dusky Skate *Pavoraja umbrosa* Last, Mallick & Yearsley, 2008

Dr Stephen Anthony Mallick is a wildlife consultant who works for the Biodiversity Conservation Branch, Department of Primary Industries and Water, Hobart, Tasmania. Previously (1997) he was on the staff of the Tasmanian Parks & Wildlife Service, Hobart. The University of Tasmania awarded his doctorate (2001). He co-wrote *A Review of the Australian Skate Genus Pavoraja* Whitley (*Rajiformes: Arhynchobatidae*) (2008).

Malm

Skate genus *Amblyraja* Malm, 1877
Hardnose Skate genus *Leucoraja* Malm, 1877

August Wilhelm Malm (1821–1882) was a Swedish zoologist who was the first director of Göteborgs Naturhistoriska Museum (1848–1882) and was titular Professor there (1861). He studied zoology at both the University of Copenhagen and at Lund in

Sweden. He worked at the Riksmuseet in Stockholm (1840) and wrote *Svenska Iglar* (1863).

Mangalore

Smoothhound sp. *Mustelus mangalorensis* Cubelio, Remya & Kurup, 2011

The smoothhound is named after the type locality, Mangalore. The fishermen of Munumbam Fisheries Harbour, Kerala, India supplied the holotype to the authors.

Manilo

Aden Gulf Torpedo *Torpedo adenensis* Carvalho, Stehmann & Manilo, 2002

Dr Leonid Georghievich Manilo (b.1953) is a zoologist whose bachelor's degree in hydrobiology and ichthyology was awarded by Kubanskiy University, Krasnodar, Russia (1975). The P.P. Shirshov Institute of Oceanology of the Russian Academy of Sciences awarded his doctorate (2001). He worked in the Southern Scientific Research Institute of Marine Fisheries and Oceanography, Kerch, Crimea (1976–1985) during which time he took part in ten expeditions to the Indian and Antarctic Oceans. Since 1986 he has worked as an ichthyologist at the National Museum of Natural History, National Academy of Sciences of Ukraine, Kiev, where he is now a senior scientist and curator of ichthyology. He was a biologist on the ninth Ukrainian Antarctic Expedition (2004–2005). Among his 71 scientific publications are *Gobies Fishes (Gobiidae, Perciformes) of North-Western Part of the Black Sea and Adjoining Estuary Ecosystems* (2008) and, as co-author, *The Longnose Chimaera Neoharriotta pinnata in the Waters of Arabian Sea* (1989) and *Taxonomic Composition, Diversity and Distribution of Coastal Fishes of the Arabian Sea* (2003).

Manjali

(See **Manjaji-Matsumoto**)

Manjaji-Matsumoto

Roughnose Stingray *Pastinachus solocirostris* Last, Manjaji & Yearsley, 2005
Hortle's Whipray *Himantura hortlei* Last, Manjaji-Matsumoto & Kailola, 2006
Black-spotted Whipray *Himantura astra* Last, Manjaji-Matsumoto & Pogonoski, 2008
Freshwater Whipray *Himantura dalyensis* Last & Manjaji-Matsumoto, 2008
Leopard Whipray *Himantura leopard* Manjaji-Matsumoto & Last, 2008
Narrowtail Stingray *Pastinachus gracilicaudus* Last & Manjaji-Matsumoto, 2010
Arabian Whipray *Himantura randalli* Last, Manjaji-Matsumoto & Moore, 2012

Dr Bernadette Mabel Manjaji-Matsumoto is a marine scientist and ichthyologist and Deputy Director of the Borneo Marine Research Institute, Universiti Malaysia Sabah. Universiti Kebangsaan Malaysia awarded her bachelor's degree (1992) and the University of Tasmania her doctorate (2004). She wrote *New Records of Elasmobranch Species from Sabah* (2002).

Margarita

Daisy Stingray *Dasyatis margarita* Günther, 1870

Margarita has several meanings including 'Pearl' and 'Daisy'. The vernacular name tells us which is intended here (see also **Margaritella**).

Margaritella

Pearl Stingray *Dasyatis margaritella* Compagno & TR Roberts, 1984

Margaritella is the diminutive of Margarita and is used to indicate this species is smaller than its relative (see **Margarita** above). The vernacular name is used not only to convey 'Little Pearl' but to reflect that this species has a pearl spine.

Maria

Ukranian Brook Lamprey *Eudontomyzon mariae* LS Berg, 1931

Maria Mikhailovna Berg née Ivanova was (1922) the second wife of the describer. He honoured her, not just for the family connection, but because it was she 'who examined many thousands of river lampreys from the mouth of the Neva and other streams, falling into the Finnish Gulf'.

Mariana

Brazilean Large-eyed Stingray *Dasyatis marianae* Gomes, Rosa & Gadig, 2000

Mariana Ramos de Oliveira Gadig Gonçalves is the third describer's daughter. She works at the epidemiology department at the Instituto Adolpho Lutz, Santos, Brazil.

Marina

River Stingray sp. *Potamotrygon marinae* Deynat, 2006

Marina Deynat is the describer's daughter.

Marini

Argentine Angelshark *Squatina argentina* Marini, 1930
Blotched Sand Skate *Psammobatis bergi* Marini, 1932
Angular Angelshark *Squatina guggenheim* Marini, 1936
Angelshark sp. *Squatina punctate* Marini, 1936

Dr Tomás Leandro Marini (b.1902) was an Argentine ichthyologist and fisheries biologist. The University of Buenos Aires awarded his doctorate (1927). He was Chief of Experimental Work in Agricultural Zoology, University of Buenos Aires (1927) then Chief of the Division of Fisheries and Pisciculture at the Argentine Department of Agriculture (1930). He was Director General of Fisheries, Buenos Aires (1967).

Markle

Nova Scotia Skate *Breviraja marklei* McEachran & Miyake, 1987

Dr Douglas Frank Markle (b.1947) is an ichthyologist who is Professor Emeritus of Fisheries, Oregon State University, which he joined (1985). Cornell University awarded his bachelor's degree (1969) and the College of William and Mary, Virginia Institute of Marine Science, Gloucester Point, Virginia, his master's (1972) and doctorate (1976). He sent the describers the type series of the species. His major collaboration has been with John Edward Olney on larval taxonomy and systematics of *Carapidae* and has carried out other systematics work was on *Gadiformes, Alepocephalidae*, and *Catostomidae* and early life history ecology of *Pleuronectidae* and *Catostomidae*. His major publications include *Audubon's Hoax: Ohio River Fish Described by Rafinesque* (1997) and, as co-author, *Systematics of Pearlfishes* (Pisces: Carapidae) (1990).

Marley

Longnose Pygmy Shark *Heteroscymnoides marleyi* Fowler, 1934

Harold Walter Bell-Marley (1872–1945) was Principal Fisheries Officer (1918–1937) at Durban, South Africa, and a naturalist with a particular interest in entomology. He was born in England and went to South Africa at the time of the Boer War, then decided to stay on. He collected continuously for c.50 years nearly everywhere south of the Zambesi River. Museums in many parts of the world have specimens sent by him. He also collected birds' eggs, and his collection, now in the Pretoria Museum, is regarded as one of the most complete ever assembled. He contracted blackwater fever (1944) while collecting in northern Zululand and died soon after his return to Durban. A mammal and three birds are also named after him. He was honoured because he had 'collected many interesting South African fishes' for Fowler.

Marshall

Western Longnose Spurdog *Squalus nasutus* Last, Marshall & WT White, 2007

Dr Lindsay J. Marshall was at the Centre for Fish and Fisheries Research, School of Biological Sciences and Biotechnology, Murdoch University, Western Australia (2006), which awarded her bachelor's degree. The University of Tasmania awarded her doctorate (2011). She co-wrote *Reproductive Biology and Diet of the Southern Fiddler Ray*, Trygonorrhina fasciata (Batoidea: Rhinobatidae), *an Important Trawl by Catch Species*. (2006). In addition to being a shark scientist she is also an artist and biological illustrator: she is presently illustrating the sharks and rays of the world for the Tree of Life Project.

Martín Salazar

Brownband Numbfish *Diplobatis guamachensis* Martín Salazar, 1957

Dr Felipe José Martín Salazar (b.1930) was a Venezuelan ichthyologist. He was head of the Fisheries and Wildlife Division, Venezuelan Ministry of Agriculture (1960). He wrote *Las Especies del Género Farlowella de Venezuela* (Piscis-Nematognalhi-Loricariidae) *con Descripción de 5 Especies y 1 Subespecie Nuevas* (1964).

Matallanas

Striped Rabbitfish *Hydrolagus matallanasi* Soto & Vooren, 2004

Jesús Matallanas García is a Spanish ichthyologist and professor in the Faculty of Biosciences at the University of Barcelona. Many of his published papers have described new species, such as *Description of* Gosztonyia antarctica, *a New Genus and Species of Zoarcidae (Teleostei: Perciformes) from the Antarctic Ocean* (2009). He was honoured for his 'extensive work and tireless dedication to ichthyology'.

Matheson

Brightspot Skate *Breviraja claramaculata* McEachran
& Matheson, 1985
Blackbelly Skate *Breviraja nigriventralis* McEachran &
Matheson, 1985
Blacknose Skate *Breviraja mouldi* McEachran &
Matheson, 1995

Dr Richard Edmond 'Eddie' Matheson Jr is an associate research scientist with the Florida Fish and
Wildlife Conservation Commission, Fish and Wildlife Research Institute, which he joined in 1987. He
graduated from the College of William and Mary,
Williamsburg, Virginia (1975) and then was at the
Department of Wildlife and Fishery Sciences, Texas
A & M University, where he gained his doctorate
(1983).

Matsubara

Pitted Stingray *Dasyatis matsubarai* Miyosi, 1939
Lamprey sp. *Lethenteron matsubarai* Vladykov & Kott,
1978

Kamohara's Sand-shark *Pseudocarcharias kamoharai*
Matsubara, 1936

Dr Kiyomatsu Matsubara (1907–1968) was originally called Kiyomatsu Sakamoto but on marrying
he took his wife's name as the family surname. He
was a Japanese herpetologist and ichthyologist. The
Imperial Fisheries Institute (now the Tokyo University of Fisheries) awarded his bachelor's degree
(1929). He was Professor of Fisheries at the Imperial
Fisheries Institute (1943–1947) and was Professor
in the Department of Fisheries, Kyoto University
(1947–1968). He wrote *Fish Morphology and Hierarchy*
(1955). He was a student of Carl Hubbs, who wrote
his obituary. Miyosi honoured him in the stingray
name as the person who collected one of the para-
types at a fish market, and 'to whom the author is
much indebted for many favours'.

Maul

Bigeye Sand Tiger *Odontaspis noronhai* Maul, 1955

Günther Edmund Maul (1909–1997) was a German
ichthyologist and taxidermist who lived and worked
most of his life in Madeira. He started work as a taxidermist at the Museu Municipal do Funchal (1930)
and went on to become Director (1940–1979) for
the rest of his working life, although he continued
his research after he retired. He opened the Museum's aquarium to the public (1959). He took part

in several expeditions to the Salvage Islands (1963),
notably with the French bathyscaphe 'Archimède'
(1966). He described several new fish species and
has several fishes and other taxa named after him.

Maurin

West African Catshark *Scyliorhinus cervigoni* Maurin &
Bonnet, 1970

Claude Maurin was Professor and Director at the
French Institute of Marine Fisheries, Nantes (1970–
1982). He was associated with the University of
Montpellier.

Mazhar

Alexandrine Torpedo *Torpedo alexandrinsis* Mazhar,
1987

Dr Fatma M. Mazhar is Professor of Vertebrates and
Embyology at the Zoology Department, Women's
College for Arts, Science and Education, Ain
Shams University, Cairo, which she joined before
1971, having previously (1964) been on the staff of
the Marine Biological Station, Al-Ghardaqa. She
co-wrote *The Elasmobranchs of the North-western Red
Sea* (1964).

McCain

McCain's Skate *Bathyraja maccaini* S Springer, 1971

Dr John Charles McCain (b.1939) was formerly a
professor and senior research scientist at the University of Petroleum and Minerals, Dhahran, Saudi
Arabia, where he was the principal investigator of
the Northern Area Marine Environmental Baseline
Study and investigated the effects of oil spills in the
Persian Gulf. Texas Christian University awarded
his bachelor's degree (1962), the College of William
and Mary, Virginia his master's (1964) and George
Washington University his doctorate (1967). He collected the holotype (1967) when aboard the marine
vessel 'Hero'. He co-wrote *The Gulf War Aftermath:
An Environmental Tragedy* (1993).

McConnaughey

Shorthead Hagfish *Eptatretus mcconnaugheyi* Wisner
& CB McMillan, 1990

Ronald R. McConnaughey is a marine technician and
diver. He was at the Scripps Institution of Oceanography, rising to Manager of Scientific Collections
and Experimental Aquariums (retired 2000). Among
his publications is *A Tropical Eastern Pacific Barna-*

cle, Megabalanus coccopoma *(Darwin), in Southern California, following El Nino 1982–83* (1987). He was honoured because he had helped develop the gear (a three-chambered trap) used by the authors to capture the holotype.

McCosker

Hagfish sp. *Myxine mccoskeri* Wisner & CB McMillan, 1995

Hagfish sp. *Eptatretus mccoskeri* CB McMillan, 1999

Galápagos Ghostshark *Hydrolagus mccoskeri* Barnett, Didier, DJ Long & Ebert, 2006

Galapagos Gray Skate *Rajella eisenhardti* DJ Long & McCosker, 1999

Hagfish sp. *Eptatretus lakeside* Mincarone & McCosker, 2004

Jaguar Catshark *Bythaelurus giddingsi* McCosker, DJ Long & CC Baldwin, 2012

Dr John Edward McCosker (b.1945) is an ichthyologist and evolutionary biologist who is Senior Scientist and First Professor of Aquatic Research at California Academy of Sciences, San Francisco. The Scripps Institution of Oceanography awarded his doctorate (1973). He was Director of the Steinhart Aquarium (1973–1994) and is an adjunct professor in marine biology at San Francisco State University. His research has included the marine life of the Galápagos, the biology of the coelacanth, the biology and systematics of snake eels and moray eels and white shark attack behaviour. He has written more than 270 articles and books including *The History of Steinhart Aquarium: A Very Fishy Tale* (1999) and, as co-author, *Great White Shark* (1991). He has appeared in a number of films, including 'Galapagos: Beyond Darwin' (1996) and 'Naked Science' (2004). Two slime eels are named after him.

McCulloch

Southern Round Skate *Irolita waitii* McCulloch, 1911

Rusty Carpetshark *Parascyllium ferrugineum* McCulloch, 1911

Ornate Angelshark *Squatina tergocellata* McCulloch, 1914

Dumb Gulper Shark *Centrophorus harrissoni* McCulloch, 1915

Longsnout Dogfish *Deania quadrispinosum* McCulloch, 1915

Wide Stingaree *Urolophus expansus* McCulloch, 1916

Greenback Stingaree *Urolophus viridis* McCulloch, 1916

Alan Riverstone McCulloch (1885–1925) was an Australian ichthyologist and field naturalist. He began his scientific career at the very young age of thirteen when he was an unpaid assistant to E.R. Waite at the Australian Museum where he was encouraged to study zoology. After three years (1901) he was employed as a 'mechanical assistant' and five years after that was appointed Curator of Fishes (1905), a post he held for the rest of his life. He also had responsibility for the crustacean collection (1905–1921). He soon began to publish scientific papers and was prolific, completing over 100, many of which he illustrated himself. He also wrote some longer works such as *Check List of Fishes and Fish-like Animals of New South Wales* (1922). He travelled and collected widely in Australasia including Queensland, New Guinea (1922), the Great Barrier Reef and many Pacific islands. Apparently overwork led to poor health and he began a sabbatical year in Hawaii but shot himself in his hotel room in Honolulu.

McEachran

Madagascar Pygmy Skate *Fenestraja maceachrani* Séret, 1989

Ray family *Urotrygonidae* McEachran, Dunn & Miyake, 1996

Pygmy Skate genus *Fenestraja* McEachran & Compagno, 1982

Pygmy Skate genus *Neoraja* McEachran & Compagno, 1982

Deepsea Skate genus *Brochiraja* Last & McEachran, 2006

Maya Skate *Leucoraja caribbaea* McEachran, 1977

Virginia Skate *Leucoraja virginica* McEachran, 1977

Tumbes Round Stingray *Urobatis tumbesensis* Chirichigno & McEachran, 1979

Onefin Skate *Gurgesiella dorsalifera* McEachran & Compagno, 1980

East African Skate *Okamejei heemstrai* McEachran & Fechhelm, 1982

Allen's Skate *Pavoraja alleni* McEachran & Fechhelm, 1982

Freckled Sand Skate *Psammobatis lentiginosa* McEachran, 1983

Shortfin Sand Skate *Psammobatis normani* McEachran, 1983

Smalltail Sand Skate *Psammobatis parvacauda* McEachran, 1983

Colombian Electric Ray *Diplobatis colombiensis* Fechhelm & McEachran, 1984

Carolina Pygmy Skate *Neoraja carolinensis* McEachran
& Stehmann, 1984

Brightspot Skate *Breviraja claramaculata* McEachran &
Matheson, 1985

Blackbelly Skate *Breviraja nigriventralis* McEachran &
Matheson, 1985

Nova Scotia Skate *Breviraja marklei* McEachran &
Miyake, 1987

Cortez' Ray *Raja cortezensis* McEachran & Miyake, 1988

Dwarf Roundray *Urotrygon nana* Miyake & McEachran,
1988

Reticulate Roundray *Urotrygon reticulate* Miyake &
McEachran, 1988

Fake Roundray *Urotrygon simulatrix* Miyake &
McEachran, 1988

Aleutian Dotted Skate *Rhinoraja longi* Raschi &
McEachran, 1991

Pale Skate *Notoraja ochroderma* McEachran & Last, 1994

Blacknose Skate *Breviraja mouldi* McEachran &
Matheson, 1995

Butterfly Skate *Bathyraja mariposa* Stevenson, Orr, Hoff
& McEachran, 2004

Enigma Skate *Brochiraja aenigma* Last & McEachran,
2006

White-lipped Skate *Brochiraja albilabiata* Last &
McEachran, 2006

Smooth Blue Skate *Brochiraja leviveneta* Last &
McEachran, 2006

Small Prickly Skate *Brochiraja microspinifera* Last &
McEachran, 2006

Ghost Skate *Notoraja hirticauda* Last & McEachran, 2006

Blue Skate *Notoraja azurea* McEachran & Last, 2008

Broken Ridge Skate *Notoraja lira* McEachran & Last, 2008

Blotched Skate *Notoraja sticta* McEachran & Last, 2008

Leopard Skate *Bathyraja panther* Orr, Stevenson, Hoff,
Spies & McEachran, 2011

Dr John D. McEachran (b.1941) is Professor of Ichthyology, Texas A & M University and Chief Curator and Curator of Fishes at the Texas Cooperative Wildlife Collection. Michigan State University awarded his bachelor's degree (1965) and the William and Mary College, Virginia his master's (1968) and doctorate (1973). During his 34-year career at Texas A & M he taught ichthyology, herpetology, conservation of marine resources, museums and their functions, systematics of vertebrates, evolutionary mechanisms of vertebrates, and biology of nekton, at both undergraduate and graduate level. His research was mainly focused on systematics and evolutionary relationships of skates and rays (*Batoidea*) and the biogeography of fishes of the Gulf of Mexico. Among his many publications he co-wrote the two-volume *Fishes of the Gulf of Mexico* (1998 and 2006) and wrote

Revision of the South American Skate Genus Sympterygia (Chondrichthyes, Rajiformes) (1982). He was honoured in the skate name for his 'major contributions to skate and ray systematics'.

McKay

Banded Numbfish *Narcine westraliensis* McKay, 1966
Circular Stingaree *Urolophus circularis* McKay, 1966
Lobed Stingaree *Urolophus lobatus* McKay, 1966

Roland J. McKay (b.1935) is an Australian marine biologist and systematic ichthyologist who worked as a senior technician of fishes at the Museum of Western Australia (1964–1972) and became Curator of Fishes, Queensland Museum, Brisbane (1972). He is now Director and owner of the Chillagoe Museum in Northern Queensland. He is a recognised expert on Sillaginid fishes and wrote the FAO Species Catalogue 14, *Sillaginid Fishes of the World (Family Sillaginidae)* (1992). He has also written and published many papers and articles, including *Pearl Perches of the World (Family Glaucosomatidae): An Annotated and Illustrated Catalogue of the Pearl Perches Known to Date* (1997).

McMillan, CB

Hagfish sp. *Myxine mcmillanae* Hensley, 1991

Hagfish sp. *Nemamyxine kreffti* CB McMillan & Wisner,
1982

Giant Hagfish *Eptatretus carlhubbsi* CB McMillan &
Wisner, 1984

Hagfish sp. *Eptatretus laurahubbsae* CB McMillan &
Wisner, 1984

Hagfish sp. *Eptatretus strahani* CB McMillan & Wisner,
1984

Hagfish sp. *Eptatretus nanii* Wisner & CB McMillan, 1988

Guadaloupe Hagfish *Eptatretus fritzi* Wisner & CB
McMillan, 1990

Shorthead Hagfish *Eptatretus mcconnaugheyi* Wisner &
CB McMillan, 1990

Cortez Hagfish *Eptatretus sinus* Wisner & CB McMillan,
1990

Hagfish sp. *Myxine debueni* Wisner & CB McMillan, 1995

Hagfish sp. *Myxine dorsum* Wisner & CB McMillan, 1995

Hagfish sp. *Myxine fernholmi* Wisner & CB McMillan,
1995

Hagfish sp. *Myxine hubbsi* Wisner & CB McMillan, 1995

Hagfish sp. *Myxine hubbsoides* Wisner & CB McMillan,
1995

Hagfish sp. *Myxine knappi* Wisner & CB McMillan, 1995

Hagfish sp. *Myxine mccoskeri* Wisner & CB McMillan,
1995

Hagfish sp. *Myxine pequenoi* Wisner & CB McMillan, 1995

Hagfish sp. *Myxine robinsorum* Wisner & CB McMillan, 1995

Hagfish sp. *Eptatretus grouseri* CB McMillan, 1999

Hagfish sp. *Eptatretus mccoskeri* CB McMillan, 1999

Hagfish sp. *Eptatretus wisneri* CB McMillan, 1999

Hagfish sp. *Eptatretus fernholmi* CB McMillan & Wisner, 2004

Hagfish sp. *Eptatretus moki* CB McMillan & Wisner, 2004

Hagfish sp. *Eptatretus walkeri* CB McMillan & Wisner, 2004

Dr Charmion B. McMillan (b.1925) was a marine biologist at the Marine Biology Research Division, Scripps Institution of Oceanography, University of California San Diego. She published widely, including papers describing many new species such as *Three New Species of Seven-gilled Hagfishes (Myxinidae, eptatretus) from the Pacific Ocean* (1984). She was honoured in the name of the hagfish for her 'fine contributions to hagfish science'.

McMillan, DA

Hagfish sp. *Eptatretus grouseri* CB McMillan, 1999

David 'Grouser' Allen McMillan (b.1957) is a Chief Engineer in the US Merchant Marines and the author's son. She honoured him in the name for 'continued encouragement of Mom's hagfish studies and for his knowledge and love of ships and the sea'.

McMillan, PJ

McMillan's Catshark *Parmaturus macmillani* Hardy, 1985

Peter John McMillan (b.1955) is a fisheries biologist who has a master's degree in zoology. He is a specialist in deep-sea species and collected the holotype and many examples of un-described or poorly known marine fish and invertebrate species from deep water off New Zealand. He has worked for the New Zealand National Institute of Water and Atmospheric Research since 1980. He is an honorary research associate of the Museum of New Zealand, Te Papa Tongarewa, Wellington. He co-wrote *Two New Species of* Coelorinchus (Teleostei, Gadiformes, Macrouridae) *from the Tasman Sea* (2009).

Mead

Blotched Catshark *Scyliorhinus meadi* S Springer, 1966

Dr Giles W. Mead Jr (1928–2003) was the first person to draw the describer's attention to a specimen of the species. His doctorate in ichthyology was awarded

by Stanford University. He initially worked at the Smithsonian for the US Fish and Wildlife Service as a laboratory director in charge of fish taxonomy. In the 1960s he was Curator of fishes at the Museum of Comparative Zoology and a professor of biology at Harvard. In the 1970s he was Director of the Los Angeles County Natural History Museum. His family was wealthy – his father was a co-founder of the chemical giant Union Carbide – and he was able to retire to the Mead family ranch and vineyard in the Napa Valley, California (1978).

Mee

Bigeye Electric Ray *Narcine oculifera* Carvalho, Compagno & Mee, 2002

Jonathan Kevin Lorne Mee (b.1960) is an American marine biologist whose bachelor's degree was awarded by San Francisco State University (1982). He then worked at the Steinhart Aquarium, San Francisco, leaving to be an aquarium curator in the Sultanate of Oman (1987–1990). He worked at the Hatfield Marine Science Center, a part of Oregon State University, which awarded his master's degree (1996). He also worked as senior aquarist at the Oregon Coast Aquarium (1991–1993). He then set up an aquarium consultancy business that operated in Asia and the Middle East (1993–1997), becoming Director of the AAA Aquarium on the Red Sea in Jeddah, Saudi Arabia. He is currently Director of Issham Aquatics, a large aquarium design, installation and operation company also in Saudi Arabia. He has authored or co-authored numerous new fish species from Omani waters including the electric ray *Narcine oculifera* and a shallow water chaetodontid fish of the genus *Chaetodon* (1989), perhaps the last new anemonefish, *Amphiprion omanensis* (1991), a new genus and species of batrachoid fish (1994), and several species of labrid and apogonid fishes (1995).

Meerdervoort

Bigeye Skate *Okamejei meerdervoortii* Bleeker, 1860

Dr Johannes Lijdius Catharinus Pompe van Meerdervoort (1829–1908) qualified at Utrecht and became a naval surgeon (1849). He lived at Dejima, a Dutch enclave in Nagasaki Harbour (1857–1863) and was invited by the Japanese authorities to teach western medicine, chemistry and photography in the Kaigun Denshujo Naval Academy. He opened the first western-style hospital and medical school in Japan (1861) before returning to the Netherlands (1863). He wrote *Five Years in Japan* (1868). He collected for

Bleeker, who wrote an article, *Fish Species of Japan, gathered in Dejima by Jhr. J. L. C. Pompe van Meerdervoort* (1859).

Meissner

Whiteleg Skate *Amblyraja taaf* Meissner, 1987
[Syn. *Raja taaf*]

Emil Emilevich Meissner (1937–2007) was a Russian ichthyologist who graduated from the Faculty of Biology, Odessa State University (1964) and worked as a senior laboratory assistant at the Odessa Branch of the Azov-Black Sea Research Institute of Marine Fisheries and Oceanography. He was the engineer and Assistant Captain on a number of scientific research vessels including the 'Marlin', 'Black Sea', 'Forest' and 'Kara-Dag' exploring the Indian and Southern Oceans (1965–1978). After 1978 he became Deputy Chief and, ultimately, Head of the laboratory concerned with fishing forecasting, upon which he wrote many papers. He described the skate in the article, *A New Species of Ray (Rajidae, Batoidei) from the Indian Ocean Sector of the Antarctic* in 'Zoologichesky Zhurnal'.

Menchaca

Freshwater Stingray sp. *Potamotrygon menchacai* Achenbach, 1967
[JS. *Potamotrygon falkneri*]

Dr Manuel J. Menchaca (1876–1969) was an Argentine politician and physician. During his tenure as Governor of Santa Fé Province, he founded the Provincial Museum of Natural Sciences 'Florentino Ameghino', which became the centre for investigating the genus Potamotrygon.

Mendoza

Mendoza's Hagfish *Eptatretus mendozai* Hensley, 1985

Luis H. Mendoza was captain of the research vessel 'Crawford', which belonged to the Department of Marine Sciences, University of Puerto Rico. The holotype was taken during a cruise of this vessel. He was honoured for his 'experiential knowledge and academic curiosity of the sea, without whose determination and nautical wisdom' the author would never have collected type.

Menezes

Hagfish sp. *Eptatretus menezesi* Mincarone, 2000

Professor Dr Naércio Aquino Menezes (b.1937) is a Brazilian zoologist who is Professor, Department of Zoology, Institute of Biosciences, University of São Paulo and at the Museu de Zoologia da Universidade de São Paulo. The University of São Paulo awarded his bachelor's degree (1962) and Harvard his PhD (1968). His specialist areas are systematics, biogeography and evolution of fishes, mainly of marine fish around the coasts of Brazil and neotropical freshwater fish. He was honoured in the hagfish name for 'his extensive contribution to Brazilian ichthyology'.

Meng

Skate sp. *Okamejei mengae* Jeong, Nakabo & Wu, 2007

South China Legskate *Anacanthobatis nanhaiensis* Meng & Li, 1981
[Syn. *Springeria nanhaiensis*]
Blackfin Gulper Shark *Centrophorus isodon* Chu, Meng & JX Liu, 1981
Spotless Catshark *Bythaelurus immaculatus* Chu & Meng, 1982
[Syn. *Halaelurus immaculatus*]
Dogfish Shark sp. *Squalus acutirostris* Chu, Meng & Li, 1984
Humpback Catshark *Apristurus gibbosus* Chu, Meng & Li, 1985
Broadmouth Catshark *Apristurus macrostomus* Meng, Chu & Li, 1985
South China Cookiecutter Shark *Isistius labialis* Meng, Chu & Li, 1985
Smalldorsal Catshark *Apristurus micropterygeus* Meng, Chu & Li, 1986

Professor Dr Qing-Wen Meng was President of Shanghai Fisheries College (now Shanghai Ocean University). She wrote *Studies on the Blood Vessels of Eye, Gas Bladder and Kidney of Snakehead* (1990). She was honoured for her 'great contributions to elasmobranch studies in China'. A fossil lamprey from the early cretaceous epoch is also named after her.

Menni

South Brazilian Skate *Dipturus mennii* Gomes & Paragó, 2001

Skate genus *Atlantoraja* Menni, 1972

Electric Ray sp. *Discopyge castelloi* Menni, Rincón & Garcia, 2008

Dr Roberto Carlos Menni is an ichthyologist and professor at the Museo de La Plata, Universidad

Nacional de La Plata, Argentina. Among his numerous publications, as both sole and joint author, is *Peces y Ambientes en la Argentina Continental* (2003). He was honoured in the name of the skate for his contributions to the study of South American rays.

Merrett

Roughskin Spurdog *Cirrhigaleus asper* Merrett, 1973

Dr Nigel Robert Merrett (b.1940) is a British zoologist and ichthyologist. Bristol University awarded his bachelor's degree (1963) and doctorate (1992) and the University of East Africa, Dar-es-Salaam his master's (1968). After a temporary job attached to the Southern Harvester Whaling Expedition in Antarctica (1962–1963), he was scientist at the East African Marine Fisheries Research Organization (1963–1968). He was ichthyologist at what is now the National Oceanographic Centre, Southampton University (1968–1989) and head of the fish group in the Zoology Department, BMNH (1989–1999). In his career he accumulated about six years of sea time on more than 50 scientific cruises, including the 1998 'Vitjaz' cruise (NV Parin deep-sea expedition). Among some 100 scientific papers he co-wrote *Deep-Sea Demersal Fish and Fisheries* (1997). The holotype of the roughskin spurdog was collected on the Royal Society Indian Ocean Deep Slope Fishing Expedition (1969) of which he was a member.

Metzelaar

Ringed Torpedo *Torpedo mackayana* Metzelaar, 1919

Dr Jan Metzelaar (1891–1929) was a Dutch ichthyologist and fisheries biologist. The University of Amsterdam awarded his doctorate (1919). He moved to the USA and worked in the Fish Division, Michigan Department of Conservation (1923–1929) and was Honorary Custodian of Michigan Fishes (1925–1929). He drowned in Lake Huron whilst engaged in fishery investigations.

Meyen

Blotched Fantail Ray *Taeniura meyeni* Müller & Henle, 1841

Dr Franz Julius Ferdinand Meyen (1804–1840) was a German surgeon, botanist and collector who was a professor in Berlin. He suggested that new cells were created through cell division rather than by the creation of new free cells. He published this theory in his work on plant anatomy, *Phytotomie* (1830). He took part in the circumnavigation (1830–1832) by the 'Prinzess Louise', an expedition that included considerable time in South America (1830–1832). A mammal and two birds are named after him.

Meyer

Port Jackson Shark *Heterodontus portusjacksoni* Meyer, 1793

Dr Friedrich Albrecht Anton Meyer (1768–1795) was a German natural historian and physician who graduated at Göttingen. He wrote *Systematisch-summarische Übersicht der Neuesten Zoologischen Entdeckungen in Neuholland und Afrika* (1793).

Michael

Michael's Epaulette Shark *Hemiscyllium michaeli* Allen & Dudgeon, 2010
[Alt. Milne Epaulette Carpetshark]

Scott W. Michael is an author and photographer who specialises in elasmobranchs and coral reef fishes. He wrote *Reef Sharks and Rays of the World* (1994). He was the first person to recognise this species' unique colour pattern and to bring the difference between this species and *H. freycineti* to the authors' attention. They also honoured him for contributing information and photographs to the first author's research on Indo-Pacific fishes.

Milius

Australian Ghost Shark *Callorhinchus milii* Bory de Saint-Vincent, 1823

Baron Pierre Bernard Milius (1773–1829) was a sailor, naturalist and civil servant who took part in an exploratory voyage (1804) of the Mascarene Islands, Indian Ocean under Nicolas Baudin, during which he became friends with Bory. He was Governor of Bourbon (now Réunion) (1818–1821) where he established a port and undertook agricultural projects. He was despotic and despised by the locals. He was later appointed Governor of French Guyana. He was present at the Battle of Navarino during the war of Greek independence.

Mincarone

Southern Sawtail Catshark *Galeus mincaronei* Soto, 2001

Hagfish sp. *Eptatretus menezesi* Mincarone, 2000
Hagfish sp. *Myxine sotoi* Mincarone, 2001
Hagfish sp. *Eptatretus lakeside* Mincarone & McCosker, 2004

Thorny-tail Skate *Dipturus diehli* Soto & Mincarone, 2001

Goliath Hagfish *Eptatretus goliath* Mincarone & Stewart, 2006

Australian Hagfish sp. *Eptatretus alastairi* Mincarone & Fernholm, 2010

Hagfish sp. *Eptatretus astrolabium* Fernholm & Mincarone, 2010

Hagfish sp. *Eptatretus gomoni* Mincarone & Fernholm, 2010

Dr Michael Maia Mincarone (b.1971) is a Brazilian ichthyologist who is a professor at Universidade Federal do Rio de Janeiro. The Universidade do Vale do Itajaí awarded his bachelor's degree in oceanography (1997) and the Pontifícia Universidade Católica do Rio Grande do Sul his doctorate in zoology (2007). His research is focused on the systematics and conservation of deep-sea fishes. Soto honoured him for his 'extensive work and tireless dedication' to the Museu Oceanográfico Univali, Brazil. He has described other genera and species of fish in addition to the cartilaginous ones above.

Miranda-Ribeiro

Spotback Skate *Atlantoraja castelnaui* Miranda-Ribeiro, 1907

Freckled Catshark *Scyliorhinus haeckelii* Miranda-Ribeiro, 1907

Dr Alípio de Miranda-Ribeiro (1874–1939) was one of the foremost Brazilian naturalists and zoologists of his era. He initially studied medicine but joined the National Museum, Rio de Janeiro (1894) before he had completed his course. He was Assistant Naturalist for the Museum (1897), Secretary, Department of Zoology (1899), and Deputy Head and Professor, Zoology Department (1910–1929). He explored the Amazon and created the Inspectorate of Fisheries (1911), the first South American oceanographical service, and was its first director (1911–1912). Ribeiro started to describe new fish species but he had competition from Goeldi who would go to a market and buy fish from the same ship, describe them and send them to the British Museum. Ribeiro became Professor of the Department of Zoology (1929), a post he held until his death. He wrote *Fauna Brasiliensis* (1907–1915). Many taxa, including seven amphibians and two birds, are named in his honour.

Misra

Broadnose Catshark *Apristurus investigatoris* Misra, 1962

Kamla Sankar Misra (1900–1969) was an Indian zoologist and ichthyologist. He wrote *The Fauna of India and the Adjacent Countries:* Pisces (1969).

Mitchill

Roughtail Stingray *Dasyatis centroura* Mitchill, 1815

Winter Skate *Leucoraja ocellata* Mitchill, 1815

Dusky Smooth-Hound *Mustelus canis* Mitchill, 1815

Cownose Ray *Rhinoptera bonasus* Mitchill, 1815

Barndoor Skate *Dipturus laevis* Mitchill, 1818

Little Skate *Leucoraja erinacea* Mitchill, 1825

Dr Samuel Latham Mitchill (1764–1831) was an American physician, naturalist and politician. His wealthy uncle paid for him to attend Edinburgh University, where he graduated (1786). He taught chemistry, botany, zoology, mineralogy and natural history at Columbia College (1792–1801), during which time he collected plants and animals with a particular emphasis on aquatic species. He went on to teach at the New York College of Physicians and Surgeons (1807–1826) and then helped organise the Rutgers Medical College, serving as its Vice-President (1827–1830). His theory of disease proved to be false but it did have the benefit of promoting better sanitation and hygiene. Concurrently he entered politics serving in the New York State Assembly (1791 and 1798), the US House of Representatives (1801–1804) and the US Senate (1804–1809 and 1810–1913). Because of his breadth of knowledge he was known as a 'living encyclopedia', and a 'stalking library'.

Mitsukuri

Shortspine Spurdog *Squalus mitsukurii* Jordan & Snyder, 1903

Spookfish *Hydrolagus mitsukurii* Jordan & Snyder, 1904

Pacific Spookfish *Rhinochimaera pacifica* Mitsukuri, 1895

Dr Kakichi Mitsukuri (1858–1909) was a Japanese zoologist who first went to the USA (1873) and achieved doctorates from Yale (1879) and Johns Hopkins University (1883). He worked for a time at the Naples Zoological Station and founded the Misaki Marine Laboratory (1887). He was a professor at the College of Science, Imperial University of Tokyo (1882–1909) and was Dean of the College of Science there (1901–1909). Among his publications is a Japanese-English dictionary. He was with Jordan and Snyder at Misaki, Japan when the holotype was taken.

Miyake

Ray family *Urotrygonidae* McEachran, Dunn & Miyake, 1996

Peruvian Skate *Bathyraja peruana* McEachran & Miyake, 1984
Nova Scotia Skate *Breviraja marklei* McEachran & Miyake, 1987
Cortez' Ray *Raja cortezensis* McEachran & Miyake, 1988
Dwarf Roundray *Urotrygon nana* Miyake & McEachran, 1988
Reticulate Roundray *Urotrygon reticulate* Miyake & McEachran, 1988
Fake Roundray *Urotrygon simulatrix* Miyake & McEachran, 1988

Dr Tsutomu 'Tom' Miyake is a biologist and ichthyologist who since 2012 is Project Professor at the Graduate School of Science and Technology, Keio University, Yokohama, Japan. Tokai University, School of Marine Sciences awarded his bachelor's degree, the University of Michigan his master's and Texas A & M University his doctorate (1988). He was (1997) at the Department of Biology, University of Dalhousie, Halifax, Nova Scotia. He was a staff scientist at the Benaroya Research Institute, Seattle (2001–2007). He co-wrote *Development of Dermal Denticles in Skates* (Chondrichthyes, Batoidea): *Patterning and Cellular Differentiation* (1999).

Miyamoto

Hyuga Fanray *Platyrhina hyugaensis* Iwatsuki, Miyamoto & Nakaya, 2011

Kei Miyamoto is an ichthyologist who was at the Division of Fisheries Science, Faculty of Agriculture, University of Miyazaki, Japan (2011) and is now (2014) at the Animal Research Department, Okinawa Churashima Research Center, Okinawa Churashima Foundation.

Miyosi

Pitted Stingray *Dasyatis matsubarai* Miyosi, 1939
Blacktip Tope *Hypogaleus hyugaensis* Miyosi, 1939

Dr Yasunori Miyosi (1909–1995) was a biologist, arachnologist, botanist and ichthyologist who was Professor Emeritus of St Catherina Women's Junior College, Hojo, Japan (1966–1984). He graduated from the Ehime Teachers' School (now Faculty of Education, Ehime University) (1929). He taught in a primary school (1929–1933), at Yawatahama Women's High School, Shikoku, Japan (1933–1938), and at Matsuyama Women's High School (now Matsuyama-Minami High School), Matsuyama (1938–1944). He started to study myriapods and moved to Tokyo (1944) to study at Tokyo Bunrika University (now Tsukuba University), but the allied attacks on Tokyo forced him to evacuate to Matsuyama where he rejoined the staff of Matsuyama Women's High School. After the war he became a teacher at Matsuyama-Kita High School (1945–1966). He submitted (1959) a thesis on arachnology to the University of Hokkaido, which awarded him a doctorate of science (1960). He switched disciplines again on his appointment as Professor at St Catherina Women's Junior College to botany, especially the flora of Ehime prefecture and adjoining areas of Shikoku.

Mochizuki

Dogfish genus *Trigonognathus* Mochizuki & Ohe, 1990

Viper Dogfish *Trigonognathus kabeyai* Mochizuki & Ohe, 1990

Dr Kenji Mochizuki is a deep-sea fish ichthyologist who was at the Department of Fisheries, University Museum, University of Tokyo (1973–1990) and at the Department of Zoology, Natural History Museum and Institute, Chiba (1991). He co-wrote *Trigonognathus kabeyai, a New Genus and Species of the Squalid Sharks from Japan* (1990). A species of sea-bass is named after him.

Mok

Hagfish sp. *Eptatretus moki* CB McMillan & Wisner, 2004

Hagfish sp. *Eptatretus chinensis* Kuo & Mok, 1994
Hagfish sp. *Eptatretus nelsoni* Kuo, Huang & Mok, 1994
Hagfish sp. *Eptatretus sheni* Kuo, Huang & Mok, 1994
Hagfish sp. *Paramyxine fernholmi* Kuo, Huang & Mok, 1994
Hagfish sp. *Paramyxine wisneri* Kuo, Huang & Mok, 1994
Hagfish sp. *Eptatretus wayuu* Mok, Saavedra-Diaz & Acero, 2001
Hagfish sp. *Myxine formosana* Mok & Kuo, 2001
Hagfish sp. *Myxine kuoi* Mok, 2002
Hagfish sp. *Eptatretus rubicundus* Kuo, Lee & Mok, 2010

Dr Michael Hin-Kiu Mok (b.1947) is a Taiwanese marine biologist and ichthyologist who was a research associate in vertebrate zoology and ichthyology at AMNH. Since his retirement he has become an adjunct professor, Department of Oceanography, National Sun Yat-sen University. The National Taiwan University awarded his bachelor's degree (1967) and the City University of New York his master's (1975) and doctorate (1978). After being a post-doctoral research fellow at the Harbor Branch Oceanographic Institution, Florida (1978–1980), he taught at Ocean College Taiwan (1980–1981). He joined National Sun Yat-Sen University, Taiwan (1981) and was Professor, Institute of Marine Biology (1983–2014), Dean of the College of Marine Sciences (1992–1998), Director of the Institute of Marine Biology (1984–1989) and Director of the Institute of Undersea Technology (1966–1999). Among his publications are *Osteology and Phylogeny of Squamipinnes* (1983) and *Study on the Accumulation of Heavy Metals in Shallow-water and Deep-sea Hagfishes* (2010), as well as a number of papers describing new hagfish. He was honoured in the hagfish name for his many 'outstanding contributions to hagfish knowledge'.

Moller

Moller's Slendertail Lantern Shark *Etmopterus molleri*
Whitley, 1939

Captain K. Moller was in command of the trawler that brought the holotype to the surface (1933). He commanded a number of trawlers in his career, including the 'Durraween' (1929) and the 'Ben Bow' (1947), and is said to have been assiduous in remarking unusual specimens and sending them to the Australian Museum in Sydney.

Møller

Jespersen's Hagfish *Myxine jespersenae* Møller, Feld,
Poulsen, Thomsen & Thormar, 2005
Strickrott's Hagfish *Eptatretus strickrotti* Møller &
Jones, 2007

Dr Peter Rask Møller (b.1969) is a Danish ichthyologist, specialising in morphological and molecular fish taxonomy and phylogeny. He is Associate Professor and Curator at the Natural History Museum of Denmark, University of Copenhagen. He has participated in many expeditions to both the Arctic and the Antarctic. He has written or co-written over 60 papers and longer works including *A Checklist of the Fish Fauna of Greenland Waters* (2010). He is a specialist in the bony fish families *Zoarcidae* and *Bythithidae* and has authored or co-authored more than 100 new species descriptions of these fishes.

Monkolprasit

Giant Freshwater Stingray *Himantura chaophraya*
Monkolprasit & TR Roberts, 1990

Dr Supap Punpoka Monkolprasit was an ichthyologist and biologist at the Faculty of Fisheries, Kasetsart University, Bangkok, where she was Dean of the Faculty and a professor. She was there from not later than 1966 to not earlier than 2002. The University of Medical Science, Bangkok awarded her bachelor's degree in pharmacy, Stanford University, California her master's in biology and the University of Michigan her doctorate in fisheries. She wrote *The Cartilagenous Fishes (class* Elasmobranchii*) Found in Thai Waters and Adjacent Areas* (1984).

Montagu

Spotted Ray *Raja montagui* Fowler, 1910

Small-eyed Ray *Raja microocellata* Montagu, 1818

Colonel George Montagu (1751–1815) was a soldier and natural history writer. He attained the rank of captain in the British army when very young and later served in the American Revolution as a lieutenant-colonel in the English militia. His military career was curtailed when he was court-martialed and cashiered for causing trouble among his brother officers by what was described as 'provocative marital skirmishing'. Montagu then devoted himself to science, particularly biology. In his own words, 'I have delighted in being an ornithologist from infancy, and, was I not bound by conjugal attachment, should like to ride my hobby to distant parts.' Montagu was among the first members of the Linnean Society. He was also an expert on shells and (1803) wrote *Testacea Britannica, a History of British Marine, Land and Freshwater Shells*. He wrote many papers on the birds of southern England, but his greatest work was the *Ornithological Dictionary or Alphabetical Synopsis of British Birds* (1802). He was renowned for his meticulous work and observations bordering on the clinical. Such observations led to a better understanding of phenomena which had previously been romanticised. For example, he said of bird song: 'males of song-birds and of many others do not in general search for the female, but, on the contrary, their business in the spring is to perch on some conspicuous spot, breathing out their full and amorous notes, which, by instinct, the female knows, and repairs to the spot to choose her mate.' He died of lockjaw (tetanus) after stepping on a rusty nail. Fowler named the spotted ray after

Montagu as a replacement name for *Raja maculata* when it was discovered to be pre-occupied by *Raja maculata* Shaw, 1804.

Montalban

Indonesian Greeneye Spurdog *Squalus montalbani* Whitley, 1931

Whitley's original description gives no explanation. However, the name is a replacement of *S. philippinus* because the name was pre-occupied, so it is probable that it honours Filipino fisheries biologist Heraclio R. Montalban. He wrote, among other works, *Pomacentridae of the Philippine Islands* (1927) as well as descriptions of new species.

Moore

Arabian Whipray *Himantura randalli* Last, Manjaji-Matsumoto & Moore, 2012

Alec B. M. Moore (b.1973) is a marine biologist, ecologist and independent elasmobranch researcher. The University of Southampton awarded his bachelor's degree (1995), the University of Warwick his master's (1999) and Bangor University his doctorate (2013), which he achieved whilst working full-time for RSK Environment Ltd, Cheshire, England, where he is principal marine environmental consultant. Previously he was with the Department of Biological Sciences, University of Warwick, England (2001) and the School of Ocean Sciences, Bangor University, Wales (2011). He rediscovered the very rare smoothtooth blacktip shark *Carcharhinus leiodon* Garrick, 1985 in Kuwait. He undertook the first focused studies of elasmobranchs in the Persian Gulf e.g. surveys of fisheries landings, biodiversity and a comprehensive literature review. He co-wrote *Elasmobranchs of the Persian (Arabian) Gulf: Ecology, Human Aspects and Research Priorities for their Improved Management* (2012) and *Sharks of the Persian (Arabian) Gulf: A First Annotated Checklist* (Chondrichthyes: Elasmobranchii) (2012).

Moresby

Dark Blind Ray *Benthobatis moresbyi* Alcock, 1898

Captain Robert Moresby (1794–1854) of the Indian Navy was a hydrographer, marine surveyor and draughtsman. He surveyed dangerous waters and reefs in the 1820s and 1830s including the most comprehensive survey of the Red Sea. Commanding 'HMS Palinurus' and based in Suez (1829–1833), he did not leave the Red Sea until the survey was com-plete. It was regarded as vital as the advent of steam power meant that the new route from England to India – by sea to Alexandria, overland to Suez, then by sea to India – was the fastest and most reliable until the opening of the Suez Canal. Moresby also surveyed off Travancore (1834–1838) from where came the holotype of the ray. His health gave way and he had to give up surveying. He transferred to the Merchant Navy and worked for P & O, commanding their steamer 'Hindostan' (1842) and later the 'Ripon', which was on the England to Alexandria run (1852). Moresby came from a distinguished naval family. His elder brother was Admiral of the Fleet Sir Fairfax Moresby, after whom Port Moresby in Papua New Guinea was named by the Admiral's son, Captain John Moresby.

Mori

Korean Lamprey *Eudontomyzon morii* LS Berg, 1931

Dr Tamezo Mori (1884–1962) was a Japanese entomologist, ichthyologist and ornithologist. He taught at Seoul Higher Common School for Keiji Imperial University, Seoul (1909–1945). He was expelled from Korea (1945) by the American authorities and became Director of Zoology at the Agricultural University Hyogo, retiring as Professor Emeritus (1961). Two birds are named after him.

Motomura

Greeneye Spurdog *Squalus chloroculus* Last, WT White & Motomura, 2007

Whitefin Swellshark *Cephaloscyllium albipinnum* Last, Motomura & WT White, 2008

Dr Hiroyuki Motomura (b.1973) is a Japanese ichthyologist who became a professor at Kagoshima University (2010). The Department of Biology, Faculty of Science, Yamagata awarded his bachelor's degree (1997) and the Graduate School of Agriculture, University of Miyazaki his master's (1999) and doctorate (2001). He wrote *Revision of the Scorpionfish Genus* Neosebastes (Scorpaeniformes: Neosebastidae), *with Descriptions of Five New Species*. (2004), *Fish Diversity of Kagoshima Prefecture Kuroshio Nurtured* (2012), *Fish Collection Building and Procedures Manual* (2013) and *Field Guide to Fishes of Yoron Island in the Middle of the Ryukyu Islands* (2014).

Mould

Blacknose Skate *Breviraja mouldi* McEachran & Matheson, 1995

B. Mould noticed that *Breviraja schroederi*, McEachran & Matheson 1985 was pre-occupied by *Breviraja schroederi*, Krefft 1968 and brought it to the describers' attention, so they re-named the skate after him. The etymology only mentions his family name and initial. We think that he is Brian Mould, a biodiversity researcher and taxonomist at the University of Nottingham, who wrote *Classification of the Recent Elasmobranchii* (1997). We have been completely unable to trace him and the University of Nottingham advised us that no one there remembers him and they have no written record that he was ever there at all.

Moura

Ghost Shark sp. *Hydrolagus lusitanicus* Moura, Figueiredo, Bordalo-Machado, Almeida & Gordo, 2005

Dr Teresa Moura (b.1980) is a Portuguese marine biologist who joined the Departamento do Mar e Recursos Marinhos, Divisão de Odelação e Gestão de Recursos da Pesca, Instituto Português do Mar e Atmosfera (2002), where she specialises in fisheries biology of demersal and deep-water resources. The University of Lisbon awarded her bachelor's degree (2003) and her doctorate (2012) and the University of Oporto her master's (2007). Among her published papers are the co-written *Molecular Barcoding of NE Atlantic Deepwater Sharks: Species Identification and Application to Fisheries Management and Conservation* (2008) and *Embryonic Development and Maternal–embryo Relationships of the Portuguese Dogfish* Centroscymnus coelolepis (2011).

Moy-Thomas

Sawfish genus *Anoxypristis* El White & Moy-Thomas, 1941

James Allan Moy-Thomas (1908–1944) whose bachelor's degree in zoology was awarded by Oxford (1930) was an expert of fossil fishes and an ardent philatelist. He was appointed University Demonstrator in the Department of Zoology and Comparative Anatomy (1933) and became a Fellow of The Queen's College (1936). He was a member of the English-Norwegian-Swedish Expedition to Spitzbergen (1939). In the Second World War he served in the Royal Air Force as a night fighter navigator/observer. He was killed in a motor accident while on duty. He wrote *Palaeozoic Fishes* (1939).

Müller

Guitarfishes *Rhinobatidae* Müller & Henle, 1837

Guitarfishes *Rhynchobatus* Müller & Henle, 1837
Mackerel Sharks *Lamnidae* Müller & Henle, 1838
Whale Sharks *Rhincodontidae* Müller & Henle, 1839
Rays *Urolophidae* Müller & Henle, 1841

Gulper Shark genus *Centrophorus* Müller & Henle, 1837
Bamboo Shark genus *Chiloscyllium* Müller & Henle, 1837
Requiem Shark genus *Galeocerdo* Müller & Henle, 1837
Nurse Shark genus *Ginglymostoma* Müller & Henle, 1837
Bamboo Shark genus *Hemiscyllium* Müller & Henle, 1837
Stingray genus *Himantura* Müller & Henle, 1837
Sawshark genus *Pristiophorus* Müller & Henle, 1837
Carpet Shark genus *Stegostoma* Müller & Henle, 1837
Fanskate genus *Sympterygia* Müller & Henle, 1837
Stingray genus *Taeniura* Müller & Henle, 1837
Reef Shark genus *Triaenodon* Müller & Henle, 1837
Stingray genus *Urolophus* Müller & Henle, 1837
Stingray genus *Urogymnus* Müller & Henle, 1837
Sliteye Shark genus *Loxodon* Müller & Henle, 1838
Fanray genus *Platyrhina* Müller & Henle, 1838
Spadenose Shark genus *Scoliodon* Müller & Henle, 1838
Fiddler Ray genus *Trygonorrhina* Müller & Henle, 1838
Sand Shark genus *Odontaspididae* Müller & Henle, 1839
Houndshark genus *Triakis* Müller & Henle, 1839
Deepwater Dogfish genus *Centroscyllium* Müller & Henle, 1841
Guitarfish genus *Rhinidae* Müller & Henle, 1841
Rio Skate genus *Rioraja* Müller & Henle, 1841
Stingaree genus *Trygonoptera* Müller & Henle, 1841

Porcupine River Stingray *Potamotrygon hystrix* Müller & Henle, 1834

Sixgill Hagfish *Eptatretus hexatrema* Müller, 1836
Ticon Cownose Ray *Rhinoptera brasiliensis* Müller, 1836
Grey Bamboo Shark *Chiloscyllium griseum* Müller & Henle, 1838
Brownbanded Bamboo Shark *Chiloscyllium punctatum* Müller & Henle, 1838
Blackspotted Catshark *Halaelurus buergeri* Müller and Henle, 1838
Dark Shyshark *Haploblepharus pictus* Müller & Henle, 1838
Spinetail Mobula *Mobula japonica* Müller & Henle, 1838

Shortfin Devil Ray *Mobula kuhlii* Müller & Henle, 1838

Leopard Catshark *Poroderma pantherinum* Müller & Henle, 1838

Narrowmouthed Catshark *Schroederichthys bivius* Müller & Henle, 1838

Spadenose Shark *Scoliodon laticaudus* Müller & Henle, 1838

Yellow-spotted Catshark *Scyliorhinus capensis* Müller & Henle, 1838

Pigeye Shark *Carcharhinus amboinensis* Müller & Henle, 1839

Spinner Shark *Carcharhinus brevipinna* Müller & Henle, 1839

Whitecheek Shark *Carcharhinus dussumieri* Müller & Henle, 1839

Silky Shark *Carcharhinus falciformis* Müller & Henle, 1839

Pondicherry Shark *Carcharhinus hemiodon* Müller & Henle, 1839

Finetooth Shark *Carcharhinus isodon* Müller & Henle, 1839

Bull Shark *Carcharhinus leucas* Müller & Henle, 1839

Blacktip Shark *Carcharhinus limbatus* Müller & Henle, 1839

Hardnose Shark *Carcharhinus macloti* Müller & Henle, 1839

Spot-tail Shark *Carcharhinus sorrah* Müller & Henle, 1839

Ganges Shark *Glyphis gangeticus* Müller & Henle, 1839

Speartooth Shark *Glyphis glyphis* Müller & Henle, 1839

Japanese Topeshark *Hemitriakis japonica* Müller & Henle, 1839

Daggernose Shark *Isogomphodon oxyrhynchus* Müller & Henle, 1839

Broadfin Shark *Lamiopsis temminckii* Müller & Henle, 1839

Barbeled Houndshark *Leptocharias smithii* Müller & Henle, 1839

Sliteye Shark *Loxodon macrorhinus* Müller & Henle, 1839

Brazillian Sharpnose Shark *Rhizoprionodon lalandii* Müller & Henle, 1839

Banded Houndshark *Triakis scyllium* Müller & Henle, 1839

Eagle Ray *Aetomylaeus milvus* Müller & Henle, 1841

Short-snouted Shovelnose Ray *Aptychotrema bougainvillii* Müller & Henle, 1841

African Softnose Skate *Bathyraja smithii* Müller & Henle, 1841

Whip Stingray *Dasyatis akajei* Müller & Henle, 1841

Bennett's Stingray *Dasyatis bennetti* Müller & Henle, 1841

Pale-edged Stingray *Dasyatis zugei* Müller & Henle, 1841

Tentacled Butterfly Ray *Gymnura tentaculata* Müller & Henle, 1841

Dwarf Whipray *Himantura walga* Müller & Henle, 1841

Cuckoo Skate *Leucoraja naevus* Müller & Henle, 1841

Blue-spotted Stingray *Neotrygon kuhlii* Müller & Henle, 1841

Ocellate Spot Skate *Okamejei kenojei* Müller & Henle, 1841

Ocellate River Stingray *Potamotrygon motoro* Müller & Henle, 1841

Large-tooth Sawfish *Pristis perotteti* Müller & Henle, 1841

Lesser Sandshark *Rhinobatos annulatus* Müller & Henle, 1841

Bluntnose Guitarfish *Rhinobatos blochii* Müller & Henle, 1841

Brazilian Guitarfish *Rhinobatos horkelii* Müller & Henle, 1841

Widenose Guitarfish *Rhinobatos obtusus* Müller & Henle, 1841

Brown Guitarfish *Rhinobatos schlegelii* Müller & Henle, 1841

Rough Cownose Ray *Rhinoptera adspersa* Müller & Henle, 1841

Flapnose Ray *Rhinoptera javanica* Müller & Henle, 1841

Rio Skate *Rioraja agassizii* Müller & Henle, 1841

Smallnose Fanskate *Sympterygia bonapartii* Müller & Henle, 1841

Blotched Fantail Ray *Taeniura meyeni* Müller & Henle, 1841

Southern Fiddler Ray *Trygonorrhina fasciata* Müller & Henle, 1841

Common Stingaree *Trygonoptera testacea* Müller & Henle, 1841

New Ireland Stingaree *Urolophus armatus* Müller & Henle, 1841

Sepia Stingaree *Urolophus aurantiacus* Müller & Henle, 1841

Striped Panray *Zanobatus schoenleinii* Müller & Henle, 1841

Lesser Guitarfish *Zapteryx brevirostris* Müller & Henle, 1841

New Zealand Rough Skate *Zearaja nasuta* Müller & Henle, 1841

Professor Johannes Peter Müller (1801–1858) was a German physiologist, ichthyologist and comparative anatomist. He entered the University of Bonn (1819), graduated and started teaching (1824), becoming Professor (1830). He was Professor of Anatomy and Physiology at Humboldt University of Berlin (1833–1858) and a member of the Royal Prussian Academy of Science, Berlin. He wrote *Handbuch der Physiologie des Menschen* (1833–1840). Among his discoveries was (1835) that caecilians are amphibians and not snakes. A reptile and an amphibian are named after him.

Munk

Munk's Devil Ray *Mobula munkiana* Notarbartolo di
Sciara, 1987

Walter Heinrich Munk (b.1917) is an Austrian-born American oceanographer who was sent to the USA before 1932, when he attended a school in upper New York State. He worked in a bank, which he hated, and studied at Columbia University for three years before resigning to go to the California Institute of Technology, which awarded his bachelor's degree in physics (1939). He went to the Scripps Institute of Oceanography (1939) and completed a master's degree in geophysics (1940) but did not obtain his doctorate from the University of California, Los Angeles until 1947 as the Second World War interrupted his progress. He became an American citizen (1939) and enlisted in the US Army as a private in the ski troops, but was later released to do research at Scripps in relation to amphibious warfare. He has been involved in a great many important research projects including the Mohole Project and analysis of the currents, diffusion and water exchanges at Bikini Atoll in the South Pacific, where the USA was testing nuclear weapons. He developed plans for a La Jolla branch of the Institute of Geophysics (1956) and oversaw its construction (1959–1963). It is now the centre of the campus of the Scripps Institute of Oceanography, La Jolla, California. He became an assistant professor at Scripps (1947) and a full professor (1954). He is now Emeritus Professor of Geography and holds the chair of oceanography at Scripps. He was a friend of the author.

Murofushi

Japanese Roughshark *Oxynotus japonicus* Yano &
Murofushi, 1985

Dr Makoto Murofushi (b.1950) is a professor in the Department of Food and Nutrition, Mishima Junior College, Nihon University, Japan. Nihon University awarded his bachelor's degree (1973) and his master's (1975). He co-wrote *Fish in Uchiura-bay, Shizuoka Prefecture-Collecting Record of Shallow Water Coastal Area Fishes* (2011).

Murray

Murray's Skate *Bathyraja murrayi* Günther, 1880

Sir John Murray (1841–1914) was a Canadian marine naturalist, often described as the founder of modern oceanography. He explored the Faroe Channel (1880–1882) and took part in and financed expeditions to Christmas Island. He was in charge of collections on the 'HMS Challenger' expedition (1874–1876) that collected the holotype and edited *Report on the Scientific Results of the Voyage of 'HMS Challenger'* (1880–1895). A motorcar killed him as he was crossing a street. A bird and a reptile are named after him.

Myagkov

Tasmanian Lantern Shark *Etmopterus tasmaniensis*
Myagkov & Pavlov, 1986
Spiny Dogfish ssp. *Squalus acanthias ponticus*
Myagkov & Kondyurin, 1986

Nikolai Alexandrovich Myagkov (1954–2007) of the All-Union Research Institute of Sea Fisheries and Oceanography, Moscow was an ichthyologist whose research focused on the taxonomy and physiology of sharks. He graduated from Kaliningrad Technical Institute of Fishing Industry with a thesis on the brain of the frilled shark *Chlamydoselachus anguineus*. Among his published works are several articles on the determinants of sharks; he co-wrote *Sharks of the World Ocean: Identification Handbook* (1986).

N

Nair

Stripenose Guitarfish *Rhinobatos variegatus* Nair & Lal Mohan, 1973

Dr N. V. Nair was Director, Central Marine Fisheries Institute, Cochin (Kochi) where he worked (1947–1976). After his retirement he worked as an Emeritus scientist at the University of Kerala. He wrote *Studies on some Postlarval Fishes of the Madras Plankton* (1952).

Nakabo

Korean Skate genus *Hongeo* Jeong & Nakabo, 2009

Skate sp. *Okamejei mengae* Jeong, Nakabo & Wu, 2007
Wu's Skate *Dipturus wuhanlingi* Jeong & Nakabo, 2008

Dr Tetsuji Nakabo (b.1949) is a Japanese ichthyologist who is a professor at the graduate course of agriculture, Kyoto University and was Director of the University's Museum. All three of his degrees were awarded by the University of Kyoto. He has published around 120 papers, mostly with a variety of co-authors, such as *Oncorhynchus Kawamurae 'Kunimasu' a Deepwater Trout, Discovered in Lake Saiko, 70 years after Extinction in the Original Habitat, Lake Tazawa* (2011). He compiled *Fishes of Japan with Pictorial Keys to the Species* (1993), followed by an English edition (2002).

Nakamura

Bigeyed Sixgill Shark *Hexanchus nakamurai* Teng, 1962

Pelagic Thresher *Alopias pelagicus* Nakamura, 1935

Dr Hiroshi Nakamura DSc was a Japanese ichthyologist. He described the thresher in *On the Two Species of the Thresher Shark from Formosan Waters* (1935), thinking it to be *H griseus* when he illustrated it the following year. He was a technician (1936–1947) at the Fisheries Experiment Station of the Taiwan Government-General during the Japanese occupation of Taiwan. He also wrote many papers from early 1930s and several longer works, such as *The Tunas and their Fisheries* (1949) and *Tuna Distribution and Migration* (1969).

Nakaya

Rasptooth Dogfish genus *Miroscyllium* Shirai & Nakaya, 1990

Japanese Catshark *Apristurus japonicus* Nakaya, 1975
Longhead Catshark *Apristurus longicephalus* Nakaya, 1975
Broadfin Sawtail Catshark *Galeus nipponensis* Nakaya, 1975
Sharpnose Stingray *Dasyatis acutirostra* Nishida & Nakaya, 1988
Izu Stingray *Dasyatis izuensis* Nishida & Nakaya, 1988
Highfin Dogfish *Centroscyllium excelsum* Shirai & Nakaya, 1990
Izu Catshark *Scyliorhinus tokubee* Shirai, Hagiwara & Nakaya, 1992
White Ghost Catshark *Apristurus aphyodes* Nakaya & Stehmann, 1998
Catshark sp. *Apristurus albisoma* Nakaya & Séret, 1999
Flaccid Catshark *Apristurus exsanguis* Sato, Nakaya & Stewart, 1999
Black Roughscale Catshark *Apristurus melanoasper* Iglésias, Nakaya & Stehmann, 2004
Ocellate Japanese Topeshark *Hemitriakis complicofasciata* Takahashi & Nakaya, 2004
Bareback Shovelnose Ray *Rhinobatos nudidorsalis* Last, Compagno & Nakaya, 2004
Roughskin Catshark *Apristurus ampliceps* Sasahara, Sato & Nakaya, 2008
Pinocchio Catshark *Apristurus australis* Sato, Nakaya & Yorozu, 2008
Hyuga Fanray *Platyrhina hyugaensis* Iwatsuki, Miyamoto & Nakaya, 2011
Yellow-spotted Fanray *Platyrhina tangi* Iwatsuki, Zhang & Nakaya, 2011
Milk-eye Catshark *Apristurus nakayai* Iglésias, 2012
Garrick's Catshark *Apristurus garricki* Sato, Stewart & Nakaya, 2013

Dr Kazuhiro Nakaya (b.1945) is a Japanese marine scientist and ichthyologist. He graduated BA from Hokkaido University (1968), which awarded his PhD (1972). He is Professor of Marine Environment and Resources at the Japanese Marine Laboratory for Biodiversity. His specialism is the taxonomy and evolution of sharks, rays and chimaeras and the fish of Lake Tanganyika. He has written widely, particularly articles and books on sharks and other fishes.

Nancy

African Dwarf Sawshark *Pristiophorus nancyae* Ebert & Cailliet, 2011

Mrs Nancy Ann Burnett née Packard (b.1943), of the Packard family of Hewlett-Packard fame, is a marine biologist who has donated generously to organisations researching the oceans, including as one of the founders of the Monterey Bay Aquarium. She has a bachelor's degree in biology from Stanford University and a master's from San Francisco State University. She has served as Chairperson of the Sea Studios Foundation and was Executive Director for 'The Shape of Life', a revolutionary eight-part television series documenting the dramatic rise of the animal kingdom by means of the breakthroughs of scientific discovery. She was honoured for her 'for her gracious support of chondrichthyan research' at the Pacific Shark Research Center at Moss Landing Marine Laboratories.

Nani

Hagfish genus *Notomyxine* Nani & Gneri, 1951

Hagfish sp. *Eptatretus nanii* Wisner & CB McMillan, 1988

Professor Alberto Nani Caputo was a zoologist who worked at the Museo Argentino de Ciencias Naturales, Buenos Aires (1940s) and was a member of the Argentine Antarctic Expedition (1942). Later in his career he was at the Department of Oceanography, University of Chile, Viña del Mar. He was honoured for 'his early work on the Myxinidae of Chile, for his considerable aid to us via correspondence with Carl L. Hubbs', and for providing specimens to the authors.

Nardo

Sandbar Shark *Carcharhinus plumbeus* Nardo, 1827

Giovanni Domenico Nardo (1802–1877) was an Italian naturalist. His uncle, an abbot, taught him taxidermy, a useful skill when he was at the University of Padua studying medicine and also organising the zoological collection. He went on to reorganise the invertebrate collection at the Imperial Natural History Museum in Vienna (1882). He wrote hundreds of scientific papers, mostly on zoology but also on medicine, social science, philology, technology and physics.

Narinari

Spotted Eagle Ray *Aetobatus narinari* Euphrasen, 1790

Narinari was a word in use in Brazil in the seventeenth and eighteenth centuries, meaning 'stingray'. Euphrasen in his description refers to Francis Willughby's book *De Historia Piscium* (1686) as his source but in fact Willughby copied both the description and the name from Georg Marcgrave (1610–1644), who was co-author with Willem Piso (1611–1678) of *Historia Naturalis Brasiliae* (1648). Euphrasen was the first person to use 'narinari' as a binomial.

Naseka

Nina's Lamprey *Lethenteron ninae* Naseka, Tuniyev & Renaud, 2009

Dr Alexander Mikhailovich Naseka is a Russian ichthyologist. He is senior researcher at the Department of Ichthyology, Zoological Institute, Russian Academy of Sciences. Among his publications is *Freshwater Fishes of Russia: Preliminary Results of the Fauna Revision // Zoological Sessions* (2001).

Navarra

Blackish Stingray *Dasyatis navarrae* Steindachner, 1892

Bruno R. A. Navarra was a scholar (variously described as a French traveller and a German soldier) who was an expert on Chinese history and culture including music. He supplied the Imperial Court Museum of Natural History (Vienna) with fish specimens collected in Shanghai. He was editor (1896–1899) of *Der Ostasiatische Lloyd*, a German language daily newspaper published in Shanghai. He wrote *China und die Chinesen* (1901) and translated (1910) Sun Tzu's *Art of War*.

Naylor

Starrynose Stingray *Pastinachus stellurostris* Last, Fahmi & Naylor, 2010

Dr Gavin J. P. Naylor (b.1960) is a British biologist who was born in Tanzania. His bachelor's degree was awarded by Durham University, England (1982) and his doctorate by the University of Maryland (1989). He was an assistant professor at Iowa State University (2001) and was an associate professor at the Department of Biological Science, Florida State University, moving to a new School of Computational Science and Information Technology (2005). He is now Professor of Biology at the College of Charleston, South Carolina. He is primarily interested in improving methods for estimating phylogenetic relationships from DNA sequence data; his empirical work has focused on estimating relationships among chondrichthyan fishes. He co-wrote *Interrelationships of Lamniform Sharks: Testing Phylogenetic Hypotheses with Sequence Data* (1997) and *Sharks and Rays of Borneo* (2010).

Nelson

Hagfish sp. *Eptatretus nelsoni* Kuo, Huang & Mok, 1994
Eastern Numbfish *Narcine nelsoni* Carvalho, 2008

Dr Gareth 'Gary' Jon Nelson (b.1937) is Emeritus Curator of Vertebrate Zoology at the American Museum of Natural History, New York City (1967–1998), having been Chairman of ichthyology (1982–1987) and of ichthyology and herpetology (1987–1993). After serving in the US Army (1955–1958) he gained a bachelor's degree at Roosevelt University, Chicago (1962). The University of Hawaii awarded his doctorate (1966), after which he did post-doctoral research at the BMNH and at the Swedish Museum of Natural History, Stockholm. He was an adjunct associate professor at Long Island University (1973–1974) and at New York University (1973–1978). In retirement he moved to Australia and is a professorial associate, School of Botany, University of Melbourne and an honorary associate in ichthyology at Museum Victoria, Melbourne. Among his more than 270 publications are *Outline of a Theory of Comparative Biology* (1974), *Identity of the Anchovy* Engraulis clarki *with Notes on the Species Groups of* Anchoa (1986) and *Cladistics and Evolutionary Models* (1989). While he was Adjunct Professor, City University of New York, he was Carvalho's PhD supervisor. Carvalho honoured him for his 'unique and meaningful contributions to both ichthyology and comparative biology in general'.

Nichols

Spot-on-spot Roundray *Urobatis concentricus* Osburn & Nichols, 1916
Roughtail Catshark *Galeus area* Nichols, 1927

John Treadwell Nichols (1883–1958) was an American ichthyologist and ornithologist. He studied vertebrate zoology at Harvard, graduating AB (1906). He then joined the AMNH (1907) as an assistant in the mammology department. He founded the journal of the American Society of Ichthyologists and Herpetologists, 'Copeia' (1913). He became first Assistant Curator (1913–1952) then Associate Curator and finally Curator of Ichthyology (1952–1958). He wrote over 1,000 articles and scientific papers and a number of books including *Fishes and Shells of the Pacific World* (1945). Two reptiles are named after him.

Nieuhof

Nieuhof's Eagle Ray *Aetomylaeus nichofii* Bloch & JGT Schneider, 1801

[Alt. Barbless Eagle Ray, Banded Eagle Ray, Syn. *Raja niehofii, Myliobatus nieuhofii*]

Johan Nieuhoff (1618–1672) was a Dutch traveller who wrote about his journeys to India, China and Brazil. His first trip was to Brazil (1640–1649), after which he joined the Dutch East India Company and lived in Batavia (Jakarta) for several years before being put in charge of an expedition to China (1654–1657). He was in India (1663) and then in Ceylon (Sri Lanka). He returned to the Netherlands (1672) before going to Madagascar where he disappeared. His *Voyages & Travels to the East Indies 1653–1670* was not published until 1988. There is no formal etymology in Bloch & Schneider but they do refer to *De Historia Piscium libri quator* Willughby, Raius & Nieuhof (1686), which is probably a reference to the first informal description of the ray.

Nina

Nina's Lamprey *Lethenteron ninae* Naseka, Tuniyev & Renaud, 2009

Dr Nina G. Bogutskaya (b.1958) is a Russian ichthyologist. Leningrad University awarded her master's degree (1981) and her doctorate (1988). She worked in the Zoological Institute in St Petersburg and lectured at the St Petersburg State University from 1997 until she moved to her present position as a researcher at the Naturhistorisches Museum, Vienna. Her main area of interest is cypriniform fishes. She was honoured in the lamprey name for her contribution to the knowledge of Eurasian freshwater fishes. Among her publications is the co-written *On the Migratory Black Sea Lamprey and the Nomenclature of the Ludoga, Peipsi and Ripus Whitefishes (Agnatha: Petromyzontidae; Teleostei: Coregonidae)* (2005).

Nishida

Deepwater Stingrays *Plesiobatidae* Nishida, 1990

Sharpnose Stingray *Dasyatis acutirostra* Nishida & Nakaya, 1988
Izu Stingray *Dasyatis izuensis* Nishida & Nakaya, 1988

Dr Kiyonori Nishida was an ichthyologist and marine zoologist at Hokkaido University, Hakodate, Japan (1990). In 2012 he was at Osaka Aquarium.

Nobili

Atlantic Torpedo *Torpedo nobiliana* Bonaparte, 1835
[Alt. Dark Electric Ray]

Leopoldo Nobili (1784–1835) was an Italian physicist who studied animal electricity and invented a number of instruments used in thermodynamics and electrochemistry. He was educated at the Military Academy of Modena and then served in Napoleon's campaign in Russia (1812) as an artillery officer, winning the Légion d'Honneur. He was later appointed Professor of Physics at the Regal Museum of Physics and Natural History in Florence.

Norman

Shortfin Sand Skate *Psammobatis normani* McEachran, 1983

Guitarfish genus *Aptychotrema* Norman, 1926

Slender Guitarfish *Rhinobatos holcorhynchus* Norman, 1922
Twin-spot Butterfly Ray *Gymnura bimaculata* Norman, 1925
Annandale's Guitarfish *Rhinobatos annandalei* Norman, 1926
Taiwan Guitarfish *Rhinobatos formosensis* Norman, 1926
Grayspotted Guitarfish *Rhinobatos leucospilus* Norman, 1926
Smoothback Guitarfish *Rhinobatos lionotus* Norman, 1926
Speckled Guitarfish *Rhinobatos ocellatus* Norman, 1926
Zanzibar Guitarfish *Rhinobatos zanzibarensis* Norman, 1926
White-spotted Guitarfish *Rhinobatos albomaculatus* Norman, 1930
Spineback Guitarfish *Rhinobatos irvinei* Norman, 1931
Bigthorn Skate *Rajella barnardi* Norman, 1935
White-dotted Skate *Bathyraja albomaculata* Norman, 1937
Graytail Skate *Bathyraja griseocauda* Norman, 1937
Patagonian Skate *Bathyraja macloviana* Norman, 1937
Multispine Skate *Bathyraja multispinis* Norman, 1937
Cuphead Skate *Bathyraja scaphiops* Norman, 1937
Antarctic Starry Skate *Amblyraja georgiana* Norman, 1938
Crying Izak *Holohalaelurus melanostigma* Norman, 1939
Bigeye Houndshark *Iago omanensis* Norman, 1939

John Roxborough Norman (1898–1944) started his working life as a bank clerk. He had rheumatic fever during his military service (WW1), which affected him ever after. He started work at the BMNH (1921) under Charles Tate Regan. He was Curator of Zoology (1939–1944) at Tring. He wrote, among other books, *A History of Fishes* (1931) and *A Draft Synopsis of the Orders, Families and Genera of Recent Fishes* (1957). He was an admirer of Albert Günther (1830–1914), who described the *Psammobatis* genus. He was honoured in the name of the sand skate as he had been the first (1937) to describe it, but misidentified it as *P. scobina*.

Noronha

Bigeye Sand Tiger *Odontaspis noronhai* Maul, 1955

Adolfo César de Noronha (1873–1963) was a naturalist and librarian who was Director of the Funchal Museum in Madeira (where the type specimen is housed). He particularly studied the ichthyology, ornithology and malacology of Madeira and worked at the Funchal Municipal Library, where he was appointed Librarian (1914) and Director (1928). He wrote *A New Species of Deep-water Shark* (Squaliobis Sarmienti) *from Madeira* (1926).

Norris

Narrowfin Smooth-Hound *Mustelus norrisi* S Springer, 1939

Dr Harry Waldo Norris (1862–1946) was a zoologist and anatomist who was Professor of Biology and Zoology and Curator of the museum at Grinnell College, Iowa (1891–1931). He became Research Professor of Zoology (1931–1941), retiring as Emeritus Professor. Iowa College (now Grinnell College) awarded his bachelor's degree (1880), his master's (1886) and his doctor of science degree (1924). His graduate work included stints at Cornell University, the University of Nebraska and at Freiburg, Germany. He wrote *On the Cranial Nerves in Elasmobranch Fishes* (1926).

Notarbartolo di Sciara

Munk's Devil Ray *Mobula munkiana* Notarbartolo di Sciara, 1987

Dr Giuseppe Notarbartolo di Sciara (b.1948) is an Italian marine ecologist and conservationist. The University of Parma awarded him bachelor's degrees in biological sciences (1974) and natural sciences (1976) and the Scripps Institution of Oceanography, University of California San Diego his doctorate in marine biology (1985). During his time there he discovered and named a new manta ray. His early work as a species-oriented marine ecologist led him towards place-based conservation. He led the creation of the Italian national cetacean stranding network, which he coordinated (1986–1990). He founded (1986) the Tethys Research Institute and directed it (1986–1997 and again since 2010). He proposed (1991) the creation of the Pelagos Sanctuary for Mediterranean Marine Mammals, established by a treaty between Italy, France and Monaco (1999) and was President of the Central Institute for Applied Marine Research (ICRAM, later merged into ISPRA), the Italian governmental body providing scientific support to national marine conservation policy (1996–2003). His current activities include being Regional Coordinator for the Mediterranean and Black Seas, IUCN WCPA – Marine (since 2000), Deputy Chairman, IUCN Species Survival Commission – Cetacean Specialist Group (since 1991), member, IUCN Shark Specialist Group (since 1993) and teaching science and policy of marine biodiversity conservation at the University 'Statale' of Milan. Among his many publications are *A Revisionary Study of the Genus Mobula Rafinesque, 1810* (Chondrichthyes: Mobulidae), *with the Description of a New Species* (1987), *Natural History of the Rays of the Genus Mobula in the Gulf of California* (1988) and *Ancient Waves, Recent Concerns: The Budding of Marine Mammal Conservation Science in Italy* (2012).

O

Oakley

Oakley's Catshark *Scyliorhinus sp.* X Undescribed

Douglas 'Doug' Oakley was a field assistant with the South Carolina Department of Natural Resources shark survey. He died of a heart attack during a research cruise.

Oda

Oda's Skate *Rhinoraja odai* Ishiyama, 1958

Mikiji Oda discovered the type specimen at the Miya fish market; it had been netted near Mikomoto-jima, off Izu Peninsula. Others have said that it is a mere Latinisation of the Japanese vernacular, Oda-ei.

Ogilby

Ratfish sp. *Hydrolagus ogilbyi* Waite, 1898

Carpetshark genus *Brachaelurus* Ogilby, 1908

Australian Spotted Catshark *Asymbolus analis* Ogilby, 1885
Australian Butterfly Ray *Gymnura australis* Ramsay & Ogilby, 1886
Common Whaler *Carcharias macrurus* Ramsay & Ogilby, 1887
Collared Carpetshark *Parascyllium collare* Ramsay & Ogilby, 1888
Prickly Dogfish *Oxynotus bruniensis* Ogilby, 1893
Thorntail Stingray *Dasyatis thetidis* Ogilby, 1899 [Alt. Black or Longtail Stingray]
Estuary Stingray *Dasyatis fluviorum* Ogilby, 1908
Bluegrey Carpetshark *Heteroscyllium colcloughi* Ogilby, 1908
Argus Skate *Dipturus polyommata* Ogilby, 1910
Purple Eagle Ray *Myliobatis hamlyni* Ogilby, 1911
Australian Cownose Ray *Rhinoptera neglecta* Ogilby, 1912
Australian Sharpnose Shark *Rhizoprionodon taylori* Ogilby, 1915

James Douglas Ogilby (1853–1925) was an Irish-born Australian ichthyologist and taxonomist, son of the famous zoologist William Ogilby. He worked at BMNH before emigrating to Australia (1884) and joining the Australian Museum, Sydney (1885). He was sacked (1890) for being drunk on the job. The contemporary report criticised his 'extreme and undiscriminating affinity for alcohol'. Though sacked as a permanent employee, he went on working on a contract basis. He worked for the Queensland Museum (1901–1904 and 1913–1920). A reptile is also named after him.

Ohe

Dogfish genus *Trigonognathus* Mochizuki & Ohe, 1990

Viper Dogfish *Trigonognathus kabeyai* Mochizuki & Ohe, 1990

Fumio Ohe is a palaeontologist and ichthyologist who was at, and is now retired from, the Department of Earth and Planetary Science, Nagoya University, Japan (2005), having previously been at the University of Tokyo. He was a member of the Tokai Fossil Society, Nagoya (1998). He co-wrote *Trigonognathus kabeyai, a New Genus and Species of the Squalid Sharks from Japan* (1990).

Olfers

Brazilian Electric Ray *Narcine brasiliensis* Olfers, 1831
Panther Electric Ray *Torpedo panther* Olfers, 1831
Variable Torpedo *Torpedo sinuspersici* Olfers, 1831

Ignaz Franz Werner Maria von Olfers (1793–1871) was a German naturalist, historian and diplomat. He went to Brazil (1816) as a diplomat. While in Brazil he described a number of new mammal species. Back in Germany he was made Director of the royal art collections (1839), was influential in the establishment of the Neues Museum, Berlin and became Director General of the Königlichen Museen. A reptile and an amphibian are named after him.

Oliva

Vladykov's Lamprey *Eudontomyzon vladykovi* Oliva & Zanandrea, 1959 [Alt. Danubian Brook Lamprey]

Professor Dr Ota Oliva (1926–1994) was a Czech ichthyologist at Charles University, Prague where he graduated (1950) and later earned his doctorate (1958). He was asked to stay on as a research assistant after his first degree and remained for the whole of his professional life, eventually becoming Professor of Zoology.

Olsen

Spreadfin Skate *Dipturus olseni* Bigelow & Schroeder, 1951

Dr Yngve H. Olsen (1905–2000) was a Danish American ichthyologist who worked (until at least 1963) at the Bingham Oceanic Laboratory, Yale University. He took part in the Vermillion Sea Expedition (1959). Gordon Arthur Riley in his reminiscences recalled that Olsen was (1937) Assistant to Albert E. Parr and preferred to work in the laboratory as he was prone to seasickness. When Parr became Director of the Laboratory and Museum and no longer had time for cruising, Olsen turned to editing and was honoured for his work as editor of the monograph series, *Fishes of the Western North Atlantic*.

Orbigny

Smooth Back River Stingray *Potamotrygon orbignyi* Castelnau, 1855

Alcide Dessalines d'Orbigny (1802–1857) was a 'learned and intrepid' traveller, collector, illustrator and naturalist. His father, Charles-Marie Dessalines d'Orbigny (1770–1856), was a ship's surgeon. He and Alcide studied shells. Alcide went to the Academy of Science, Paris to pursue his methodical paintings and classification of natural history specimens. The Muséum National d'Histoire Naturelle, Paris sent him to South America (1826), where the Spanish briefly imprisoned him, mistaking his compass and barometer for 'instruments of espionage'. After prison he lived for a year with the Guarani Indians, learning their language. He spent five years in Argentina, then travelled north along the Chilean and Peruvian coasts, before moving into Bolivia and later returning to France (1834). Once home he donated thousands of specimens of all kinds to the Muséum National d'Histoire Naturelle. His fossil collection led him to determine that there were many geological layers, revealing that they must have been laid down over millions of years. This was the first time such an idea had been put forward. He wrote *Dictionnaire Universel d'Histoire Naturelle*. Ten birds, five reptiles, two mammals and an amphibian are named after him.

Oregon

Hooktail Skate *Dipturus oregoni* Bigelow & Schroeder, 1958

The name is after the fishery explorations of the US Fish and Wildlife Service vessel 'Oregon' in the Gulf of Mexico and the Caribbean Sea.

ORI

Black-leg Skate *Anacanthobatis ori* Wallace, 1967

The skate is named after the acronym formed by the Oceanographic Research Institute, Durban, South Africa for whom the species was collected and described.

Orr

Butterfly Skate *Bathyraja mariposa* Stevenson, Orr, Hoff & McEachran, 2004
Leopard Skate *Bathyraja panther* Orr, Stevenson, Hoff, Spies & McEachran, 2011

Dr James Wilder Orr (b.1958) is (since 1995) a research fishery biologist at the Alaska Fisheries Science Center, Seattle, USA and also an affiliate associate professor at the College of Ocean and Fishery Sciences, University of Washington, Seattle (1999). He is also Museum Research Associate, Burke Museum of Natural History and Culture (2008). Wheaton College, Wheaton, Illinois awarded his bachelor's degree (1980), Auburn University, Auburn, Alabama his master's (1987) and the University of Washington, Seattle his doctorate (1995). He co-wrote *Recent Contributions to the Knowledge of the Skates of* Alaska (2005).

Osburn

Spot-on-spot Roundray *Urobatis concentricus* Osburn & Nichols, 1916

Dr Raymond Carroll Osburn (1872–1955) was an American zoologist whose bachelor's degree (1898) and master's (1900) were both awarded by Ohio State University while his doctorate was awarded by Columbia University (1906). He worked at Barnard College and at Connecticut College for Woman (1907–1917). He was Professor of Zoology and Entomology, Ohio State University (1917–1942). He also acted as summer Director of the F. T. Stone Laboratory in Lake Erie from 1925.

Owston

Goblin Shark *Mitsukurina owstoni* Jordan, 1898
Owston's Chimaera *Chimaera owstoni* Tanaka (I), 1905
Roughskin Dogfish *Centroscymnus owstoni* Garman, 1906

Alan Owston (1853–1915) was an Englishman who was a collector of Asian wildlife, as well as a businessman and yachtsman. He left for the Orient when still quite young. He married Shimada Rei Jkao (c.1880) in Japan and they had one child, Susie. He later married Kame (Edith) Miyahara (c.1893) having eight children by that marriage. Owston's most active collecting period was in the early twentieth century. He died of lung cancer in Yokohama, Japan. Eleven birds, an amphibian and a mammal are named after him. He obtained the type specimen of the goblin shark from a fisherman.

Ozouf-Costaz

Kerguelen Sandpaper Skate *Bathyraja irrasa* Hureau & Ozouf-Costaz, 1980

Dr Catherine Ozouf-Costaz (b.1951) is a research engineer at CNRS (Centre national de la recherche scientifique) whose main areas of interest are Antarctic notothenioïds and cichlids of the African great lakes. She has spent her entire professional life at the Muséum Nationale d'Histoire Naturelle, Paris in various teams. She works in Antarctica, which she first visited in 1974, as an ichthyologist and cytogeneticist. She co-wrote Barbapellis pterygalces, *New Genus and New Species of a Singular Eelpout* (Zoarcidae,: Teleostei) *from the Antarctic Deep Waters* (2011) and *Estimation Préliminaire de la Variabilité Chromosomique parmi quelques Espèces de* Cichlidae *des Grands Lacs Africains* (2003).

P

Palmer

Painted Electric Ray *Diplobatis pictus* Palmer, 1950

Geoffrey Palmer was originally employed at the BMNH as an assistant to the well-known ichthyologist J. R. Norman (q.v.). He wrote *New Records, and One New Species, of Teleost Fishes from the Gilbert Islands* (1970), which was based on a collection made (1962) by Mrs Jane Cooper, a resident of Betio, Tarawa in the Gilbert Islands.

Paragó

South Brazilian Skate *Dipturus mennii* UL Gomes & Paragó, 2001

Cristina Paragó is a biologist who was at the Museu Nacional, Universidade Federal do Rio de Janeiro (2001). She co-wrote *Anatomical Study on the Prebranchial Region of* Sphyrna lewini *(Griffith & Smith) and* Rhizoprionodon lalandii *(Valenciennes) (Elasmobranchii, Carcharhiniformes) Related with the Cephalofoil in Sphyrna Rafinesque* (1997).

Parin

Pocket Shark *Mollisquama parini* Dolganov, 1984
Dwarf False Catshark *Planonasus parini* Weigmann, Stehmann & Thiel, 2013

Smalldisk Torpedo *Torpedo microdiscus* Parin & Kotlyar, 1985
Semipelagic Torpedo *Torpedo semipelagica* Parin & Kotlyar, 1985
Smalleye Lantern Shark *Etmopterus litvinovi* Parin & Kotlyar, 1990

Professor Nikolai Vasilyevich Parin (1932–2012) of the P. P. Shirov Institute of Oceanology, Russian Academy of Sciences was an oceanographer and ichthyologist who was an expert on flying fishes. He graduated from the Institute of Fisheries (1955) and later was awarded a PhD (1961) and a DSc (1967). He became Chief of the Laboratory of Oceanic Ichthyofauna (1973) and studied pelagic fishes for the rest of his life, taking part in 20 cruises and notching up 8 years of sea time. During his career he published (1958–2012) numerous articles and papers including *Ichthyofauna of the Epipelagic Zone* (1970) and co-wrote *Biology of the Nazca and Sala-y-*

Gómez Submarine Ridges, an Outpost of the Indo-West Pacific Fauna in the Eastern Pacific Ocean: Composition and Distribution of the Fauna, its Communities and History (1997). He was honoured in the name of the false catshark as he was chief scientist of the 'memorable' Cruise 17 of the research vessel 'Vityaz' (1988–1989), when 13 ichthyologists from 8 countries were assembled. During this cruise the type was collected. The pocket shark was named for Parin because he was an authority on the fauna of the Nazca submarine ridge (east of Chile), where the only known specimen was collected. In addition to these 2 sharks, he is remembered in the names of 27 other fishes.

Paucke

Freshwater Stingray sp. *Potamotrygon pauckei* Castex, 1963
[JS. *Potamotrygon motoro*]

Florian Paucke (1719–1780) was a German-Austrian Jesuit monk and ethologist who was an accomplished artist. He left behind hundreds of drawings of fauna, flora and people when he was accused by the Spanish authorities of being a British spy and expelled from Argentina (1767).

Paul

Smooth Deep-sea Skate *Brochiraja asperula* Garrick & Paul, 1974
Prickly Deep-sea Skate *Brochiraja spinifera* Garrick & Paul, 1974
New Zealand Smooth Skate *Dipturus innominatus* Garrick & Paul, 1974

Lawrence 'Larry' James Paul BSc (b.1939) is an ichthyologist who is with the Fisheries Division, Marine Department, Wellington, New Zealand (1966). He wrote *New Zealand Fishes: Identification, Natural History & Fisheries* (2000).

Pavlenko

Golden Skate *Bathyraja smirnovi* Soldatov & Pavlenko, 1915

Mikhail Nikolaevich Pavlenko (1886–1919) was a Russian ichthyologist who co-authored a number of scientific papers including descriptions of new fish, such as *Description of a New Species of Family Rajidae from Peter the Great Bay and from Okhotsk Sea* (1915) and *Fish of Peter the Great Bay* (1910). The University of Kazan, where his collection is housed, published the latter paper.

Pavlov

Tasmanian Lantern Shark *Etmopterus tasmaniensis* Myagkov & Pavlov, 1986

N. A. Pavlov is a Russian ichthyologist who collected the type specimen of this species.

Pel

Lusitanian Cownose Ray *Rhinoptera peli* Bleeker, 1863 [Syn. *Rhinoptera marginata*]

Hendrik Severinus Pel (1818–1876) was the Dutch Governor of the Gold Coast (Ghana) (c.1840–1850). He was also an amateur naturalist and trained taxidermist, and acted as such for the Leiden Museum, to which he sent shipments of animal specimens. He was honoured as the person whose 'enlightened zeal' led to the deposition of natural history specimens at the Leiden Museum, including the type of this species. He is remembered in the names of other taxa, including two mammals and two birds.

Pellegrin

Bluespotted Bamboo Shark *Chiloscyllium caerulopunctatum* Pellegrin, 1914

Dr Jacques Pellegrin (1873–1944) was a French zoologist. He was an ichthyologist at the Muséum National d'Histoire Naturelle, Paris where he became Assistant Curator of Zoology (1894) when Vaillant (q.v.) relinquished the post. During this time he continued to study and was awarded his MD (1899) and his doctorate (1904), becoming Assistant Professor (1908). He undertook a number of overseas trips collecting for the Museum and was appointed as Deputy Director (1937) and Curator of Herpetology and Ichthyology. He published more than 600 books and scientific papers and discovered over 350 species new to science. A number of other fishes are named after him.

Penggali

Indonesian Shovelnose Ray *Rhinobatos penggali* Last, WT White & Fahmi, 2006

Penggali is the Indonesian for 'shovel' and refers to the shape of this species' head.

Penrith

Taillight Shark genus *Euprotomicroides* Hulley & Penrith, 1966

Taillight Shark *Euprotomicroides zantedeschia* Hulley & Penrith, 1966

Dr Michael John Penrith was an ichthyologist and biologist at the Oceanic Research Unit, University of Cape Town, South Africa (1964–1971), at the State Museum, Windhoek, Southwest Africa (Namibia) (1972) and at the National Zoological Gardens, Pretoria (1998) in charge of the aquarium and reptile park. He wrote *Earliest Description and Name for the Whale Shark* (1972). His wife, Professor Marie-Louise Penrith (b.1942), is also a biologist and ichthyologist.

Pequeño

Hagfish sp. *Myxine pequenoi* Wisner & CB McMillan, 1995

Dr German Enrique Pequeño Reyes (b.1941) was Professor of Zoology at and Director of the Instituto de Zoología Ernst F. Kilian, Universidad Austral de Chile (1973–2007). The University of Chile awarded his bachelor's degree (1967) and Oregon State University his doctorate (1984). He was an assistant in the ichthyology department of the National Museum of Natural History (1965–1970). He undertook post-doctoral studies in Barcelona (1986 and 1988) and in London (1988). He received his professional diploma as Profesor de Estado en Biología y Ciencias at the Universidad de Chile, Santiago (1968–1972). Interestingly he is a descendant of the Chilean hero Bernardo O'Higgins. He has described a number of fish species and three fishes are named after him, including the Hagfish sp. He was honoured in the name of that hagfish for his work on Chilean fishes and for providing the authors with the holotype. Among his many publications are *Sinopsis de Macrouriformes de Chile (Pisces: Teleostomi)* (1971) and *Las Colecciones de Animales Chilenos y el Problema de su Ordenación* (1979).

Pérez

Caribbean Reef Shark *Carcharhinus perezii* Poey, 1876

Professor Laureano Pérez Arcas (1824–1894) was a Spanish entomologist and malacologist who became Professor of Zoology at the University of Madrid and Poey's friend and companion. He was co-founder of the Spanish Society of Natural History. Poey used Arcas' book, *Elementos de Zoología* when at the University of Havana.

Péron

Eastern Australian Sawshark *Pristiophorus peroniensis* Yearsley, Last & WT White, 2008

Broadnose Sevengill Shark *Notorynchus cepedianus* Péron, 1807

Tiger Shark *Galeocerdo cuvier* Péron & Lesueur, 1822

François Péron (1775–1810) was a French voyager and naturalist. Originally intending to become a priest, he was a reluctant army volunteer (1792) against Prussia, was wounded and taken prisoner (1793), being repatriated over a year later (1794) and invalided out as he had lost an eye. He became a town clerk, then gained a scholarship to study medicine in Paris. After an unhappy love affair he became an anthropological observer on Baudin's scientific expedition with the ships 'Géographe' and 'Naturaliste' (1800–1804), which visited New Holland, Maria Island, Van Diemen's Land (Tasmania) and Timor, Indonesia. He was constantly clashing with Baudin but was soon the only zoologist left on the expedition and with Lesueur (q.v.) collected more than 100,000 zoological specimens. As well as collecting, Péron conducted pioneering experiments on seawater temperatures at depth. He died at the age of 35 of tuberculosis. The Peron Peninsula, Western Australia is named after him as are two mammals, six birds, three reptiles and two amphibians.

Perrottet

Large-tooth Sawfish *Pristis perotteti* Müller & Henle, 1841

Georges (Guerrard or Gustave) Samuel Perrottet (1793–1867/1870) was a Swiss-born French botanist and horticulturalist. He was a gardener at the Jardin des Plantes, Paris. He became the naturalist on the 'Rhône' during an expedition (1819–1821) to Réunion, Java, and the Philippines and was sent to Cayenne (French Guiana) to introduce plants that were considered useful. While there Perrottet made large mineralogical and botanical collections, then returned to France. He made a number of voyages to Africa and South America (1822–1832), including one circumnavigation, and wrote *Souvenirs d'un Voyage Autour du Monde* (1831). He explored in Senegambia (1824–1829) and was an administrator at a government trading post. He also explored Cape Verde (1829) before returning to France. He co-wrote a work on the plant life of that part of Africa, *Florae Senegambiae Tentamen* (1830–1833). He became correspondent of the MNHN, Paris (1832) and was then assigned to a botanical garden in Pondicherry (India) (1834–1839), after which he returned to France and cultivated silk worms. He returned to Pondicherry as a botanist and there lived out his life (1843–1870). He supplied a sample of this fish. Four reptiles are named after him.

Perry (Gilbert)

Dwarf Lantern Shark *Etmopterus perryi* S Springer & Burgess, 1985

Perry Webster Gilbert (1912–2000) was an American biologist with a particular interest in sharks. He studied for his first degree at Dartmouth College, graduating in zoology (1934), and was given an instructorship there. He entered the programme at Cornell University (1936), which awarded his PhD (1940) in comparative vertebrate anatomy. He was an instructor there until becoming Assistant Professor (1943–1946), gaining tenure (1946–1952), then full Professor (1952–1978) until retirement after which he continued to research as Professor Emeritus. His many dissections of sharks continued to fuel an interest that he pursued whenever time (and sabbaticals) allowed, including undertaking research trips to the Bahamas, Tahiti, Australia, Belize, South Africa and Japan as well as within the US. He published on virtually every aspect of shark biology, writing over 150 papers and editing two books including *Sharks, Skates and Rays* (1967). He was honoured for his contributions to the knowledge of elasmobranch reproduction and other aspects of shark biology.

Peters

Black-spotted Torpedo *Torpedo fuscomaculata* Peters, 1855
Cobbler Wobbegong *Sutorectus tentaculatus* Peters, 1864
Longsnout Butterfly Ray *Gymnura crebripunctata* Peters, 1869

Wilhelm Karl Hartwig Peters (1815–1883) was a German zoologist and explorer. He was Assistant to the anatomist Müller (q.v.) who with Humbolt encouraged Peters to travel. He made a major expedition to Africa (1842–1847), travelling from Angola to Mozambique. He returned to Berlin with very important zoological collections. For many years (from 1858) he was Curator of the Berlin Zoology Museum. He was elected as a corresponding member of the Russian Academy of Sciences (1876). He described 122 new genera and nearly 650 new species. Thirty-nine reptiles, 23 mammals, 18 amphibians and 2 birds are named after him.

Petit

Madagascar Guitarfish *Rhinobatos petiti* Chabanaud, 1929

Professor Georges Jean-Jacques Petit (1892–1973)

was a French marine biologist at the Muséum National d'Histoire Naturelle. He became Director of the marine research stations at Banyuls-sur-mer and Villefranche-sur-mer. He collected the holotype.

Philip

Skate sp. *Okamejei philipi* Lloyd, 1906
[Syn. *Raja philipi*]

The description has no etymology and we have been completely unsuccessful at identifying 'Philip'. If any reader knows the answer, we would love to hear from them.

Philippi

Guitarfish sp. *Tarsistes philippii* Jordan, 1919

Raspthorn Sand Skate *Psammobatis scobina* Philippi
    {Krumweide}, 1857
Chilean Angelshark *Squatina armata* Philippi
    (Krumweide), 1887
Skate sp. *Dipturus flavirostris* Philippi (Krumweide), 1892
Chilean Devil Ray *Mobula tarapacana* Philippi
    (Krumweide), 1892
Chilean Eagle Ray *Myliobatis chilensis* Philippi
    (Krumweide), 1893
Chilean Roundray *Urobatis marmoratus* Philippi
    (Krumweide), 1893
Magellan Skate *Bathyraja magellanica* Philippi
    {Krumweide}, 1902

Professor Dr Rodulfo Amando (sometimes Rudolph Amandus) Philippi (1808–1904) {Krumweide}* was a German-born Chilean naturalist principally interested in palaeontology and zoology. He was educated in Berlin, qualifying as a physician with further study in zoology. He was then Professor of Natural History and Geography at the Polytechnic of Kassel. Believing he was mortally ill, he moved to southern Italy, duly recovered and began working there. His brother worked for the Chilean government and invited him to join him (1851). He became a professor of zoology and botany and Director of the Museo Nacional de Historia Natural, Santiago (1853–1883). He organized over 60 expeditions within Chile, published 456 scientific papers and described more than 6,000 plants before retiring (1896). He proposed a genus for the guitarfish (known only from a dried head) using the name *Rhynchobatis* that Jordan mistakenly thought was preoccupied. He described seven species of *Elasmobranchii* and is commemorated in the names of numerous plants, an amphibian, a mammal and three reptiles.
*{Krumweide}, derived from his maternal line, is used to distinguish him in scientific descriptions from his zoologist grandson with exactly the same name.

Phillipps

Oval Electric Ray *Typhlonarke tarakea* Phillipps, 1929
Northern Spiny Dogfish *Squalus griffini* Phillipps, 1931
Spotted Estuary Smooth-Hound *Mustelus lenticulatus*
    Phillipps, 1932

William John Phillipps (1893–1967) was a New Zealand ichthyologist, ornithologist and ethnologist who worked at the Dominion Museum (1917–1958). He wrote several books on New Zealand fishes and on Maori material culture.

Pietschmann

Spotless Smooth-Hound *Mustelus griseus*
    Pietschmann, 1908
Blackspot Shark *Carcharhinus sealei* Pietschmann,
    1913
Prickly Shark *Echinorhinus cookie* Pietschmann, 1928

Dr Viktor Pietschmann (1881–1956) was an Austrian ichthyologist who worked as Curator of the Fish Collection at the Vienna Museum of Natural History (1919–1946). He graduated from the University of Vienna (1904). He made several collecting trips, notably to Greenland, Armenia, Hawaii, Romania, Poland and the Middle East. He was a member of the Nazi party (from 1932) so was forced to resign (1946).

Pinocchio

Pinocchio Catshark *Apristurus australis* Sato, Nakaya & Yorozu, 2008

Pinocchio is a fictional character created by Carlo Collodi in his children's novel, *The Adventures of Pinocchio* (1883). One of his main features was his nose that grew longer and longer when he was under stress and particularly when he was telling lies!

Planer

European Brook Lamprey *Lampetra planeri* Bloch, 1784

Professor Dr Johann Jacob Planer (1743–1789) was a German physician and botanist. He gave Bloch the holotype. A tree genus *Planera* is named after him.

Plunket

Lord Plunket's Shark *Proscymnodon plunketi* Waite, 1910

Sir William Lee Plunket, fifth Baron Plunket of Newton, Count Cork (1864–1920) was an Anglo-Irishman who was the sixteenth Governor of New Zealand (1904–1910). He served as a diplomat in Rome and Constantinople (Istanbul) (1889–1894). The Plunket Society in New Zealand is named after his wife, Victoria. He was honoured for his interest in the Canterbury Museum and 'gratefully remembering His Excellency's kindness when, as his guest, [Waite] accompanied him on his cruise to the southern islands of New Zealand in 1907.'

Poeppig

Filetail Fanskate *Sympterygia lima* Poeppig, 1835

Professor Eduard Friedrich Poeppig (1798–1868) was a German naturalist, collector and explorer. He studied medicine and natural science at Leipzig University, leaving to undertake an expedition to Cuba and the USA (1823–1826). He then went to Chile and remained in South America until 1832. When he returned to Germany, he became Professor of Zoology, Leipzig University. He wrote *Reise nach Chili, Peru, und auf dem Amazonen-Flusse* (1835). A mammal, a bird, an amphibian and a reptile are named after him.

Poey

Cuban Legskate *Cruriraja poeyi* Bigelow & Schroeder, 1948

Blacknose Shark *Carcharhinus acronotus* Poey, 1860
Oceanic Whitetip Shark *Carcharhinus longimanus* Poey, 1861
Caribbean Lantern Shark *Etmopterus hillianus* Poey, 1861
Caribbean Sharpnose Shark *Rhizoprionodon porosus* Poey, 1861
Night Shark *Carcharhinus signatus* Poey, 1868
Lemon Shark *Negaprion brevirostris* Poey, 1868
Caribbean Reef Shark *Carcharhinus perezii* Poey, 1876

Professor Felipe Poey y Aloy (1799–1891) was a Cuban zoologist, naturalist, and artist. He was brought up in France (1804–1807) and later Spain. He qualified as a lawyer in Madrid, but his ideas were too liberal for the age, and he was forced to return to Cuba (1823). He returned to France (1825) and was one of the founders of the Société Entomologique de France (1832), finally returning to Cuba (1833).

He concentrated on natural history, describing 85 species of Cuban fish. His *Memorias sobre la Historia Natural de la Isla de Cuba, Acompañadas de Sumarios Latinos y Extractos en Francés* (1858) depicts mainly fishes and snails but also some mammals, hymenoptera, and lepidoptera; the drawings were all done by Poey. He founded the Museum of Natural History, Havana (1839) and became its first director. He also was appointed the first Professor of Zoology and Comparative Anatomy at the University of Havana (1842). The Museum merged (1849) with the University of Havana and was later named in his honour. Poey, who also supplied Cuvier and Valenciennes in Paris with Cuban fish specimens, and prepared many of the Museum's exhibits, especially of fishes. He is commemorated in the scientific names of other taxa, especially fish and a mammal.

Pogonoski

Bighead Spurdog *Squalus bucephalus* Last, Séret & Pogonoski, 2007
Bighead Catshark *Apristurus bucephalus* WT White, Last & Pogonoski, 2008
Whitefin Chimaera *Chimaera argiloba* Last, WT White & Pogonoski, 2008
Deepwater Skate *Dipturus acrobelus* Last, WT White & Pogonoski, 2008
Pale Tropical Skate *Dipturus apricus* Last, WT White & Pogonoski, 2008
Heald's Skate *Dipturus healdi* Last, WT White & Pogonoski, 2008
Blacktip Skate *Dipturus melanospilus* Last, WT White & Pogonoski, 2008
Queensland Deepwater Skate *Dipturus queenslandicus* Last, WT White & Pogonoski, 2008
Black-spotted Whipray *Himantura astra* Last, Manjaji-Matsumoto & Pogonoski, 2008
Network Wobbegong *Orectolobus reticulatus* Last, Pogonoski & WT White, 2008
Indonesian Wobbegong *Orectolobus leptolineatus* Last, Pogonoski & WT White, 2010

John James Pogonoski is an ichthyologist with CSIRO, Hobart, Tasmania. He previously (2004) worked at the Australian Museum, Sydney. He co-wrote: *Revision of the Genus Parequula (Pisces: Gerreidae) with a New Species from Southwestern Australia* (2012).

Poll

Spotted Skate *Raja straeleni* Poll, 1951
African Lantern Shark *Etmopterus polli* Bigelow, Schroeder, & S Springer, 1953

African Sawtail Catshark *Galeus polli* Cadenat, 1959

Dr Max Fernand Leon Poll (1908–1991) was a Belgian ichthyologist, and 'connoisseur of the fish fauna'. He worked in the Congo and in the Musée Royal du Congo Belge, Tervuren, and was Professor at the Université Libre de Bruxelles. He led expeditions to Lake Tanganyika (1946–1947) and a Congo River survey (1953). Cadenat named the catshark after him as Poll had first drawn it to his attention, having discovered it and sent it to the Harvard Museum of Comparative Zoology. Many fishes, an amphibian and a mammal are named after him.

Port Jackson

Port Jackson Shark *Heterodontus portusjacksoni* Meyer, 1793

As the name implies, it is named after a place – Port Jackson, Sydney, New South Wales, near Botany Bay, where the type specimen was collected.

Potter

Non-parasitic Lamprey *Mordacia praecox* Potter, 1968

Professor Dr Ian C. Potter of the School of Biological Sciences and Biotechnology, Murdoch University, Perth, Australia is considered to be one of the world's most knowledgeable researchers into lamprey biology. He was formerly Director of the Centre for Fish and Fisheries Research there. He described the lamprey in *Mordacia Praecox, n. sp., a Nonparasitic Lamprey (Petromyzonidae), from New South Wales, Australia* (1968). He also co-wrote the five-volume *The Biology of Lampreys* (Volume 1, 1971).

Poulsen

Jespersen's Hagfish *Myxine jespersenae* Møller, Feld, Poulsen, Thomsen & Thormar, 2005

Mrs Idahella Hyldgaard Bacher née Poulsen (b.1979) is a Danish biologist and parasitologist whose master's degree was awarded by the University of Copenhagen. She now works as a consultant for Danish Science Factory (a private Danish non-governmental organisation). Among her papers is *Prevalence of Gastrointestinal Nematodes in Growing Pigs in Kabale District in Uganda* (2011).

Powell

Indian Ringed Skate *Okamejei powelli* Alcock, 1898

Lieutenant Frederick Thomas Powell of the Indian Navy was one of Robert Moresby's officers on 'HMS Palinurus' (1829–1833) and on 'HMS Benares' surveying the Chagos Archipelago (1837–1838). While Moresby was on sick leave, Powell commanded 'HMS Benares' in undertaking other Indian coastal surveys. He served in the Indian Navy for more than 30 years and was a post captain commanding the 'Oriental' on the Suez-India leg of the same journey from England as the 'Ripon'. (See **Moresby**)

Pozzi

Southern Thorny Skate *Amblyraja doellojuradoi* Pozzi, 1935

Aurelio Juan Santiago Pozzi (1895–1959) was an ichthyologist at the Museo Argentino de Ciencias Naturales de Buenos Aires (1912–1946).

Prabhu

Quilon Electric Ray *Heteronarce prabhui* Talwar, 1981

Madhav Sudhakar Prabhu (b.1922) was Director of the Central Marine Fisheries Research Institute (1970), Mandapam Camp, India. The holotype was taken nearby. He wrote *Mackerel and Oil Sardine Tagging Programme, 1966–67 to 1968–69* (1970).

Prahl

Gorgona Guitarfish *Rhinobatos prahli* Acero & Franke, 1995

Henry von Prahl (1948–1989) was a pioneering Colombian marine biologist who studied Gorgona Island (the type locality) and whose family emigrated from Germany (1953). He was killed by a bomb, which exploded on Avianca Airlines Flight 203 over Bogotá.

Priapus

Phallic Catshark *Galeus priapus* Séret & Last, 2008

Priapus was a Greek fertility god. The description relates this to the male shark's very long claspers.

Princeps

Great Lantern Shark *Etmopterus princeps* Collett, 1904

Princeps is Latin for 'chief' and is probably just an allusion to the fact that this species is the largest of its genus.

Puckridge

Ningaloo Maskray *Neotrygon ningalooensis* Last, WT
White & Puckridge, 2010

Dr Melody Puckridge (b.1979) is a phylogeneticist. Southern Cross University, Australia awarded her honours degree (2001) and the University of Tasmania her master's (2006) and doctorate on fish phylogenetics (2013). She co-wrote *Cryptic Diversity in Flathead Fishes* (Scorpaeniformes: Platycephalidae) *across the Indo-West Pacific Uncovered by DNA Barcoding* (2013) and *Phylogeography of the Indo-West Pacific Maskrays* (Dasyatidae, Neotrygon): *A Complex Example of Chondrichthyan Radiation in the Cenozoic* (2013).

Q

Quaranta

White-spot Chimaera *Hydrolagus alphus* Quaranta,
Didier, DJ Long & Ebert, 2006

Dr Kimberly Lisa Quaranta (b.1978) was a post-
graduate research worker at California State
University, Moss Landing Marine Laboratory (2004–
2009), where her doctorate was awarded. California
Lutheran University awarded her bachelor's degree
in science.

Quattrini

Hagfish sp. *Eptatretus lopheliae* Fernholm & Quattrini,
2008

Dr Andrea M. Quattrini is a marine biologist. Millersville University awarded her bachelor's degree
(1999), the University of North Carolina her master's (2002) and Temple University her doctorate
(2013). She was Research Assistant (2000–2002) then
Research Associate (2002–2009) at UNCW Center
for Marine Science. She has written or co-written
a number of papers, including *Megafaunal Habitat
Associations at a Deep-sea Coral Mound off North Carolina* (2012). She has also taken part in numerous
research cruises (2000–2011).

Quattro

Carolina Hammerhead *Sphyrna gilberti* Quattro,
Driggers, Grady, Ulrich & MA Roberts, 2013

Dr Joseph M. Quattro is an American
biologist and ichthyologist who is a professor in the Marine Science Program and
Department of Biological Sciences, University of
South Carolina. Frostburg State University, Maryland awarded his master's degree in fisheries
management (1985) and Rutgers University awarded
his doctorate in ecology and evolution (1991). He
was a Sloan Post-doctoral Fellow at Stanford University (1991–1994). His research interests include
the population genetics, evolutionary relationships
and demography of rare, threatened and/or endangered species of freshwater and marine fishes. As
part of this research programme, Dr Quattro has

contributed to the conservation and management
of species of anadromous (sturgeons), freshwater
(pygmy sunfishes, centrarchids) and marine fishes
(sharks, rays and teleosts) and sea turtles. He is a
co-author of *Freshwater Fishes of South Carolina* (2009),
which includes a chapter on the conservation of rare,
threatened and endangered fishes. Similarly, he has
contributed to reviews on the status of endangered
species, fisheries management at local state and
federal levels and is currently involved in several
projects focusing on the identification of eggs from
commercially important marine teleosts.

Quekett

Flapnose Houndshark *Scylliogaleus quecketti*
Boulenger, 1902

John Frederick Whitlie Quekett (b.1849) FZS was
Curator of the Durban Museum (1895–1909). He
was appointed to organise the collection of the Natal
Society (1886). He sent the holotype to Boulenger
who named the species after him but misspelled his
name, as he did with a frog species also named after
him.

Quoy

Galapagos Bullhead Shark *Heterodontus quoyi*
Fréminville, 1840

Blacktip Reef Shark *Carcharhinus melanopterus* Quoy
& Gaimard, 1824
Pygmy Shark *Euprotomicrus bispinatus* Quoy &
Gaimard, 1824
Indian Speckled Carpetshark *Hemiscyllium freycineti*
Quoy & Gaimard, 1824
Cookiecutter Shark *Isistius brasiliensis* Quoy &
Gaimard, 1824

Jean René Constant Quoy (1790–1869) was a French
naval surgeon and zoologist who named and
described many species, often with Joseph Paul
Gaimard. He took part in a number of voyages of
discovery, including a circumnavigation aboard
the 'Astrolabe' (1826–1829) with Jules Dumont
d'Urville. He became chief medical officer of the
naval hospital at Toulon (1835). Fréminville named
the Bullhead Shark 'for his friendship, untiring zeal
and wide knowledge of zoology'. He is commemorated in the names of other taxa, including several
fishes, five birds, two amphibians and a reptile.

R

Radcliffe

Pygmy Ribbontail Catshark *Eridacnis Radcliffe* HM
Smith, 1913

Catshark genus *Pentanchus* HM Smith & Radcliffe, 1912
Deepsea Dogfish Shark genus *Squaliolus* HM Smith &
Radcliffe, 1912
Carpetshark genus *Cirrhoscyllium* HM Smith &
Radcliffe, 1913

Aguja Skate *Bathyraja aguja* Kendall & Radcliffe, 1912
Arrowhead Dogfish *Deania profundorum* HM Smith &
Radcliffe, 1912
Short-tail Lantern Shark *Etmopterus brachyurus* HM
Smith & Radcliffe, 1912
Philippine Chimaera *Hydrolagus deani* HM Smith &
Radcliffe, 1912
Onefin Catshark *Pentanchus profundicolus* HM Smith
& Radcliffe, 1912
Spined Pygmy Shark *Squaliolus laticaudus* HM Smith
& Radcliffe, 1912
Barbelthroat Carpetshark *Cirrhoscyllium expolitum*
HM Smith & Radcliffe, 1913

Lewis Radcliffe (1880–1950) was an American naturalist particularly interested in malacology and
ichthyology. He was Assistant Naturalist to Hugh
McCormick Smith (q.v.) on the Philippines Expedition (1907–1910). He later became Scientific Assistant
then Deputy Commissioner of the US Bureau of
Fisheries (to 1932). He was Director of the Oyster
Institute of North America when he died. He was
a member of the Bureau of Fisheries team that collected type from the steamer 'Albatross'.

Rafinesque

Thresher Shark genus *Alopias* Rafinesque, 1810
Great White Shark genus *Carcharias* Rafinesque, 1810
Kitefin Shark genus *Dalatias* Rafinesque, 1810
Skate genus *Dasyatis* Rafinesque, 1810
Skate genus *Dipturus* Rafinesque, 1810
Lantern Shark genus *Etmopterus* Rafinesque, 1810
Catshark genus *Galeus* Rafinesque, 1810
Sharpnose Shark genus *Heptranchias* Rafinesque, 1810
Sixgill Shark genus *Hexanchus* Rafinesque, 1810
Mackerel Shark genus *Isurus* Rafinesque, 1810
Eagle Ray genus *Mobula* Rafinesque, 1810

Rough Shark genus *Oxynotus* Rafinesque, 1810
Hammerhead Shark genus *Sphyrna* Rafinesque, 1810

Sand Tiger Shark *Carcharias Taurus* Rafinesque, 1810
Blackmouth Catshark *Galeus melastomus* Rafinesque,
1810
Shortfin Mako *Isurus oxyrinchus* Rafinesque, 1810
Little Gulper Shark *Squalus uyato* Rafinesque, 1810

Professor Constantine Samuel Rafinesque-Schmaltz
(1783–1840) was born in Constantinople of a French
father and a German mother. He was sent to live in
Tuscany to escape the turmoil of the French Revolution. His father was a merchant who died in
Philadelphia (1793), leaving the family very badly
off. Despite being unable to attend a university,
Rafinesque was a highly gifted individual and
his accomplishments included being a botanist, a
geologist, a historian, poet, philosopher, philolo-
gist, economist, merchant, manufacturer, professor,
architect, author and editor. He was apprenticed
(1802) to a merchant house in Philadelphia, and
for the next two years he roamed the fields and
woods and made collections of plants and animals.
He was in Sicily (1805–1815) as secretary to the US
Consul and carried on a lucrative trade in commod-
ities. He scoured the island for plants and collected
previously unrecorded fishes from the stalls of the
Palermo market. He sailed for New York (1815) but
was shipwrecked in Long Island Sound, losing all
his unpublished manuscripts and collections. He
sailed down the Ohio River (1818) and conducted
a comprehensive survey of the fish species there,
published as *Ichthyologia Ohiensis* (1820). He was
Professor of Botany and Natural Science, University
of Transylvania (Lexington, Kentucky) (1819–1826).
He returned to Philadelphia (1826) with 40 crates of
specimens. He had a remarkable gift for inventing
scientific names, some 6,700 in botany alone. He
died in poverty but was later re-interred in Lexington. A mammal and an amphibian are named after
him.

Ragno

Long-tailed River Stingray *Plesiotrygon nana* Carvalho
& Ragno, 2011

Maira Portella Ragno is a biologist at the Department of Zoology, Institute of Bio Sciences, University
of São Paulo, Brazil. She wrote *Distribution and
Morphology of Lateral Line Canals in Rays and its Systematic Relevance* (Chondrichthyes: Elasmobranchii:
Batoidea) (2014) as the thesis for her master's degree.

Raleigh

Pacific Longnose Chimaera *Harriotta raleighana*
Goode & Bean, 1895

Sir Walter Raleigh (1554–1618) was an English aristocrat, explorer, writer, poet, soldier and spy who popularised the use of tobacco in England. So much has been written about him that no more is needed here.

Ramalheira

Whitespotted Bullhead Shark *Heterodontus*
ramalheira JLB Smith, 1949
[Alt Mozambique Bullhead Shark, Syn *Gyropleurodus*
ramalheira]

João Ramalheira was the captain of the trawler that collected the holotype and, according to the etymology, 'has brought in much valuable scientific material'.

Ramsay

Australian Butterfly Ray *Gymnura australis* Ramsay &
Ogilby, 1886
Common Whaler *Carcharias macrurus* Ramsay &
Ogilby, 1887
Collared Carpetshark *Parascyllium collare* Ramsay &
Ogilby, 1888

Edward Pearson Ramsay (1842–1916) was an Australian naturalist, oologist, ornithologist and particularly a marine zoologist. He attended medical school at the University of Sydney (1863–1865) but did not graduate. Ramsay became Curator of the Australian Museum (1874–1894); he resigned owing to ill health. When he attended an international fisheries exhibition in London (1883) he met Francis Day, who had amassed a collection of fishes from Malaysia and India, which he purchased and took back to Sydney. After resigning he still served as 'consulting ornithologist' (1894–1909). Eleven birds, a mammal and two reptiles are named after him.

Rancurel

Devil Ray *Mobula rancureli* Cadenat, 1959
Cyrano Spurdog *Squalus rancureli* Fourmanoir &
Rivaton, 1979

Sicklefin Devil Ray *Mobula coilloti* Cadenat &
Rancurel, 1960

Dr Paul G. Rancurel was a zoologist and biologist who worked for ORSTOM. He was at the Centre de

recherches océanographiques, Abidjan, Ivory Coast (1967) and later in Noumea (New Caledonia) and Marseille. He wrote *Topographie Générale du Plateau Continental de la Côte d'Ivoire et du Libéria* (1968). Fourmanoir wrote in his etymology: 'Nous dédions ce nouveau Squale à P. RANCUREL océanographe O.R.S.T.O.M., qui est le premier à l'avoir capturé et à avoir signalé ses caractères principaux'. He collected the holotype and was the first to note its principal characteristics.

Randall

Slender Weasel Shark *Paragaleus randalli* Compagno,
Krupp & KE Carpenter, 1996
Arabian Whipray *Himantura randalli* Last, Manjaji-
Matsumoto & Moore, 2012

Spotted Guitarfish *Rhinobatos punctifer* Compagno &
Randall, 1987
Elat Electric Ray *Heteronarce bentuviai* Baranes &
Randall, 1989
Salalah Guitarfish *Rhinobatos salalah* Randall &
Compagno, 1995
Atz's Numbfish *Narcine atzi* Carvalho & Randall, 2003
Steven's Swellshark *Cephaloscyllium stevensi* E Clark & ·
Randall, 2011

Dr John 'Jack' Ernest Randall Jr (b.1924) is one of the most respected contemporary ichthyologists and a world authority on coral reef fishes. He has described over 600 species and has written 11 books and over 670 scientific and popular articles. Jack has participated in numerous scientific expeditions and has made countless SCUBA dives to collect specimens. He has developed excellent photographic methods, both underwater and of newly collected specimens. His photographs have been published widely. The University of California, Los Angeles awarded his bachelor's degree (1950) and the University of Hawai'i-Manoa his doctorate (1950). He was a research fellow at the Bishop Museum, Honolulu (1955–1957). He worked at the Marine Laboratory, University of Miami (1957–1961) and then as Professor of Zoology (1961–1965) and Director of the Institute of Marine Biology, University of Puerto Rico (1965). He returned to Hawaii as Director of the Oceanic Institute (1965) and became Senior Ichthyologist at the Bishop Museum (1967). He wrote *Coastal Fishes of Oman* (1995). He collected the first specimen of the slender weasel shark but misidentified it as *Hypogaleus hyugaensis* in his book, *Sharks of Arabia* (1986). He was among the first to photograph the Arabian whipray and he was honoured in its name for this and for his 'legendary' work on the taxonomy of Indo-Pacific fishes.

Ranzani

Smalltail Shark *Carcharhinus porosus* Ranzani, 1839

Camillo Ranzani (1775–1841) was an Italian priest and naturalist. He was Director of the Museum of Natural History, Bologna (1803–1841), attached to the University of Bologna. He wrote *Elementi di Zoologia* (1819–1825). Other marine organisms are named after him.

Raoul

Kermadec Spiny Dogfish *Squalus raoulensis* Duffy & Last, 2007

This species is named after its type locality, Raoul Island.

Raschi

Aleutian Dotted Skate *Rhinoraja longi* Raschi & McEachran, 1991

Dr William 'Bill' Glen Raschi (b.1950) is a biologist who was an assistant curator at the Virginia Institute of Marine Science (1979) and was a member of the faculty at the Department of Biology, Bucknell University, Lewisburg, Pennsylvania (1996). He was President of the American Society of Ichthyologists and Herpetologists (1996). He was arrested and charged (2012) with a number of offences connected with consuming alcohol whilst driving. He was convicted and, having a number of similar previous convictions, was sentenced to a prison term of up to 4½ years (2013): he is now in the Livingston County Jail. He co-wrote *Hydrodynamic Aspects of Shark Scales* (1986).

Regan

Izak Catshark *Holohalaelurus regani* Gilchrist, 1922

Wobbegong genus *Eucrossorhinus* Regan, 1908
Carpet Shark genus *Heteroscyllium* Regan, 1908
Sleeper Ray genus *Heteronarce* Regan, 1921

Eyespot Skate *Atlantoraja cyclophora* Regan, 1903
Tiger Catshark *Halaelurus natalensis* Regan, 1904

Shortnose Velvet Dogfish *Centroscymnus cryptacanthus* Regan, 1906
Japanese Wobbegong *Orectolobus japonicus* Regan, 1906

Sixgill Sawshark *Pliotrema warreni* Regan, 1906
Largespine Velvet Dogfish *Proscymnodon macracanthus* Regan, 1906
Australian Angelshark *Squatina australis* Regan, 1906
Clouded Angelshark *Squatina nebulosa* Regan, 1906
African Angelshark *Squatina africana* Regan, 1908
Carpathian Lamprey *Eudontomyzon danfordi* Regan, 1911
Cape Hagfish *Myxine capensis* Regan, 1913
Hagfish sp. *Myxine paucidens* Regan, 1913
Balloon Swellshark *Cephaloscyllium sufflans* Regan, 1921
Natal Electric Ray *Heteronarce garmani* Regan, 1921
Speckled Ray *Raja polystigma* Regan, 1923

Charles Tate Regan (1878–1943) was a British ichthyologist. He was educated at Queen's College, Cambridge before joining (1901) the staff of the BMNH, becoming Keeper of Zoology (1921) and later Director of the Museum (1927–1938). He described many South African fishes.

Reighard

Northern Brook Lamprey *Ichthyomyzon fossor* Reighard & Cummins, 1916

Professor Jacob Ellsworth Reighard (1861–1942) was an American zoologist. He studied biological science at the University of Michigan (1878–1882), where he then worked as Instructor in Zoology (1886) and Acting Assistant Professor (1887–1888). He became Assistant Professor (1889–1891), full Professor of Animal Morphology (1892), Professor of Zoology (1895) and Director of the Biological Station and Professor Emeritus (1927–1942). He was a co-founder of the University of Michigan Biological Station on Douglas Lake (1909) and its first director. He also directed the scientific work of the Michigan Fish Commission (1890–1895).

Reinhardt

Black Dogfish *Centroscyllium fabricii* Reinhardt, 1825

Dr Johannes Christopher Hagemann Reinhardt (1778–1845) was a Norwegian zoologist. He was Professor of Zoology (1814) at the University of Copenhagen. He studied theology, zoology, botany, mineralogy and anatomy in Copenhagen, Freiberg, Göttingen and Paris. His doctorate was an honorary PhD, awarded by the University of Copenhagen (1836). When the Danish Royal Museum of Natural History was founded (1805), he was invited to be its 'inspector' of newly purchased collections of the Society for Natural History. He accepted but only returned to Copenhagen (late 1806) after studying the museums in Paris and attending Cuvier's lectures. He started lecturing in the Museum (1809) and was eventually employed by the University (1813), being appointed a 'professor extraordinarius' (1814). He became a member of the Royal Danish Academy of Sciences and Letters (1821), a full professor of the University (1830) and a titular councillor of state (1839). A bird is named after him.

Reissner

Far Eastern Brook Lamprey *Lethenteron reissneri* Dybowski, 1869

We believe this honours Ernst Reissner (1824–1878), who was an Estonian (Baltic German) anatomist. The University of Dorpat awarded his medical degree (1851) and he became Professor of Anatomy there (1855) until retiring early in poor health (1875). He is also commemorated in the name of an anatomical detail: 'Reissner's membrane', part of the cochlea in the inner ear.

Remya

Smoothhound sp. *Mustelus mangalorensis* Cubelio, Remya & Kurup, 2011

Reg Remya is no longer involved in ichthyological research: we hear she is now working in the Kerala Revenue Department.

Renaud

Macedonia Brook Lamprey *Eudontomyzon hellenicus* Vladykov, Renaud, Kott & Economidis, 1982
Nina's Lamprey *Lethenteron ninae* Naseka, Tuniyev & Renaud, 2009
Epirus Brook Lamprey *Eudontomyzon graecus* Renaud & Economidis, 2010

Dr Claude B. Renaud (b.1955) is a Canadian research scientist (ichthyology) at the Canadian Museum of Nature. The University of Ottawa awarded his doctorate (1989). His main interests are lamprey taxonomy and systematics, functional morphology and the conservation of freshwater fishes. He has written widely, including *A New Nonparasitic Lamprey Species from the North-eastern Black Sea Basin* (2009), describing the lamprey, *Conservation Status of Northern Hemisphere Lampreys* (1997) and the FAO Species Catalogue *Lampreys of the World* (2011).

Richardson, J

Western Brook Lamprey *Lampetra richardsoni* Vladykov & Follett, 1965

Skate genus *Dentiraja* J Richardson, 1845

Pacific Lamprey *Entosphenus tridentatus* J Richardson, 1836
Atlantic Sharpnose Shark *Rhizoprionodon terraenovae* J Richardson, 1836
Tasmanian Numbfish *Narcine tasmaniensis* J Richardson, 1841
Speckled Carpetshark *Hemiscyllium trispeculare* J Richardson, 1843
Thornback skate *Dentiraja lemprieri* J Richardson, 1845
Shortheaded Lamprey *Mordacia mordax* J Richardson, 1846
Chinese Numbfish *Narcine lingual* J Richardson, 1846
Ringstreaked Guitarfish *Rhinobatos hynnicephalus* J Richardson, 1846

Sir John Richardson (**1787–1865**) was a Scottish naval surgeon, naturalist and Arctic explorer, knighted (1846), who assisted Swainson. Edinburgh University awarded his MD (1807) and he joined the navy on leaving. He was a friend of Sir John Franklin, to whom he was also related by marriage, and took part in Franklin's expeditions (1819–1822 and 1825–1827). He also participated (1847) in the vain search for Franklin and his colleagues; their fate (the ships were icebound near King William Island) was not discovered until Rae's expedition (1853–1854). In September 2014 a Canadian expedition located and filmed the wreck of one of the sunken vessels, which was remarkably well preserved. At the time of writing it had not been established whether it is HMS Erebus or HMS Terror – they were both bomb vessels of very similar specification. Among other works Richardson wrote *Icones Piscium* (1843), *Catalogue of Apodal Fish in the British Museum* (1856) and the second edition of Yarrell's *History of British Fishes* (1860). The Richardson Mountains in Canada are also named after him as are eight birds, five mammals and four reptiles.

Richardson, LR

Richardson's Ray *Bathyraja richardsoni* Garrick, 1961

Hagfish sp. *Neomyxine biniplicata* LR Richardson & Jowett, 1951
Hagfish sp. *Nemamyxine elongate* LR Richardson, 1958

Professor Laurence 'Larry' Robert Richardson (1911–1988) was a British zoologist. His master's degree and doctorate were awarded by McGill University, Montreal, Canada. He was working in New Zealand after the Second World War and became a professor in the Department of Zoology, Victoria University of Wellington. He fell out with his colleagues and was reprimanded. In a fit of anger he resigned (1964) and when he tried to withdraw his resignation, found that he could not. He left New Zealand and settled in Australia. He became a Fellow of the Royal Society of New Zealand (1959). He was honoured in the ray name 'for his extensive contribution to deep water research in New Zealand, and especially in Cook Strait where the type specimen was taken'. He is also honoured in the cephalopod name *Megalocranchia richardsoni*.

Richardson, RE

Blacktip Sawtail Catshark *Galeus sauteri* Jordan & RE Richardson, 1909
Yellow-spotted Skate *Okamejei hollandi* Jordan & RE Richardson, 1909

Robert Earl Richardson (1877–1935) was an American ichthyologist. He co-wrote *Check-list of Species of Fishes Known from the Philippine Archipelago* (1910).

Riera

Slender Electric Ray *Narcine rierai* Lloris & Rucabado, 1991

Ignacio Riera Julia was chief of the Spanish fisheries office and adviser to the General Direction of International Fisheries Relations, Ministry of Agriculture and Fisheries, Mahé, Seychelles (1989). He was a good friend of Lloris.

Rincón

Brazilian Blind Electric Ray *Benthobatis kreffti* Rincón, Stehmann & Vooren, 2001
Electric Ray sp. *Discopyge castelloi* Menni, Rincón & García, 2008

Dr Getúlio Rincón Filho (b.1972) is a Brazilian ichthyologist, biologist and taxonomist at the Conselho Nacional de Pesca e Aquicultura, Brasilia and is based at the Universidade Estadula Paulista, São Paulo. The Universidade Federal de Rio Grande awarded his bachelor's degree (1993) and his master's (1997) and the Universidade Estadula Paulista his doctorate (2006). He wrote *Taxonomia, Alimentação e Reprodução da Raia Elétrica* Benthobatis sp. (Torpediniformes: Narcinidae) *no sul do Brasil* (1997).

Risso

Smalltooth Sand Tiger *Odontaspis ferox* Risso, 1810
Marbled Electric Ray *Torpedo marmorata* Risso, 1810
Blackspotted Smooth-Hound *Mustelus punctulatus* Risso, 1827
Little Sleeper Shark *Somniosus rostratus* Risso, 1827
Longnose Spurdog *Squalus blainville* Risso, 1827

Professor Giovanni Antonio Risso (1777–1845) (aka Joseph Antoine Risso) was an Italian naturalist. He was (until 1826) an apothecary in his native town of Nice (then in Italy). He later became Professor of Botany and Chemistry at the University of Nizza (Nice). He had an outstanding knowledge of ichthyology and published *Ichthyologie de Nice* (1810). Other works included *Histoire Naturelle de l'Europe Méridionale* (1826). A mammal is named after him.

Ritter

Whitefin Dogfish *Centroscyllium ritteri* Jordan & HW Fowler, 1903

Dr William Emerson Ritter (1856–1944) was an American marine biologist and tunicatologist. He financed his own education by taking a series of teaching jobs and eventually obtained a bachelor's degree from the University of California, Berkeley (1888); he was able to move to Harvard for his master's (1891) and doctorate (1893). He worked at the University of California, Berkeley (1891–1924), being a full professor (1902) and retiring as Emeritus Professor. He was a member of the Harriman Alaska Expedition (1899). He met the newspaper proprietor, E. W. Scripps (1903) and their friendship and partnership led to what is now the Scripps Institution of Oceanography, of which he became the first director until retiring (1922). He wrote *Organization in Scientific Research* (1905). He was honoured for his work on the tunicates and enteropneusta (acorn worms) of the Pacific Ocean.

Rivaton

Longnose Houndshark *Iago garricki* Fourmanoir & Rivaton, 1979
Cyrano Spurdog *Squalus rancureli* Fourmanoir & Rivaton, 1979

Jacques Rivaton is a French zoologist who worked at the marine biology laboratory in Noumea, New Caledonia. He co-wrote Pisces, Pleuronectiformes: *Flatfishes from the Waters around New Caledonia – A Revision of the Genus* Engyprosopon (1993).

River

Rivero's Catshark *Apristurus riveri* Bigelow & Schroeder, 1944
[Alt Broadgill Catshark]

(See **Howell-Rivero**)

Roberts, JD

Roberts Bigmouth Skate *Amblyraja robertsi* Hulley, 1970

Dr J. Douglas 'JD' Roberts (d.1982) was a British-born South African businessman who was Chairman of Murray & Roberts, South Africa's leading engineering, contracting and construction services company. In honouring him Hulley wrote '...who, by his kind generosity, made the study of the "Walther Herwig" material in Hamburg possible'. Roberts was generous in sponsoring research and providing funds so that important collections stayed in South Africa, such as the Harald Pager collection of San rock paintings.

Roberts, MA

Carolina Hammerhead *Sphyrna gilberti* Quattro, Driggers, Grady, Ulrich & MA Roberts, 2013

Mark A. Roberts is an American biologist in the Department of Biological Sciences, Marine Science Program, University of South Carolina, Columbia. He co-wrote *Population Genetic Structure and Taxonomy of the Southeastern United States Estuarine Grass Shrimp* (Palaemonetes spp.) (2012).

Roberts, TR

Freshwater Ray genus *Makararaja* TR Roberts, 2007

White-rimmed Whipray *Himantura signifier* Compagno & TR Roberts, 1982
Pearl Stingray *Dasyatis margaritella* Compagno & TR Roberts, 1984
Mekong Stingray *Dasyatis laosensis* TR Roberts & Karnasuta, 1987
Giant Freshwater Stingray *Himantura chaophraya* Monkolprasit & TR Roberts, 1990
Roughback Whipray *Himantura kittipongi* Vidthayanon & TR Roberts, 2005
Freshwater Ray sp. *Makararaja chindwinensis* TR Roberts, 2007

Dr Tyson Royal Roberts (b.1940) is an American naturalist and evolutionary biologist who has devoted most of his life to studying biodiversity, particularly that of fishes, throughout the tropics. Stanford University awarded his bachelor's degree (1961) and his doctorate (1968). Although for many years associated with the Department of Ichthyology, California Academy of Sciences, San Francisco, he never held a position or an honorary appointment there. He was at the Museum of Comparative Zoology, Harvard University (1969–1975) as Associate Professor of Biology and Associate Curator of Fishes. While there he submitted a five-year plan for total renovation of the MCZ fish collection to the National Science Foundation that was implemented by others after he left. He has been actively involved in building research collections of tropical marine and especially freshwater fishes including elasmobranchs at several institutions. He is currently Research Associate of the Smithsonian Tropical Research Institute, Panama and Advisor and Research Associate of the Institute for Molecular Biosciences, Mahidol University, Thailand. His publications include three faunal monographs, *Ichthyological Survey of the Fly River of Papua New Guinea* (1975), *Fishes of the Lower Rapids of the Congo River* (1976) and *Freshwater Fishes of Western Borneo (Kalimantan Barat, Indonesia)* (1989). He has described more than 30 genera and nearly 200 species of fish. He perhaps is best known for monographs on distribution of the freshwater fishes of Africa (1975) and on the oceanic oarfishes (*Regalecidae*) (2012), and for assessments of negative environmental impacts of hydropower dams in the Mekong basin.

Robins

Hagfish sp. *Myxine robinsorum* Wisner & CB McMillan, 1995
West Indian Lantern Shark *Etmopterus robinsi* Schofield & Burgess, 1997

Dr Charles Richard 'Dick' Robins (b.1928) was a systematic ichthyologist who studied ornithology at Cornell University, gaining a batchelor's degree (1949) and it was at Cornell that he became fascinated by ichthyology. He did not do a master's degree but instead went straight for a doctorate at Cornell (awarded 1954). He served in the US Army (1954–1956) and on being discharged joined the Marine Laboratory at the University of Miami where he did much to develop its ichthyology collection, becoming Professor, Marine Biology and Fisheries Division. He became Faculty Curator Emeritus, Division of Ichthyology, University of Kansas (1994). His wife Mrs Catherine Robins née Hale is also an ichthyologist, and the hagfish commemorates them both for their contributions to marine fauna of the tropical western Atlantic. He co-wrote *A Field Guide to Atlantic Coast Fishes of North America* (1986).

Rochebrune

Lesser Guinean Devil Ray *Mobula rochebrunei* Vaillant, 1879

Dr Alphonse Trémeau de Rochebrune (1836–1912) was a French military doctor who took a zoological doctorate (1874) and became a botanist, malacologist and ichthyologist. He was in Senegal (1875–1883) and wrote *Fauna de Sénégal* (1883). He worked at the Muséum National d'Histoire Naturelle, Paris (1884–1911).

Rogers

Roger's Roundray *Urotrygon rogersi* Jordan & Starks, 1895

Dr George Warren Rogers was a physician from Vermont who lived in Mazatlán, Sinaloa, Mexico. He was described in the etymology as 'a scholarly physician', who assisted the authors from his home in Mazatlán, Sinaloa, Mexico, near which is the type locality.

Rondelet

Rondelet's Ray *Raja rondeleti* Bougis, 1959

Dr Guillaume Rondelet (1507–1566) was a French physician, naturalist, botanist and zoologist. He qualified at the University of Montpellier as a physician (1537) but was not a success as a doctor and was always short of money. Cardinal François de Tournon employed him as his personal physician (1538) and with such a powerful patron his fortunes revived. He became Regius Professor of Medicine at the University of Montpellier (1545) and Chancellor of the University (1556–1566). He is satirised as 'Rondibilis' in *La vie de Gargantua et Pantagruel*, written by his great friend François Rabelais. He is also noteworthy for having expelled Nostradamus from the University for slandering doctors and being an apothecary. He wrote *Libri de Piscibus Marinis in quibus Verae Piscium Effigies Expressae sunt* (1554). He recognized the distinctiveness of this ray (1554), but assumed it was a juvenile *Raja fullonica*.

Rosa

Rosa's Roundray *Heliotrygon rosai* Carvalho & Lovejoy, 2011

Freshwater Stingray genus *Plesiotrygon* Rosa, Castello & Thorson, 1987

Long-tailed River Stingray *Plesiotrygon iwamae* Rosa, Castello & Thorson, 1987

Brazilean Large-eyed Stingray *Dasyatis marianae* Gomes, Rosa & Gadig, 2000

Freshwater Ray sp. *Potamotrygon boesemani* Rosa, Carvalho & Almeida Wanderley, 2008

Dr Ricardo de Souza Rosa (b.1954) is a Brazilian ichthyologist who is a professor at the Department of Systematics and Ecology, Federal University of Paraíba, Brazil. He was a volunteer (1972–1976) in the ichthyology department of the Zoology Museum of the University of São Paulo, which awarded his bachelor's degree in biology (1976). Virginia Institute of Marine Science, College of William and Mary, Virginia awarded his doctorate (1985), after which he carried out post-doctoral research at the University of Alberta, Canada (1990). He wrote *A Systematic Revision of the South American Freshwater Stingrays (Chondrichthyes: Potamotrygonidae)* (1985) for which he received the Rodolpho von Ihering Award presented by the Brazilian Zoology Society, and was honoured in the name of the roundray for this 'excellent' revision that 'represents a landmark in our understanding of the taxonomy and diversity of this family'. In addition to his contribution to the knowledge of freshwater stingrays, he also worked on the taxonomy of other marine and freshwater neotropical fish groups, as well as on the biology of tropical sharks. His major findings on South American stingrays are included in a book, *Rayas de Agua Dulce* (Potamotrygonidae) *de Suramérica*, Parte I edited by Carlos A. Lasso *et al.* (2013).

Roux

Skate sp. *Raja rouxi* Capapé, 1977

Dr Charles Roux (b.1920) is a marine zoologist who became Professor and Deputy Director, Muséum National d'Histoire Naturelle, Paris (1987). His doctorate was awarded in 1973. He has written a number of papers and books, including co-writing *Ocean Dwellers* (1982).

Rucabado

Slender Electric Ray *Narcine rierai* Lloris & Rucabado, 1991

Dr Jaume Rucabado Aguilar (1946–1999) was a Spanish ichthyologist at the Instituto de Ciencias del Mar, Barcelona. He co-wrote *Ictiofauna del Canal Beagle (Tierra del Fuego): Aspectos Ecológicos y Análisis Biogeográfico* (1991).

Ruocco

Shortnose Eagle Ray *Myliobatis ridens* Ruocco, Lucifora, Díaz de Astarloa, Mabragaña & Delpiani, 2012

Dr Natalia Lorena Ruocco is an Argentine ichthyologist and biologist whose doctorate was from the Universidad Nacional de Mar del Plata (2012). She wrote *Ecology and Conservation of Stingrays (Chondrichthyes, Myliobatiformes) of Buenos Aires and the Uruguayan Coastal Ecosystem* (2012).

Rüppell

Carpet Shark genus *Nebrius* Rüppell, 1837

Silvertip Shark *Carcharhinus albimarginatus* Rüppell, 1837
Sicklefin Lemon Shark *Negaprion acutidens* Rüppell, 1837

Milk Shark *Rhizoprionodon acutus* Rüppell, 1837
Great Hammerhead *Sphyrna mokarran* Rüppell, 1837
Whitetip Reef Shark *Triaenodon obesus* Rüppell, 1837

Wilhelm Peter Eduard Simon Rüppell (1794–1884) was a German naturalist and explorer. He went to Egypt and ascended the Nile as far as Aswan (1817). He later made two extended expeditions to northern and eastern Africa, Sudan (1821–1827) and Ethiopia (1830–1834). Although he brought back large zoological and ethnographical collections, his expeditions impoverished him. He wrote *Reisen in Nubien, Kordofan und dem Petraischen Arabien* (1829), *Systematische Übersicht der Vogel Nord-ost-Afrikas* (1845) and *Reise in Abyssynien* (1838–1840). He also collected in the broadest sense and presented his collection of coins and rare manuscripts to the Historical Museum in Frankfurt (his home town). Eleven birds, five mammals, an amphibian and two reptiles are named after him.

S

Saavedra-Diaz

Hagfish sp. *Eptatretus wayuu* Mok, Saavedra-Diaz & Acero, 2001

Lina Maria Saavedra-Diaz (b.1975) is a Colombian marine biologist, based at the Universidad del Magdalena, Colombia. She studied for her doctorate (2006–2010) at the University of New Hampshire, USA. She was one of the collectors of the holotype (1998) and described the hagfish in the paper, *Two New Species of Eptatretus and Quadratus (Myxinidae, Myxiniformes) from the Caribbean Coast of Colombia* (2001).

Sabaj Pérez

Tiger Ray sp. *Potamotrygon tigrina* Carvalho, Sabaj Pérez & Lovejoy, 2011

Dr Mark Henry Sabaj Pérez (b.1969) is an ichthyologist and collection manager of fishes at the Academy of Natural Sciences of Philadelphia (2000–present). The University of Richmond, Virginia awarded his bachelor's degree (1990) and master's (1992). The University of Illinois, Urbana-Champaign awarded his doctorate (2002), with a thesis on the taxonomy of the neotropical thorny catfishes (*Siluriformes: Doradidae*) and revision of genus *Leptodoras*. He has co-authored 43 peer-reviewed publications including the descriptions of 22 new species and one new subspecies of freshwater fish from Asia and South America, one new genus of Doradidae, and two new species of North American crayfishes. He has joined or led 30 expeditions to 12 countries in Asia, Europe and North and South America.

Sadowsky

Brazillian Skate *Rajella sadowskii* G Krefft & Stehmann, 1974

Polkadot Catshark *Scyliorhinus besnardi* S Springer & Sadowsky, 1970

Victor Sadowsky (1909–1990) was a Latvian-born Brazilian ichthyologist at the Instituto Oceanográfico da Universidade de São Paulo. He first arrived in Brazil (1949) and settled (1951) at Cananeia (São Paulo State), where the local museum is named in his honour. He wrote *First Record of Broad-snouted Seven-gilled Shark from Cananéia, Coast of Brazil* (1969)

and has 'considerably added to the knowledge of southwestern Atlantic elasmobranchs'.

Sæmundsson

Iceland Catshark *Apristurus laurussonii* Sæmundsson, 1922

Dr Bjarni Sæmundsson (1867–1940) was an Icelandic naturalist and their pioneering ichthyologist. He regarded himself as a mediator between Icelandic fisherman and foreign scientists as he had travelled among Icelandic fishing communities (1880s) to pick up their practical knowledge. A research vessel is named after him. He wrote *Icelandic Malacostraca* (1937).

Sainsbury

Goldeneye Shovelnose Ray *Rhinobatos sainsburyi* Last, 2004

Dr Keith John Sainsbury (b.1952) is a New Zealand ichthyologist, marine ecologist and mathematical modeler with a research focus on the assessment, ecology, economics, exploitation and conservation of marine resources and ecosystems. The University of Canterbury, New Zealand awarded both his bachelor's degree (1972) and doctorate (1977). He worked for over 20 years for CSIRO in Australia, as a senior principal research scientist, engaged in fishery research including several ecological surveys of the marine ecosystems off northern Australia and southern Indonesia (late 1970s–early 1980s). These resulted in the discovery of many new species, including the shovelnose ray, for which he was honoured for 'planning and managing the initial trawl surveys of the continental shelf of northwestern Australia, and for recognising the need to document the fauna before addressing more management-oriented questions, and for pioneering research that provided an excellent biological baseline'. He became self-employed (2006) and is the Director of SainSolutions Pty Ltd, which provides scientific advice for marine resource management. He is also Professor of Marine System Science, Institute of Marine and Antarctic Science, University of Tasmania. Since 1995 he has been closely associated with the work of the Food and Agriculture Organisation of the United Nations (FAO). His work on the science and sustainable management of fisheries has been recognized by the awards of the Japan Prize (2004) and the Swedish Seafood Award (2012). He wrote *Best Practice Reference Points for Australian Fisheries* (2008) and co-wrote *Continental Shelf Fishes of Northern and North-western Australia* (1985).

Saint-Hilaire

Bull Ray *Pteromylaeus bovinus* Geoffroy Saint-Hilaire,
1817

Blackchin Guitarfish *Rhinobatos cemiculus* Geoffroy
Saint-Hilaire, 1817

Lusitanian Cownose Ray *Rhinoptera marginata*
Geoffroy Saint-Hilaire, 1817

Round Santail Stingray *Taeniura grabata* Geoffroy
Saint-Hilaire, 1817

Étienne Geoffroy Saint-Hilaire (1772–1844) was a
naturalist. He originally trained for the Church but
abandoned theology to become Professor of Zoology
(1793), when the Jardin du Roi became the Muséum
National d'Histoire Naturelle. He expounded the
theory that all animals conform to a single plan of
structure. This was strongly opposed by Cuvier,
who had been his friend, and a widely publicised
debate between them took place (1830). Despite their
differences, the two men did not become enemies.
Geoffroy gave one of the orations at Cuvier's funeral
(1832). Modern developmental biologists have con-
firmed some of Geoffroy's ideas. He wrote *Philosophie
Anatomique* (1818–1822). Fifteen mammals, five birds
and a reptile are named after him.

Santos

Colares Stingray *Dasyatis colarensis* Santos, Gomes &
Charvet-Almeida, 2004

Groovebelly Stingray *Dasyatis hypostigma* Santos &
Carvalho, 2004

Dr Hugo Ricardo Secioso Santos (b.1966) is a
zoologist, biologist and ichthyologist at the Institute
of Biology, Department of Zoology, Universidade
do Estado do Rio de Janeiro, where he is Curator
of biological collections (vertebrates), and which
awarded his bachelor's degree (1990). His master's degree (1997) and doctorate (2007) were both
awarded by the Museu Nacional/Universidade
Federal do Rio de Janeiro. He is interested in the
taxonomy of stingrays (family *Dasyatidae*) and
recently his study group has focused on the anatomy
and taxonomy of elasmobranch fishes, from which
they published *Guia para Identificação de Tubarões e
Raias do Rio de Janeiro* (2010). He has also written a
number of papers such as *A New Species of Whiptail
Stingray of the Genus* Dasyatis *Rafinesque from the
Southwestern Atlantic Ocean* (2004) and he co-wrote
Description of Females of the Stingray Dasyatis colarensis *Santos, Gomes & Charvet-Almeida, 2004*
(Chondrichthyes: Myliobatiformes: Dasyatidae)
(2007).

Sasahara

Roughskin Catshark *Apristurus ampliceps* Sasahara,
Sato & Nakaya, 2008

Ryohei Sasahara is an ichthyologist at the Graduate
School of Fisheries Science, Hokkaido University,
Hakodate, Japan. He has co-written a number of
papers such as *Shallow-sea Fishes of the Shiretoko Peninsula, Hokkaido, Japan* (2007).

Sato

Flaccid Catshark *Apristurus exsanguis* Sato, Nakaya &
Stewart, 1999

Roughskin Catshark *Apristurus ampliceps* Sasahara,
Sato & Nakaya, 2008

Pinocchio Catshark *Apristurus australis* Sato, Nakaya &
Yorozu, 2008

Garrick's Catshark *Apristurus garricki* Sato, Stewart &
Nakaya, 2013

Dr Keiichi Sato is an ichthyologist and marine sci-
entist who is Curator, Zoological Department,
Okinawa Churashima Research Center, Okinawa
Churashima Foundation, Japan. Hokkaido Univer-
sity, Hakodate awarded his doctorate. He co-wrote
*Feeding of the Megamouth Shark (Pisces: Lamniformes:
Megachasmidae) Predicted by its Hyoid Arch: A Biome-
chanical Approach* (2011).

Sauter

Blacktip Sawtail Catshark *Galeus sauteri* Jordan & RE
Richardson, 1909

Dr Hans Sauter (1871–1943) was a German entomologist who became interested in herpetology. He
studied biology at the Universities of Munich and
Tübingen. He was in Formosa (Taiwan) under Japanese occupation, collecting insects (1902–1904). He
was in Tokyo (1905) before returning to Taiwan for
the rest of his life. He worked for a British trading
company but entomology was his passion. Although
Japan and Germany were enemies in the First World
War, he kept his job and continued collecting, albeit
under observation. He was said to be the first person
to offer private piano lessons in Taiwan, and gave
German and English lessons. He collected the holotype from a fish market in Taiwan. Three reptiles and
an amphibian are also named after him.

Sauvage

Marbled Whipray *Himantura oxyrhyncha* Sauvage,
1878

Dr Henri-Émile Sauvage (1844–1917) was a French palaeontologist, herpetologist and ichthyologist. He was also a populariser of science and often chose to publish articles in 'Popular Science Monthly' such as *Amphibious Fishes* (1876). Two reptiles and an amphibian are named after him.

Say

Bluntnose Stingray *Dasyatis say* Lesueur, 1817

Thomas Say (1787–1834) was a self-taught American naturalist whose primary interest was entomology. He described over 1,000 new species of beetle and over 400 new insects of other orders. He became a charter member and founder of the Academy of Natural Sciences of Philadelphia (where he became friends with Lesueur) (1812) and was appointed chief zoologist with Major Stephen H. Long's expeditions, which explored the Rocky Mountains (1819–1820). He lived at the utopian village of 'New Harmony' in Indiana (1826–1834). Say wrote *American Entomology, or Descriptions of the Insects of North America* (1824–1828) and *American Conchology* (1830–1834). Say's death was one of the things that prompted Lesueur to return to France. A mammal, a bird genus and a bird species and a reptile are named after him.

Schaaf-Da Silva

Broad-snout Lanternshark *Etmopterus burgessi* Schaaf-Da Silva & Ebert, 2006
Spotted Swellshark *Cephaloscyllium maculatum* Schaaf-Da Silva & Ebert, 2008
Leopard-spotted Swellshark *Cephaloscyllium pardelotum* Schaaf-Da Silva & Ebert, 2008

Jayna Ann Schaaf-Da Silva (b.1978) is an environmental marine biologist. Humboldt State University awarded her BSc degree (2001). She participated in a research experience for undergraduates program at the University of Wisconsin-Milwaukee (2000) and presented her first paper from her research on zebra mussels in Lake Michigan (2001). Jayna worked as a fisheries technician and commercial scuba diver at two Alaska salmon hatcheries before starting her graduate program. She worked as a marine biologist at the Pacific Shark Research Centre, Moss Landing Marine Laboratories, California (2005–2008). During that time she published three papers and numerous scientific illustrations and was awarded the Ernst Mayr Scholarship in Animal Systematics from Harvard University (2006). San Jose State University awarded her master's degree in marine science (2007). She wrote *A Taxonomic Revision of North Pacific Swell Sharks, Genus Cephaloscyllium* (2007). She now works as an environmental scientist for the California Department of Fish and Wildlife. She leads a large-scale, fisheries-dependent sampling program in Central California to monitor sportfishing effort and recreational marine finfish catches. She has been with the Department for 11 years and finds it highly enjoyable, winning a DFW Silver Accomplishment award (2009) and three regional awards (2010–2012).

Schinz

Puffadder Shyshark *Haploblepharus edwardsii* Schinz, 1822

Dr Heinrich Rudolf Schinz (1777–1861) was a Swiss physician, ornithologist, herpetologist and naturalist. He studied medicine in Würzburg and Jena then practised in Zurich (1798–1804), becoming a teacher at the medical institute (1804–1833). He became a professor of natural history at the University of Zurich (1833) as well as Curator of the Zurich Natural History Society. He wrote several full-length works including *Europäsche Fauna* (1840). A bird and a reptile are named after him.

Schlegel

Brown Guitarfish *Rhinobatos schlegelii* Müller & Henle, 1841

Japanese Butterfly Ray *Gymnura japonica* Temminck & Schlegel, 1850
Japanese Sleeper Ray *Narke japonica* Temminck & Schlegel, 1850

Hermann Schlegel (1804–1884) was a German-born zoologist who spent much of his life in the Netherlands. He was the first person (1844) to use trinomials to describe separate races. Schlegel made a trip on foot through large parts of Germany and Austria (1824–1825). While he was in Vienna (1825) he received a letter through Johan Natterer from Jacob Coenraad Temminck, who was looking for a researcher to explore parts of Indonesia. Schlegel went to Leiden, and worked so hard and well for Temminck that the latter decided that Schelgel should stay at Leiden since he was too valuable to risk on an overseas assignment. He succeeded Temminck as Director at the Museum (1858). Schlegel produced many publications, some co-written with Temminck, including *Fauna Japonica – Aves* and *Kritische Übersicht der Europäischen Vögel*. He married twice. He was honoured in the name of the guitarfish 'for his unhesitating support of the authors' work'. Sixteen birds, two amphibians, two mammals and ten reptiles are named after him.

Schmarda

Chupare Whipray *Himantura schmardae* Werner, 1904

Dr Ludwig Karl Schmarda (1819–1908) was an Austrian physician, naturalist and traveller. After qualifying as a physician in Vienna (1843) he became an army surgeon. He taught at the Joanneum, Graz (1848–1850). He was Professor of Natural History, Karl-Franzens-Universität, Graz (1850–1852) and Professor of Zoology and Director of the Zoological Museum, Univerzita Karlova v Praze (Charles University, Prague) (1852–1862). He undertook a private circumnavigation of the world (1853–1857) by 'hitch-hiking' on various sailing ships; it was a rough adventure as he suffered scurvy on the voyage from South Africa and Australia, lost his collections in a fire in Chile and was robbed in Panama. He was Professor of Zoology, Universität Wien (1862–1883). After retirement he travelled in Spain and North Africa (1884–1887). He wrote *Die Geographische Verbreitung der Thiere* (1853). He collected the holotype of the whipray. An amphibian is named after him.

Schmidt

Browneye Skate *Okamejei schmidti* Ishiyama, 1958
[Syn. *Raja schmidti*]
Lantern Shark sp. *Etmopterus schmidti* Dolganov, 1986

Professor Petr Yulevich Schmidt (1872–1949) was an ichthyologist at the Zoological Museum and Institute in St Petersburg. He wrote *Fishes of Eastern Seas of the Russian Empire – Scientific results of the Korea–Sakhalin Expedition of the Emperor Russian Geographical Society 1900–1901* (1904) and *Fishes of the Riu-Kiu Islands* (1931), in which he misidentified the skate as *Raja fusca*. Many of the species collected on the scientific expeditions were not completely 'written up', and some were mislabelled or misidentified. A reptile is named after him.

Schmitt

Narrownose Smooth-Hound *Mustelus schmitti*
S Springer, 1939

Dr Waldo LaSalle Schmitt (1887–1977) was an expert on crustaceans who worked at the Smithsonian (1915–1957). He retired as Head Curator of Zoology but continued his association with the Smithsonian as Honorary Research Associate (1957–1977).

His doctorate was awarded by George Washington University (1922) and the University of Southern California awarded him an honorary doctorate (1948). He had been on the staff of the US Bureau of Fisheries and was naturalist on board the 'Albatross' (1911–1914). He continued to take part in expeditions until 1963 to Marguerite Bay and the Weddell Sea. He collected the holotype.

Schnakenbeck

Sicklefin Chimaera *Neoharriotta pinnata*
Schnakenbeck, 1931

Professor Dr Werner Schnakenbeck (1887–1971) was a German zoologist and ichthyologist. He studied at the University of Freiburg and then worked at the zoological gardens in Halle (1919) whilst studying for his doctorate (1920). He was a research assistant at the German Scientific Commission for Marine Research, Biology Institute, Heligoland (1921) and at the Department of Fisheries Biology, Zoological Museum, Hamburg (1923) and in charge of it (1931–1957). He also taught on the subject of fisheries at the University of Hamburg (1936–1945). He wrote *Deutsche Fischerei in Nordsee und Nordmeer* (1947).

Schneider, AF

Hagfish sp. *Eptatretus bischoffii* AF Schneider, 1880

Dr Anton Friedrich Schneider (1831–1890) was a German zoologist and comparative anatomist. He studied mathematics, natural sciences and zoology at the University of Bonn (1849–1851) and at the University of Berlin (1851–1854), where he received his doctorate. He went on an expedition to Norway (1855) but the vessel suffered a collision, causing it to explode and sink. He was in Naples and Messina (1856–1857). He was a private tutor at the University of Berlin (1859–1869) and a curator at the Zoological Museum, where he worked particularly on the nematode collections; among his published works is *Monographie der Nematoden* (1866). He was Professor and Rector at the University of Giessen (1869–1881) and Professor of Zoology and Comparative Anatomy at the University of Breslau (1881–1890). He founded the journal, *Zoologische Beiträge* (1883).

Schneider, JGT

Guitarfish genus *Rhina* Bloch & JGT Schneider, 1801

Longheaded Eagle Ray *Aetobatus flagellum* Bloch & JGT Schneider, 1801

Nieuhof's Eagle Ray *Aetomylaeus nichofii* Bloch & JGT
Schneider, 1801
[Alt. Barbless Eagle Ray, Banded Eagle Ray, Syn. *Raja
niehofii, Myliobatus nieuhofii*]
Blind Shark *Brachaelurus waddi* Bloch & JGT
Schneider, 1801
Gulper Shark *Centrophorus granulosus* Bloch & JGT
Schneider, 1801
Longnose *Dasyatis guttata* Bloch & JGT Schneider, 1801
Smooth Butterfly Ray *Gymnura micrura* Bloch & JGT
Schneider, 1801
Scaly Whipray *Himantura imbricate* Bloch & JGT
Schneider, 1801
Blackspotted Numbfish *Narcine timlei* Bloch & JGT
Schneider, 1801
Numbray *Narke dipterygia* Bloch & JGT Schneider, 1801
Fanray *Platyrhina sinensis* Bloch & JGT Schneider,
1801
Bowmouth Guitarfish *Rhina ancylostoma* Bloch & JGT
Schneider, 1801
Smoothnose Wedgefish *Rhynchobatus laevis* Bloch &
JGT Schneider, 1801
Greenland Shark *Somniosus microcephalus* Bloch &
JGT Schneider, 1801
Porcupine Ray *Urogymnus asperrimus* Bloch & JGT
Schneider, 1801

Johann Gottlob Theaenus Schneider (1750–1822)
was a German scholar in the days when scholars
were expected to be polymaths, conversant with
everything from the natural sciences to dead languages. He became (1811) Professor of Ancient
Languages and Eloquence at Breslau (now Wrocław,
Poland). He corrected, expanded and republished
Bloch's work on fish, *Systema Ichthyologiae Iconi-
bus Illustratum* (1801). He also published editions
of works by classical authors, such as Aelian's *De
Natura Animalium*. Six reptiles, an amphibian and a
mammal are named after him.

Schoenlein

Striped Panray *Zanobatus schoenleinii* Müller & Henle,
1841

Dr Johan Lukas Schönlein (1793–1864) was a naturalist and professor of medicine. He studied at
Landshut, Jena, Göttingen and Würzburg and taught
at Würzburg and Zurich until he was appointed
to Berlin (1839) where, among his duties, he was
physician to King Frederick William IV of Prussia.
The specimen, which was in the Berlin Anatomical
Museum, was supplied to the authors by Schönlein
who was a friend and colleague of the junior author.

Schofield

West Indian Lantern Shark *Etmopterus robinsi*
Schofield & Burgess, 1997

Dr Pamela J. Schofield is a research fisheries biologist at the US Geological Survey, Florida (her home
state) where she has worked since 2000. The University of Florida awarded her bachelor's and master's
degrees and the University of Southern Mississippi
her doctorate. In her laboratory, Dr Schofield and
her graduate students study non-native fishes from
marine and freshwater environments. They investigate factors such as influences of habitat variables
on fish distribution and abundance, tolerances of
species to environmental variables such as oxygen
and salinity, and behaviour. Two major objectives
are: 1) to determine how the introduction of non-
native fishes affect native fauna and environments as
well as 2) the development of techniques to control
non-native fishes or mitigate their impacts. She has
written numerous journal articles, book chapters,
etc and wrote *Geographic Extent and Chronology of the
Invasion of Non-native Lionfish* (Pterois volitans [*Linnaeus 1758*] and P. miles [*Bennett 1828*]) in the Western
North Atlantic and Caribbean Sea (2009). She co-wrote
*Etmopterus robinsi: A New Species of Deep-water
Dogshark* (Elasmobranchii, Etmopteridae) *from the
Caribbean and Western Atlantic with a Redescription of
E. hillianus* (1997).

Schroeder

Catshark genus *Schroederichthys* S Springer, 1966

Rosette River Stingray *Potamotrygon schroederi*
Fernández-Yépez, 1958
[Alt. Flower Ray]
Bahamas Sawshark *Pristiophorus schroederi*
S Springer & Bullis, 1960
Whitemouth Skate *Bathyraja schroederi* G Krefft, 1968

Ray genus *Breviraja* Bigelow & Schroeder, 1948
Catshark genus *Cephalurus* Bigelow & Schroeder, 1941
Skate genus *Cruriraja* Bigelow & Schroeder, 1948
Electric Ray genus *Diplobatis* Bigelow & Schroeder,
1948
Chimaera Genus *Neoharriotta* Bigelow & Schroeder,
1950
Skate genus *Springeria* Bigelow & Schroeder, 1951
Skate genus *Pseudoraja* Bigelow & Schroeder, 1954

Southern Stingray *Dasyatis americana* Hildebrand &
Schroeder, 1928

Broadgill Catshark *Apristurus riveri* Bigelow &
   Schroeder, 1944
Cuban Ribbontail Catshark *Eridacnis barbouri* Bigelow
   & Schroeder, 1944
Pacific Sleeper Shark *Somniosus pacificus* Bigelow &
   Schroeder, 1944
Lightnose Skate *Breviraja colesi* Bigelow & Schroeder,
   1948
Atlantic Legskate *Cruriraja atlantis* Bigelow &
   Schroeder, 1948
Cuban Legskate *Cruriraja poeyi* Bigelow & Schroeder,
   1948
Shorttail Skate *Amblyraja jenseni* Bigelow &
   Schroeder, 1950
Spinose Skate *Breviraja spinosa* Bigelow & Schroeder,
   1950
Blackfin Pygmy Skate *Fenestraja atripinna* Bigelow &
   Schroeder, 1950
Cuban Pygmy Skate *Fenestraja cubensis* Bigelow &
   Schroeder, 1950
Gulf of Mexico Pygmy Skate *Fenestraja
   sinusmexicanus* Bigelow & Schroeder, 1950
Yucatán Skate *Leucoraja yucatanensis* Bigelow &
   Schroeder, 1950
Leaf-nose Legskate *Anacanthobatis folirostris* Bigelow
   & Schroeder, 1951
Spreadfin Skate *Dipturus olseni* Bigelow & Schroeder,
   1951
Prickly Brown Ray *Dipturus teevani* Bigelow &
   Schroeder, 1951
Ratfish sp. *Hydrolagus alberti* Bigelow & Schroeder, 1951
Speckled Skate *Leucoraja lentiginosa* Bigelow &
   Schroeder, 1951
Gulf Hagfish *Eptatretus springeri* Bigelow & Schroeder,
   1952
African Lantern Shark *Etmopterus polli* Bigelow,
   Schroeder, & S Springer, 1953
Fringefin Lantern Shark *Etmopterus schultzi* Bigelow,
   Schroeder, & S Springer, 1953
Green Lantern Shark *Etmopterus virens* Bigelow,
   Schroeder & S Springer, 1953
Fanfin skate *Pseudoraja fischeri* Bigelow & Schroeder,
   1954
Sooty Ray *Rajella fuliginea* Bigelow & Schroeder, 1954
Lined Lantern Shark *Etmopterus bullisi* Bigelow &
   Schroeder, 1957
Rough Legskate *Cruriraja rugosa* Bigelow &
   Schroeder, 1958
Hookskate *Dactylobatus clarkia* Bigelow & Schroeder,
   1958
San Blas Skate *Dipturus garricki* Bigelow & Schroeder,
   1958
Hooktail Skate *Dipturus oregoni* Bigelow & Schroeder,
   1958

American Legskate *Anacanthobatis americanus*
   Bigelow & Schroeder, 1962
Longnose Legskate *Anacanthobatis longirostris*
   Bigelow & Schroeder, 1962
Broadfoot Legskate *Cruriraja cadenati* Bigelow &
   Schroeder, 1962
Bullis Skate *Dipturus bullisi* Bigelow & Schroeder, 1962
Plain Pygmy Skate *Fenestraja ishiyamai* Bigelow &
   Schroeder, 1962
Atlantic Pygmy Skate *Gurgesiella atlantica* Bigelow &
   Schroeder, 1962
Purplebelly Skate *Rajella purpuriventralis* Bigelow &
   Schroeder, 1962
Finspot Ray *Raja cervigoni* Bigelow & Schroeder, 1964
Bahama Skate *Raja bahamensis* Bigelow & Schroeder,
   1965

William Charles Schroeder (1894–1977) was an
American oceanographer and ichthyologist. He
joined the Woods Hole Oceanographic Institution
(1932), initially as a business manager in connection
with a ship they had acquired, the 'Atlantis'. He
was also an associate curator of ichthyology at the
Harvard Museum of Comparative Zoology (1937)
and went with the 'Atlantis' on a collecting trip to
waters off Central and South America. He had a life-
long collaboration with H.B. Bigelow and together
they greatly enhanced and expanded knowledge
of the fishes of the North Atlantic. Together they
described 42 new fish species. They wrote *Fishes of
the Western North Atlantic* (1953) and *Fishes of the Gulf
of Maine* (1953). Krefft honoured him in the name of
the Whitemouth Skate 'for his outstanding contribution to the study of western Atlantic elasmobranchs'.
A reptile is named after him.

Schuhmacher

Schuhmacher's stingray *Potamotrygon schuhmacheri*
   Castex, 1964

Roberto Schuhmacher (1947–1964) was a high school
student of Castex, who died tragically young in an
accident.

Schultz

Dwarf Sawtail Catshark *Galeus schultzi* S Springer,
   1979
Fringefin Lantern Shark *Etmopterus schultzi* Bigelow,
   Schroeder, & S Springer, 1953

Venezuela Round Stingray *Urotrygon venezuelae*
   Schultz, 1949

Dr Leonard Peter Schultz (1901–1986) was an ichthyologist and field naturalist who was Assistant Curator of fishes at the Smithsonian (1936–1938), then Curator (1938–1968) and Senior Zoologist (1965–1968). Albion College awarded his bachelor's degree (1924), the University of Michigan, where he taught (1925–1927), his master's (1926) and the University of Washington, where he taught (1928–1936), his doctorate (1932). After retiring, he retained his connection with the Smithsonian as Zoologist Emeritus (1968–1986).

Scott

Magpie Fiddler Ray *Trygonorrhina melaleuca* Scott, 1954

Spotted Stingaree *Urolophus gigas* Scott, 1954

Trevor Dennison Scott (b.1929) is an Australian ichthyologist at the South Australian Museum, Adelaide, which he joined (1951) as Assistant in marine zoology, later becoming Curator of fishes. He transferred (1963) to the South Australian Department of Education. He was President of the Royal Society of South Australia (1987). He wrote *The Sharks and Rays of South Australia* (1955) and co-wrote *The Marine and Freshwater Fishes of South Australia* (1962).

Seale

Blackspot Shark *Carcharhinus sealei* Pietschmann, 1913

Pink Whipray *Himantura fai* Jordan & Seale, 1906

Alvin Seale (1871–1958) was an ichthyologist and designer of aquaria. He rode his bicycle from Indiana to California (1892) to study under Jordan at Stanford, from which he only graduated (1905) as his studies were interrupted by expeditions to Alaska (1896–1898) and Hawaii (1900–1902) as a field naturalist for the Bishop Museum, where he then worked as Curator of Fishes (1902–1904). He was again in Alaska (1906) and the Philippines (1907–1917) as chief of the Division of Fishes, Philippine Bureau of Science. He was ichthyologist at the Harvard Museum of Comparative Anatomy (1917–1920) and then retired to a ranch in California. He was persuaded to help the California Academy of Sciences with the planning and building of the aquarium near the Golden Gate and came out of retirement (1921). He was the first superintendent of the Aquarium (1923–1941). He described the blackspot shark (1910) but used a preoccupied name, *C. borneensis*.

Sellos

Chimaera sp. *Chimaera opalescens* Luchetti, Iglésias & Sellos, 2011

Professor Dr Daniel Y. Sellos is a French biologist at the Station de Biologie Marine et Marinarium de Concarneau. The University of Paris awarded his bachelor's (1974) and master's (1976) degrees in biochemistry and his doctorate (1979). He was a researcher at the Collège de France (1975–1979), then research associate (1980–1984) then research fellow (1985–2005) at CNRS. He has been Deputy Director of the Marine Biological Station of Concarneau and professor at the National Museum of Natural History (since 2005). He co-wrote *Discovery of a Normal Hermaphroditic Chondrichthyan Species:* Apristurus longicephalus *(2005)* and *Molecular Phylogeny and Node Time Estimation of Bioluminescent Lantern Sharks (Elasmobranchii: Etmopteridae) (2010)*.

Séret

West African Pygmy Skate *Neoraja africana* Stehmann & Séret, 1983

Broadheaded Catshark *Bythaelurus clevai* Séret, 1987

Madagascar Skate *Dipturus crosnieri* Séret, 1989

Madagascar Pygmy Skate *Fenestraja maceachrani* Séret, 1989

Kanakorum Catshark *Aulohalaelurus kanakorum* Séret, 1990

Catshark sp. *Apristurus albisoma* Nakaya & Séret, 1999

Tailspot Lantern Shark *Etmopterus caudistigmus* Last, Burgess & Séret, 2002

Pink Lantern Shark *Etmopterus dianthus* Last, Burgess & Séret, 2002

Lined Lantern Shark *Etmopterus dislineatus* Last, Burgess & Séret, 2002

Blackmouth Lantern Shark *Etmopterus evansi* Last, Burgess & Séret, 2002

Pygmy Lantern Shark *Etmopterus fusus* Last, Burgess & Séret, 2002

False Lantern Shark *Etmopterus pseudosqualiolus* Last, Burgess & Séret, 2002

Pointy-nosed Blue Chimaera *Hydrolagus trolli* Didier & Séret, 2002

Madagascar Numbfish *Narcine insolita* Carvalho, Séret & Compagno, 2002

Last's Numbfish *Narcine lasti* Carvalho & Séret, 2002

Chesterfield Island Stingaree *Urolophus deforgesi* Séret & Last, 2003

New Caledonian Stingaree *Urolophus neocaledoniensis* Séret & Last, 2003

Butterfly Stingaree *Urolophus papilio* Séret & Last, 2003

Coral Sea Stingaree *Urolophus piperatus* Séret & Last, 2003

Bighead Spurdog *Squalus bucephalus* Last, Séret & Pogonoski, 2007

White-tip Catshark *Parmaturus albimarginatus* Séret & Last, 2007

White-clasper Catshark *Parmaturus albipenis* Séret & Last, 2007

Beige Catshark *Parmaturus bigus* Séret & Last, 2007

Velvet Catshark *Parmaturus lanatus* Séret & Last, 2007

Cook's Swellshark *Cephaloscyllium cooki* Last, Séret & WT White, 2008

Painted Swellshark *Cephaloscyllium pictum* Last, Séret & WT White, 2008

Flagtail Swellshark *Cephaloscyllium signourum* Last, Séret & WT White, 2008

Speckled Swellshark *Cephaloscyllium speccum* Last, Séret & WT White, 2008

Western Legskate *Sinobatis bulbicauda* Last & Séret, 2008

Blue Legskate *Sinobatis caerulea* Last & Séret, 2008

Eastern Leg Skate *Sinobatis filicauda* Last & Séret, 2008

Weng's Skate *Dipturus wengi* Séret & Last, 2008

Phallic Catshark *Galeus priapus* Séret & Last, 2008

Sawback Skate *Leucoraja pristispina* Last, Stehmann & Séret, 2008

Iberian Pygmy Skate *Neoraja iberica* Stehmann, Séret, Costa & Baro, 2008

Sapphire Skate *Notoraja sapphire* Séret & Last, 2009

New Zealand Eureka Skate *Brochiraja heuresa* Last & Séret, 2012

Deep-sea Skate sp. *Brochiraja vittacauda* Last & Séret, 2012

Deepwater Skate sp. *Notoraja alisae* Séret & Last, 2012

Deepwater Skate sp. *Notoraja fijiensis* Séret & Last, 2012

Deepwater Skate sp. *Notoraja inusitata* Séret & Last, 2012

Deepwater Skate sp. *Notoraja longiventralis* Séret & Last, 2012

Dr Bernard Séret (b.1949) is an ichthyologist and marine biologist who is a senior scientist at IRD (Institut de Recherche pour le Développement) and is currently hosted by the Department of Systematics and Evolution, Muséum National d'Histoire naturelle, Paris. His current researches concern the taxonomy, eco-biology, fisheries and conservation of chondrichthyan fishes (sharks, rays and chimaeras). So far he has described more than 40 new species and published more than 120 scientific papers. He took part in the elaboration of the FAO international plan of action for the conservation and management of shark populations and elaborated (with F. Serena)

the conservation plan for Mediterranean Sea cartilaginous fishes. He is the French representative to the ICES working group on elasmobranch fishes and the expert for the French ministry of ecology for the international conventions (e.g. CITES, CMS, OSPAR). He is a member of the IUCN Shark Specialist Group and the scientific chair of the European Elasmobranch Association. He is involved in the European programmes MADE (Mitigating adverse ecological impacts of open ocean fisheries) and CPOA-Shark (Provision of scientific advice for the purpose of the implementation of the CPOA Sharks). He is also leader of a programme concerning the sharks of the Scattered Islands (SW Indian Ocean). He wrote *Guide d'identification des Principales Espèces de Requins et de Raies de l'Atlantique Tropical Oriental, à l'usage des Enquêteurs des Pêches* (2006) and *Guide des Requins, des Raies, et des Chimères des Pêches Françaises* (2010).

Seychelles

Seychelles Gulper Shark *Centrophorus seychellorum* Baranes, 2003

This species is named after the Seychellois people whom the researchers described as the 'helpful and always smiling inhabitants of the paradise island of the Republic of Seychelles'.

Shaw

Eastern Shovelnose Ray *Aptychotrema rostrata* Shaw, 1794

Sharpwing Eagle Ray *Aetobatus guttatus* Shaw, 1804

Longtail Butterfly Ray *Gymnura poecilura* Shaw, 1804

Darkfinned Numbfish *Narcine maculate* Shaw, 1804

Shaw's Shovelnose Guitarfish *Rhinobatos thouiniana* Shaw, 1804

Dr George Kearsley Shaw (1751–1813) was an English physician, botanist and zoologist. Magdalen Hall, Oxford awarded his MSc (1772) following which he practised as a physician. He lectured on botany at Oxford (1786–1791), then was Assistant Keeper (1791–1807) and Keeper (1807–1813) at the Natural History Section, British Museum. He was a co-founder of the Linnean Society (1788). He described numerous previously undescribed reptiles and amphibians as well as fishes. His works include *General Zoology* (1800–1812). Two reptiles and a mammal are named after him.

Sheiko

Rasptooth Dogfish *Miroscyllium sheikoi* Dolganov, 1986

Boris Anatolievich Sheiko (b.1957) is a Russian ichthyologist who graduated from Rostov State University. He was a junior research fellow, Pacific Research Institute of Fisheries and Oceanography, Vladivostok (1980–1982), a research fellow, Azov Research Institute of Fisheries (Rostov on Don) (1983–1987), a post-graduate student, Zoological Institute, St Petersburg (1987–1990), research fellow, Kamchatka Institute of Ecology, Petropavlovsk-Kamchatsky (1991–2000) and a junior research fellow, Zoological Institute, Russian Academy of Sciences, St Petersburg (2000) and, since 2012, a research fellow. He is also a veteran of many oceanographic expeditions to the northwestern Pacific Ocean and northern Bering Sea and his major scientific interests include species composition, vertical distribution and dispersion of marine fishes in the North Pacific. He is also interested in the taxonomy of fishes of families *Cottidae, Agonidae, Cyclopteridae, Zoarcidae, Stichaeidae* and the nomenclature of Latin names of fishes. He wrote *Ichthyofauna of Peter the Great Bay (Sea of Japan): Species Composition, Ichthyocenes, Zoogeography* (1983), *A Catalog of Fishes of the Family* Agonidae (Scorpaeniformes: Cottoidei) (1993) and *Alectrias markevichi sp. nov. – A New Species of Cockscombs* (Perciformes: Stichaeidae: Alectriinae) *from the Sublittoral of the Sea of Japan and Adjacent Waters* (2012). He has co-written a number of papers with Fedorov (q.v.) after whom Dolganov has also named a skate species.

Shen S-C

Hagfish sp. *Eptatretus sheni* Kuo, Huang & Mok, 1994

Taiwan Angelshark *Squatina Formosa* S-C Shen & Ting, 1972
Hagfish sp. *Eptatretus cheni* S-C Shen & Tao, 1975
Hagfish sp. *Eptatretus taiwanae* S-C Shen & Tao, 1975

Shih-Chieh Shen (alternative transliteration Shen Chia-jui) (1902–1975) was a marine biologist who studied at the University of London (1932–1934) and the BMNH (1935), before returning to China to work for the Chinese Academy of Sciences, Peking (Beijing). He was also at the National Taiwan University. He was honoured for his contributions to the knowledge of Taiwanese fishes.

Shen K-N

Fine-spotted Leopard Whipray *Himantura tutul* Borsa, Durand, K-N Shen, Arlyza, Solihin & Berrebi, 2013

Dr Kang-Ning Shen (b.1972) is a Taiwanese evolutionary biologist at the Department of Environmental Biology and Fisheries Science, College of Ocean Science and Resource, National Taiwan Ocean University, Keelung, where he is an assistant research fellow. The National Sun Yat-Sen University awarded his bachelor's degree (1995), the National Taiwan University his master's (1997) and doctorate (2007). He co-wrote *Species Boundaries in the* Himantura uarnak *Species Complex (Myliobatiformes: Dasyatidae)* (2013).

Sherwood

Sherwood Dogfish *Scymnodalatias sherwoodi* Archey, 1921

C. W. Sherwood discovered the holotype lying washed up on the beach at New Brighton, New Zealand (1920) and presented it to the Canterbury Museum.

Shirai

Rasptooth Dogfish genus *Miroscyllium* Shirai & Nakaya, 1990

Highfin Dogfish *Centroscyllium excelsum* Shirai & Nakaya, 1990
Izu Catshark *Scyliorhinus tokubee* Shirai, Hagiwara & Nakaya, 1992
Blurred Lantern Shark *Etmopterus bigelowi* Shirai & Tachikawa, 1993

Dr Shigeru Shirai is a Japanese ichthyologist and fisheries scientist who specialises in sharks. The University of Tokyo awarded his doctorate. He works at the Department of Aqua Bioscience and Industry, Tokyo University of Agriculture. He wrote *Squalean Phylogeny* (1992) and is editor of the Japanese journal, *Ichthyological Research*.

Shuntov

Longnose Deep-sea Skate *Bathyraja shuntovi* Dolganov, 1985

Dr Vyacheslav Petrovich Shuntov (b.1937) is an ichthyologist with an interest in marine ornithology. Kazan State University awarded his bachelor's degree (1959); his doctorate was awarded in 1973. He worked at the Pacific Scientific Research Fisheries Centre, Vladivostok (1959–1999). He became a professor (1983) and an academician of the Russian Academy of Natural Sciences (1995). He was Deputy Editor-in-Chief of the *Russian Journal of Marine Biology* (2000).

Siboga

Pale Catshark *Apristurus sibogae* Weber, 1913
Siboga Skate *Fenestraja sibogae* Weber, 1913

These species are named after the 'Siboga', the ship used by Weber who led the Siboga Expedition (1899–1900) during which the holotype was collected.

Silas

Indian Swellshark *Cephaloscyllium silasi* Talwar, 1974
[Syn. *Hexatrematobatis longirostris, Hexatrygon longirostra, H. brevirostra, H. taiwanensis, H. yangi*]

Dr Eric Godwin Silas (b.1928) is an Indian ichthyologist and fisheries scientist. He holds six degrees, including two doctorates, all awarded by Madras University (Chennai). He worked for the Zoological Survey of India, Calcutta (1949–1955) and at the Scripps Institution of Oceanography, La Jolla, California (1955–1956). He concentrated on Indian freshwater fishes (1956–1958) and since 1959 on sea fishes, mainly Indian Ocean pelagic fishes. He was based at the Central Marine Fisheries Research Institute, Cochin (1959–1993), being its Director from 1975. He is a former Vice-Chancellor, Kerala Agricultural University, Cochin. Among his many publications is *Fishes from the High Range of Travancore* (1951). Talwar's etymology says that his 'excellent publications on the ichthyofauna of the continental shelf of the south-west coast of India have added much to our knowledge of the fauna of this region'.

Smale

Whitetip Weasel Shark *Paragaleus leucolomatus* Compagno & Smale, 1985

Dr Malcolm John Smale (b.1950) is a British-born South African ichthyologist and diver scientist. The University of Natal awarded his bachelor's degree (1974) and his master's (1978) and Rhodes University, Grahamstown his doctorate (1984). He worked at the Oceanographic Research Institute, Durban, South Africa (1974–1978) and moved to the Port Elizabeth Museum, South Africa, first as a contract researcher (1978–1987) and subsequently as a permanent specialist scientist and Curator of Fish Otolith and Cephalopod Beak Collections. He was an honorary associate lecturer at the Department of Zoology, University of Port Elizabeth (1990–2004) and since 2007 has been an honorary research associate at the Department of Zoology, Nelson Mandela Metropolitan University. Among his many publications are *Otolith Atlas of Southern African Marine Fishes* (1995).

His current research focuses on prey identification of marine apex predators, and elasmobranch biology and ecology.

Smirnov

Golden Skate *Bathyraja smirnovi* Soldatov & Pavlenko, 1915
[Alt. Deepwater Stingray, Giant Stingaree, Syn. *Urolophus marmoratus, Urotrygon daviesi*]

Mr Smirnov was an inspector of fisheries who collected fishes from the Okhtosk Sea whence came the holotype.

Smith, A

Barbeled Houndshark *Leptocharias smithii* Müller & Henle, 1839
African Softnose Skate *Bathyraja smithii* Müller & Henle, 1841

Megalodon genus *Carcharodon* A Smith, 1838
Houndshark genus *Leptocharias* A Smith, 1838
Catshark genus *Poroderma* A Smith, 1838
Whale Shark genus *Rhincodon* A Smith, 1828
Catshark genus *Schroederichthys* A Smith, 1838

Blue Stingray *Dasyatis chrysonota* A Smith, 1828
Braun's Whale Shark *Rhincodon typus* A Smith, 1828
[Alt. Whale Shark]
Sharptooth Houndshark *Triakis megalopterus* A Smith, 1839

Dr Sir Andrew Smith (1797–1872) was a Scotsman who joined the Army Medical Service (1819) after graduating from Edinburgh University. He was a zoologist and herpetologist famous for his scrupulous accuracy. He was in the Cape Colony, South Africa (1820–1837), and became the first superintendent of the South African Museum of Natural History, Cape Town (1825). He led the first scientific expedition into the South African interior (1834–1836) and wrote *Report of the Expedition for Exploring Central Africa* (1836), the year he met Darwin, who later sponsored his Fellowship of the Royal Society. However, Smith ceased his natural history collecting and study after returning to Britain as Principal Medical Officer at Fort Pitt, Chatham (1841) and Director General of Army Medical Services (1853). *The Times* newspaper accused him of incompetence in organising medical services during the Crimean War but he was cleared by an enquiry. He retired on grounds of ill health (1858). Much of his private collection was given to Edinburgh University and is now in the Royal Museum of Scotland. He wrote the 28-part *Illustrations of the Zoology of South Africa* (1838–1850). He collected many South African sharks and the skate holotype and coined many of the shark names later formally described by Müller and Henle. Twelve birds, nine reptiles, two amphibians and four mammals are named after him.

Smith, CH

Bancroft's Numbfish *Narcine bancroftii* Griffith & CH Smith, 1834
[Alt. Brazilian/Lesser Electric Ray, Syn. *Narcine umbrosa*]
Scalloped Hammerhead *Sphyrna lewini* Griffith & CH Smith, 1834

Lieutenant-Colonel Charles Hamilton Smith (1776–1859) was an English artist, antiquarian, soldier, spy and naturalist. He was born in Austrian-controlled Flanders (now in Belgium) and studied at the Austrian military academy (1787). He served in the British army until 1820, much of the time in Britain. He pointed out that red was a bad colour for uniforms and that grey or green meant a less distinctive target, which would reduce casualties (1800). It took well over half a century for the War Office in London to agree. It is estimated that he produced over 38,000 drawings; most of his non-military illustrations have been lost but his notebooks of his observations as a naturalist have survived. He wrote *The Natural History of the Human Species* (1848).

Smith, HM

Carpetshark genus *Cirrhoscyllium* HM Smith & Radcliffe, 1913
Catshark genus *Pentanchus* HM Smith & Radcliffe, 1912
Deepsea Dogfish Sharks genus *Squaliolus* HM Smith & Radcliffe, 1912
Ribbontail Catshark genus *Eridacnis* HM Smith, 1913

Arrowhead Dogfish *Deania profundorum* HM Smith & Radcliffe, 1912
Short-tail Lantern Shark *Etmopterus brachyurus* HM Smith & Radcliffe,1912
Philippine Chimaera *Hydrolagus deani* HM Smith & Radcliffe, 1912
Onefin Catshark *Pentanchus profundicolus* HM Smith & Radcliffe, 1912
Spined Pygmy Shark *Squaliolus laticaudus* HM Smith & Radcliffe, 1912
Barbelthroat Carpetshark *Cirrhoscyllium expolitum* HM Smith & Radcliffe, 1913

Pygmy Ribbontail Catshark *Eridacnis Radcliffe* HM Smith, 1913

Dr Hugh McCormick Smith (1865–1941) was an American physician and ichthyologist. He graduated MD from Georgetown University (1888). He led an expedition that explored in the Philippines (1907–1910) aboard 'USS Albatross' and was assisted by Lewis Radcliffe (q.v.). He worked for the US Bureau of Fisheries (1886–1922), first as an assistant then directing their scientific research centre (1897–1903) and also directing the Marine Biology Laboratory at Massachusetts (1902–1903). Concurrently he taught medicine at Georgetown University (1888–1902). He was Deputy Commissioner of the US Bureau of Fisheries (1903–1913) and then Commissioner (1913–1922); on being pressured into resigning, he became a fisheries advisor in Siam (Thailand) (1923–1934). He was Curator of Zoology at the Smithsonian for the rest of his life (1935–1941). He published widely on fishes, including *The Salmon Fishery of Penobscot Bay and River in 1895–96*, and also wrote popular articles. Three birds are named after him.

Smith, JA

Thorny Freshwater Stingray *Dasyatis ukpam* JA Smith, 1863

Dr John Alexander Smith was a physician by training but an archaeologist, zoologist, botanist and antiquarian by preference. He was associated with the National Museum of Antiquities in Edinburgh. He took part in a number of excavations, including several during the construction of the North British Railway (1846), during which a number of ancient human remains were uncovered. He was Secretary of the Society of Antiquaries of Scotland, before which he read a paper entitled *On the Remains of the Rein-deer in Scotland* (1869). A mammal is named after him.

Smith, JLB

Houndshark genus *Hypogaleus* JLB Smith, 1957

Whitespotted Bullhead Shark *Heterodontus ramalheira* JLB Smith, 1949
[Alt Mozambique Bullhead Shark, Syn *Gyropleurodus ramalheira*]
Brown Shyshark *Haploblepharus fuscus* JLB Smith, 1950
African Ribbontail Catshark *Eridacnis sinuans* JLB Smith, 1957

Whitespotted Smooth-Hound *Mustelus palumbes* JLB Smith, 1957

Slime Skate *Dipturus pullopunctatus* JLB Smith, 1964

Dr James Leonard Brierley Smith (1897–1968) was a South African chemist and ichthyologist most famous for identifying a stuffed fish as a coelacanth, then thought to have been extinct for millions of years. The University of the Cape of Good Hope awarded his bachelor's degree (1916) and Stellenbosch University his master's (1918). Cambridge awarded his doctorate (1922), after which he worked at Rhodes University, Grahamstown, South Africa as senior lecturer and head of department and Professor of Ichthyology (1947). With his wife (below) he wrote *Sea Fishes of Southern Africa* (1949). After a long illness he committed suicide by taking cyanide.

Smith, MM

Sixgill Stingray *Hexatrygon bickelli* Heemstra & MM Smith, 1980

Professor Margaret Mary Smith née Macdonald (1916–1987) was a South African ichthyologist. She was the second wife of James Leonard Brierley Smith (above), whom she married (1938). She had a bachelor's degree from Rhodes University, Grahamstown, South Africa and became her husband's research assistant and illustrator. After his death (1968) she persuaded Rhodes University to establish the J.L.B. Smith Institute of Ichthyology and became its first director. She was appointed full professor (1981) and retired (1982). The Institute was renamed the South African Institute for Aquatic Biodiversity (2000) and the library was named after her (2010).

Snodgrass

Galapagos Shark *Carcharhinus galapagensis* Snodgrass & Heller, 1905

Dr Robert Evans Snodgrass (1875–1962) was an American entomologist and artist. At the age of 15 he was an ornithologist and a convinced disciple of Darwin, which caused trouble at home and, to his great delight, got him expelled from Sunday school. He entered Stanford University (1895), where he was converted to entomology before graduating with a bachelor's degree in zoology (1901). His doctorate was honorary and given by Eberhard-Karls-Universität, Tübingen (1953). Whilst still a student he went with Jordan to the Pribilof Islands and with Heller to the Galapagos Islands. He taught at the State College of Washington (1901–1903) and then returned to Stanford (1903–1905), but left under

a cloud for causing the demise of a mulberry tree. He worked for an advertising agency in their art department (1905–1906) but the San Francisco earthquake (1906) brought that to an end. Eventually he was employed by the Bureau of Entomology, US Department of Agriculture in Washington DC (1906–1909). When refused a pay rise, he packed and went to New York to work as an artist and cartoonist. He worked in entomology in Indianapolis (1915–1917). He rejoined the Department of Agriculture (1917) and stayed there until he retired (1945) but continued working and researching at the Smithsonian. He wrote *Principles of Insect Morphology* (1935).

Snyder

Shark genus *Deania* Jordan & Snyder, 1902

Silver Chimaera *Chimaera phantasma* Jordan & Snyder, 1900

Hagfish sp. *Myxine garmani* Jordan & Snyder, 1901

Blackbelly Lantern Shark *Etmopterus lucifer* Jordan & Snyder, 1902

Shortspine Spurdog *Squalus mitsukurii* Jordan & Snyder, 1903

Gecko Catshark *Galeus eastmani* Jordan & Snyder, 1904

Spookfish *Hydrolagus mitsukurii* Jordan & Snyder, 1904

John Otterbein Snyder (1867–1943) was an American zoologist and ichthyologist. Stanford University awarded both his bachelor's degree (1897) and his master's (1899). He was an instructor and professor at Stanford University (1899–1943). He took part in the 'USS Albatross' expeditions (1900s) and organised the fish collection at the US National Museum (1925). When he was Assistant Professor of Zoology at Stanford, he published *Notes on the Fishes of the Streams Flowing into San Francisco Bay* in *Report of the Commissioner of Fisheries to the Secretary of Commerce and Labor for the Fiscal Year Ending June 30, 1904* (1905).

Soldatov

Golden Skate *Bathyraja smirnovi* Soldatov & Pavlenko, 1915

Vladimir Konstantinovich Soldatov (1875–1941) was a Russian ichthyologist who was Professor at the Moscow Technical Institute of Fishing Industry and Fish Farming (1919–1941). He wrote *Fishes and Commercial Fishing* (1928). A seamount in the Pacific is named Soldatov after him.

Solihin

Fine-spotted Leopard Whipray *Himantura tutul* Borsa,
Durand, K-N Shen, Arlyza, Solihin & Berrebi, 2013

Dr Dedy Duryadi Solihin is a biologist who is a
researcher and a member of the Faculty of Mathematics and Natural Sciences, Bogor Agriculture
University, Bogor, Indonesia.

Soto

Hagfish sp. *Myxine sotoi* Mincarone, 2001

Thorny-tail Skate *Dipturus diehli* Soto & Mincarone,
2001
Southern Sawtail Catshark *Galeus mincaronei* Soto, 2001
Lizard Catshark *Schroederichthys saurisqualus* Soto, 2001
Striped Rabbitfish *Hydrolagus matallanasi* Soto &
Vooren, 2004

Dr Jules Marcelo Rosa Soto is a Brazilian ichthyologist and geographer who is Curator at the Museu
Oceanográfico do Vale do Itajaí, Universidade do
Vale do Itajaí. He wrote *Fauna Microbiana Ocorrente
na Cavidade Bucal da Piranha* Serrasalmus spilopleura
(Characidae) *no Município de Uruguaiana, Rio Grande
do Sul, Brasil* (2001). He was honoured in the name
of the hagfish for his work on Brazilian marine fauna
and for encouraging Mincarone to study hagfishes.

Spies

Leopard Skate *Bathyraja panther* Orr, Stevenson, Hoff,
Spies & McEachran, 2011

Ingrid Brigette Spies (b.1974) is an American biologist
and geneticist who is a PhD student at the University of Washington. She is currently employed at the
Alaska Fisheries Science Center, Seattle. Swartmore
Colledge, Pennsylvania awarded her bachelor's
degree (1996) and the University of Washington,
Seattle her master's (2002).

Springer, S

Skate genus *Springeria* Bigelow & Schroeder, 1951

Gulf Hagfish *Eptatretus springeri* Bigelow & Schroeder,
1952
Springer's Sawtail Catshark *Galeus springeri*
Konstantinou & Cozzi, 1998
Broadnose Wedgefish *Rhynchobatus springeri*
Compagno & Last, 2010

Catshark genus *Schroederichthys* S Springer, 1966

Houndshark genus *Iago* Compagno & S Springer,
1971

Narrowfin Smooth-Hound *Mustelus norrisi* S Springer,
1939
Narrownose Smooth-Hound *Mustelus schmitti*
S Springer, 1939
Scalloped Bonnethead *Sphyrna corona* S Springer,
1940
Scoophead Shark *Sphyrna media* S Springer, 1940
Bignose Shark *Carcharhinus altimus* S Springer, 1950
Green Lantern Shark *Etmopterus virens* Bigelow,
Schroeder & S Springer, 1953
African Lantern Shark *Etmopterus polli* Bigelow,
Schroeder, & S Springer, 1953
Fringefin Lantern Shark *Etmopterus schultzi* Bigelow,
Schroeder, & S Springer, 1953
Bahamas Sawshark *Pristiophorus schroederi*
S Springer & Bullis, 1960
Smalleye Smooth-Hound *Mustelus higmani*
S Springer & RH Lowe, 1963
South China Cookiecutter Shark *Isistius plutodus*
Garrick & S Springer, 1964
Longfin Sawtail Catshark *Galeus cadenati* S Springer,
1966
Peppered Catshark *Galeus piperatus* S Springer &
MH Wagner, 1966
Narrowtail Catshark *Schroederichthys maculatus*
S Springer, 1966
Slender Catshark *Schroederichthys tenuis* S Springer,
1966
White-saddled Catshark *Scyliorhinus hesperius*
S Springer, 1966
Blotched Catshark *Scyliorhinus meadi* S Springer, 1966
Harlequin Catshark *Ctenacis fehlmanni* S Springer,
1968
Polkadot Catshark *Scyliorhinus besnardi* S Springer
& Sadowsky, 1970 McCain's Skate *Bathyraja
maccaini* S Springer, 1971
New Zealand Catshark *Bythaelurus dawsoni*
S Springer, 1971
Speckled Catshark *Halaelurus boesemani* S Springer &
D'Aubrey, 1972
Hoary Catshark *Apristurus canutus* S Springer &
Heemstra, 1979
Ghost Catshark *Apristurus manis* S Springer, 1979
Smallfin Catshark *Apristurus parvipinnis* S Springer &
Heemstra, 1979
Panama Ghost Catshark *Apristurus stenseni*
S Springer, 1979
Antilles Catshark *Galeus antillensis* S Springer, 1979
Dwarf Sawtail Catshark *Galeus schultzi* S Springer,
1979

Campeche Catshark *Parmaturus campechiensis*
S Springer, 1979
Cylindrical Lantern Shark *Etmopterus carteri*
S Springer & Burgess, 1985
Dwarf Lantern Shark *Etmopterus perryi* S Springer &
Burgess, 1985

Stewart 'Stew' Springer (1906–1991) was a field naturalist who dropped out of Butler College (1929) but was awarded a baccalaureate by George Washington University (1964), by which time he was world-renowned as an expert on both the taxonomy and behaviour of sharks. He was originally interested in herpetology and collected, identified and described the Plateau Striped Whiptail *Cnemidophorus velox* (1928). During and after the Second World War he worked variously for the Office of Strategic Services (now CIA) and for the US Navy on shark repellents and on survival manuals. After a short period in the business of commercial shark fishing, he worked for the US Fish and Wildlife Service (1950–1971). After retirement from Government service (1971) he worked for the Mote Marine Laboratory, Sarasota, Florida (1971–1979). He wrote more than 80 papers on sharks and rays, including the extensive *A Revision of the Catsharks, Family Scyliorhinidae* (1979). Konstantinou and Cozzi honoured him as an 'outstanding leader in shark taxonomy …for his work with the family Scyliorhinidae'.

Springer, VG

Roughbelly Skate *Dipturus springeri* Wallace, 1967

Grey Sharpnose Shark *Rhizoprionodon oligolinx* VG
Springer, 1964

Dr Victor Gruschka Springer (b.1928) is an American ichthyologist. He is Senior Scientist Emeritus, Division of Fishes at the Smithsonian, which he originally joined (1961). He specialised in anatomy, classification and fish distribution. Emory University awarded his bachelor's degree (1948). He originally intended to be a physician but hated the sight of blood and switched to marine biology (1948) at the University of Miami, which awarded his master's (1954). His academic career was interrupted by service in the US Army (1950–1952), including in Korea during the Korean War, but he was able to spend a lot of time collecting. The University of Texas awarded his doctorate (1957). Like many naturalists he is a keen philatelist, collecting and publishing on stamps with a fish or fishing theme. He wrote *Sharks in Question: The Smithsonian Answer Book* (1989).

Starks

Giant Electric Ray *Narcine entemedor* Jordan & Starks,
1895
Roger's Roundray *Urotrygon rogersi* Jordan & Starks,
1895

Edwin Chapin Starks (1867–1932) was an American ichthyologist who was an authority on the osteology of fish. He was educated at Stanford University, where he started a course in zoology (1893). He went on many expeditions including with Jordan to Mazatlán (1894), and to Panama with the Hopkins Expedition (1895). He was also on the Harriman Expedition to Alaska (1899) and the Stanford Expedition to Brazil (1911). He was a fieldwork assistant for the US Bureau of Biological Survey (1897–1899) and Curator of the Museum and Assistant Professor of Zoology, University of Washington (1899–1901). Stanford appointed him as Curator of Zoology (1901) and Assistant Professor (1927). He retired as Professor Emeritus (1932).

Stauch

Smooth Freshwater Stingray *Dasyatis garouaensis*
Stauch & Blanc, 1962

Alfred Stauch (1921–1993) was a French oceanographer and ichthyologist. He co-wrote a number of articles such as *Les Poissons du Bassins du Tchad ed du Bassin Adjacent du Mayo-Kabbi. Études Systematiques et Biologiques* (1964), as well as describing a number of marine organisms.

Stehmann

African Pygmy Skate *Neoraja stehmanni* Hulley, 1972

Soft Skate genus *Malacoraja* Stehmann, 1970
Skate genus *Rajella* Stehmann, 1970
Shark genus *Planonasus* Weigmann, Stehmann &
Thiel, 2013
Leg Skate genus *Indobatis* Weigmann, Stehmann &
Thiel, 2014

Whitedappled Skate *Leucoraja leucosticta* Stehmann,
1971
Brazilian Skate *Rajella sadowskii* G Krefft & Stehmann,
1974
Thintail Skate *Dipturus leptocaudus* G Krefft &
Stehmann, 1975
Roughskin Skate *Dipturus trachydermus* G Krefft &
Stehmann, 1975
Blue Ray *Neoraja caerulea* Stehmann, 1976
Krefft's Skate *Malacoraja kreffti* Stehmann, 1978

Bigelow's Ray *Rajella bigelowi* Stehmann, 1978

West African Pygmy Skate *Neoraja africana* Stehmann & Séret, 1983

Carolina Pygmy Skate *Neoraja carolinensis* McEachran & Stehmann, 1984

Butterfly Skate *Bathyraja papilionifera* Stehmann, 1985

Dark-belly Skate *Bathyraja meridionalis* Stehmann, 1987

Velvet Skate *Insentiraja subtilispinosa* Stehmann, 1989

Ratfish sp. *Hydrolagus pallidus* Hardy & Stehmann, 1990

Paddle-nose Chimaera *Rhinochimaera africana* Compagno, Stehmann & Ebert, 1990

West African Skate *Bathyraja hesperafricana* Stehmann, 1995

Tigertail Skate *Leucoraja compagnoi* Stehmann, 1995

Arabian Sicklefin Chimaera *Neoharriotta pumila* Didier & Stehmann, 1996

White Ghost Catshark *Apristurus aphyodes* Nakaya & Stehmann, 1998

Brazilian Blind Electric Ray *Benthobatis kreffti* Rincón, Stehmann & Vooren, 2001

Aden Gulf Torpedo *Torpedo adenensis* Carvalho, Stehmann & Manilo, 2002

Black Roughscale Catshark *Apristurus melanoasper* Iglésias, Nakaya & Stehmann, 2004

Abyssal Skate *Bathyraja ishiharai* Stehmann, 2005

Tuna's Skate *Bathyraja tunae* Stehmann, 2005

Sawback Skate *Leucoraja pristispina* Last, Stehmann & Séret, 2008

Iberian Pygmy Skate *Neoraja iberica* Stehmann, Séret, Costa & Baro, 2008

Challenger Skate *Rajella challengeri* Last & Stehmann, 2008

Dwarf False Catshark *Planonasus parini* Weigmann, Stehmann & Thiel, 2013

Sparsely-thorned Skate *Rajella paucispinosa* Weigmann, Stehmann & Thiel, 2014

Dr Matthias F. W. Stehmann (b.1943) is a German ichthyologist whose university career at Kiel University (1962–1969) resulted in degrees covering marine sciences, zoology and limnology and culminated with a doctorate in marine sciences. He was a research scientist at the Federal Institute for Fisheries, Hamburg (1969–2002), during which time he went on many research expeditions from the Arctic to the Antarctic as well as the whole Atlantic basin. Having retired (2002) he set up and runs his own ICHTHYS research laboratory in Hamburg. He was a contributor on batoid fishes to a number of fundamental faunal checklists and handbooks,

notably *Check-list of the Fishes of the NE Atlantic and Mediterranean* (1973), *Fishes of the NE Atlantic and Mediterranean* (1984) and *Check-list of the Fishes of the Eastern Tropical Atlantic* (1990), all published by UNESCO. He has produced a very large number of publications on chondrichthyan fishes, too many to mention here, both as a sole and as a joint author, including *Batoid Fishes – Technical Terms and Principal Measurements, General Remarks, Key with Picture Guide to Families, List of Species* (1978), *Bathyraja meridionalis sp. n. (Pisces, Elasmobranchii, Rajidae), a New Deep-water Skate from the Eastern Slope of Subantarctic South Georgia Island* (1987), *First and New Records of Skates (Chondrichthyes, Rajiformes, Rajidae) from the West African Continental Slope (Morocco to South Africa), with Descriptions of Two New Species* (1995) and *Proposal of a Maturity Stages Scale for Oviparous and Viviparous Cartilaginous Fishes (Pisces, Chondrichthyes)* (2002) – of all of which he is the sole author – and, jointly with D. Ebert, *Sharks, Batoids, and Chimaeras of the North Atlantic. FAO Species Catalogue for Fishery Purposes* (2013).

Steindachner

Pacific Cownose Ray *Rhinoptera steindachneri* Evermann & OP Jenkins, 1891

Spotted Houndshark *Triakis maculate* Kner & Steindachner, 1867

Marbled Stingray *Dasyatis marmorata* Steindachner, 1892

Blackish Stingray *Dasyatis navarrae* Steindachner, 1892

Chinese Stingray *Dasyatis sinensis* Steindachner, 1892

Irrawaddy River Shark *Glyphis siamensis* Steindachner, 1896

Torpedo sp. *Torpedo suessii* Steindachner, 1898

Franz Steindachner (1834–1919) was an Austrian zoologist who specialised in herpetology and ichthyology. He originally planned to become a lawyer, but became interested in fossil fish and (1860) joined the Naturhistorisches Museum in Vienna, becoming a curator (1861) and Head of the Zoology Department (1874). He went on to become Director of the Vienna Museum (1898–1919). Unlike many museum curators he also travelled actively and collected in the Americas including the Galapagos Islands, Africa and the Middle East. His major work was to write up the amphibian and reptile sections of the published results of the circumnavigation of the globe by the Austrian frigate 'Novara'. He was honoured in the name of the cownosed ray for his 'valuable services to American ichthyology'. Seven amphibians, ten reptiles and two birds are named after him.

Steiner

Turkish Brook Lamprey *Lampetra lanceolata* Kux & Steiner, 1972

Dr Hans Martin Steiner (b.1938) is an Austrian zoologist, mammalogist and herpetologist who was interested in natural history as a child even before visiting a biological station on Lake Neusiedl (1954), where for the first time he got serious answers to his many questions and so started his scientific education. He became head of this station in 1961. He studied zoology and palaeontology at the University of Vienna from 1957, culminating in his doctorate (1966). He joined the staff of the Hochschule für Bodenkultur, now the University of Natural Resources and Life Sciences, Vienna, where he worked (1966–2004), being Professor (1977), Head of the Zoological Institute (1979) and full Professor (1981–2004) before retiring as Emeritus Professor (2004). He was a visiting professor in both Kyrgistan and Thailand and since retiring is still working and publishing on birds. The majority of his work outside Austria concerned southern European countries and the Middle East. He co-wrote a paper on the Persian Brook Salamander (1970). A reptile is named after him.

Stensen

Panama Ghost Catshark *Apristurus stenseni* S Springer, 1979

Dr Neils Stensen (variously rendered Nicolas Steno, Nicolaus Stenonis or Nicolaus Stenonius) (1638–1686) was a Danish geologist, anatomist and author. He was born Niels Stensen but, as Linnaeus did later, Latinised his name. He went to Leiden (1660) to study medicine (1660). After a short period in Paris and Montpellier, Stensen went to Florence (1665), where he studied anatomy. He was the first person to realise that what looked like sharks' teeth embedded in rocks were in fact fossilised sharks' teeth. From that discovery he was led to formulate his most important contribution to geology, Steno's Law of Superposition. He was the Royal Anatomist in Copenhagen (1672–1674). He converted to Roman Catholicism and abandoned science (1667), and was later ordained as a priest (1675). He became a bishop (1677) and spent the rest of his life ministering to the minority Catholic populations in Denmark, Norway and northern Germany. He was beatified (1987), the first step on the road to sainthood. The etymology notes that Stensen's 'scientifically accurate work on elasmobranch anatomy was highly influential in the beginnings of elasmobranch systematics.' A mammal is named after him.

Stevens

White Spotted Gummy Shark *Mustelus stevensi* WT White & Last, 2008

Stevens' Swellshark *Cephaloscyllium stevensi* E Clark & Randall, 2011

Banded Sand Catshark *Atelomycterus fasciatus* Compagno & Stevens, 1993

Slender Sawtail Catshark *Galeus gracilis* Compagno & Stevens, 1993

Deepwater Sicklefin Houndshark *Hemitriakis abdita* Compagno & Stevens, 1993

Sicklefin Houndshark *Hemitriakis falcate* Compagno & Stevens, 1993

Blotched Catshark *Asymbolus funebris* Compagno, Stevens & Last, 1999

Dwarf Catshark *Asymbolus parvus* Compagno, Stevens & Last, 1999

Variegated Catshark *Asymbolus submaculatus* Compagno, Stevens & Last, 1999

Southern Mandarin Dogfish *Cirrhigaleus australis* WT White, Last & Stevens, 2007

Indonesian Speckled Catshark *Halaelurus maculosus* WT White, Last & Stevens, 2007

Rusty Catshark *Halaelurus sellus* WT White, Last & Stevens, 2007

Eastern Highfin Spurdog *Squalus albifrons* Last, WT White & Stevens, 2007

Western Highfin Spurdog *Squalus altipinnis* Last, WT White & Stevens, 2007

Edmund's Spurdog *Squalus edmundsi* Last, WT White & Stevens, 2007

Eastern Longnose Spurdog *Squalus grahami* Last, WT White & Stevens, 2007

Bartail Spurdog *Squalus notocaudatus* Last, WT White & Stevens, 2007

Sombre Catshark *Bythaelurus incanus* Last & Stevens, 2008

Elongate Carpetshark *Parascyllium elongatum* Last & Stevens, 2008

Dr John Donald Stevens (b.1947) is a biologist and ichthyologist who was a senior principal research scientist with CSIRO Marine and Atmospheric Research, originally in Sydney (1979–1984) and subsequently in Hobart, Tasmania (1984–2011). Both his bachelor's degree (1970) and his doctorate (1976) were awarded by London University. He spent a year on Aldabra Atoll in the Seychelles carrying out his post-doctoral research on reef sharks. He has published more than 100 scientific papers and reports on sharks, contributed to and edited several books on sharks as well as co-authoring *Sharks and Rays of Australia* (2009). Clark, in honouring him in the name of the swellshark, described this work as the 'foundation for research that led to the descriptions of 37 new chondrichthyan fishes, including 11 species of Cephaloscyllium.' White honoured him because he had 'dedicated a lifetime to researching sharks around the world, and ... contributed greatly to our knowledge of sharks and rays in Australia.'

Stevenson

Butterfly Skate *Bathyraja mariposa* Stevenson, Orr, Hoff & McEachran, 2004
Leopard Skate *Bathyraja panther* Orr, Stevenson, Hoff, Spies & McEachran, 2011

Dr Duane E. Stevenson is an ichthyologist and fishery biologist who joined the staff of the Alaska Fisheries Science Center, Seattle (2001). Washington & Jefferson College, Pennsylvania awarded his bachelor's degree (1993), the University of Charleston, South Caroline his master's (1996) and the University of Washington, Seattle his doctorate (2002). He became an affiliate assistant professor at the School of Aquatic and Fishery Sciences, University of Washington (2007) and a research associate at the Burke Museum of Natural History and Culture, Seattle (2008). He co-wrote *Field Guide to Sharks, Skates, and Ratfish of Alaska* (2007).

Stewart

Flaccid Catshark *Apristurus exsanguis* Sato, Nakaya & Stewart, 1999
Goliath Hagfish *Eptatretus goliath* Mincarone & Stewart, 2006
Garrick's Catshark *Apristurus garricki* Sato, Stewart & Nakaya, 2013

Andrew L. Stewart became collection manager of fishes (1991) at the Museum of New Zealand Te Papa Tongarewa, Wellington, which he joined as a science technician (1982). He was a member of the NORFANZ survey along the Norfolk and Lord Howe Rises (2003) and the IPY/CAMP Expedition to Antarctica (2008). He co-wrote *New Zealand Fish, a Complete Guide* (1989).

Storm

Norwegian Skate *Dipturus nidarosiensis* Storm, 1881

Vilhelm Ferdinand Johan Storm (1835–1913) was a curator at the Videnskapsselskaps Natural History Collections, Trondheim, Norway (1856–1913). He worked on the fauna of the Trondheim fjord, including at the Trondheim Biological Station (1900). He left school in Trondheim (1852) and studied zoology, entomology, botany and taxidermy in Christiania (Oslo) (1853–1855). He wrote *Throndhjemsfjordens Fiske* (1884).

Stout

Pacific Hagfish *Eptatretus stoutii* Lockington, 1878

Dr Arthur B. Stout (1814–1898) was a prominent surgeon. He was corresponding secretary of the California Academy of Sciences, of which he was a member (1853–1898) and Curator of Ethnology and Osteology (1881). He wrote *Contributions to the History of the Aleutian Islands*.

Straelen

Spotted Skate *Raja straeleni* Poll, 1951

Professor Victor Émile van Straelen (1889–1964) was a Belgian geologist and naturalist. He was Director, Belgian Royal Institute of Natural Science, Brussels (1925–1954). He travelled widely, especially in Indonesia and the Belgian Congo (DRC) and became (1933) President, Institute of National Parks of the Congo. He was the first president (1959–1964) of the Darwin Foundation. He was also president of the non-profit organization (Mbizi) that sponsored the expedition that collected the skate holotype; he accompanied one of the expedition's trawler cruises off the mouth of the Congo. An amphibian is named after him.

Strahan

Hagfish sp. *Eptatretus strahani* CB McMillan & Wisner, 1984
Hooded Carpetshark *Hemiscyllium strahani* Whitley, 1967

Longfinned Hagfish *Eptatretus longipinnis* Strahan, 1975

Dr Ronald Strahan (1922–2010) was an Australian zoologist, ichthyologist and research scientist who was Director of Taronga Park Zoo, Sydney (1967–1974) when this shark was described; the holotype had been swimming in the zoo's aquarium since 1960. The University of New South Wales awarded his honorary doctorate (1999). He served in the Australian army in the Second World War in a mobile entomological research unit, tasked with identifying insect vectors for tropical diseases, particularly malaria and dengue fever, which caused more casualties than combat with the Japanese. After being demobilised he eventually completed his degree in zoology at the University of Western Australia (1947). He went to study in Europe at Oxford and then to lecture on zoology at the University of Hong Kong, returning to Australia (1961) as senior lecturer, University of New South Wales. He became a research fellow at the Australian Museum, Sydney (1974) and later head of its National Photographic Index of Australian Wildlife. He wrote *Taronga Zoo and Aquarium* (1974).

Straube

Lantern Shark sp. *Etmopterus viator* Straube, 2011

Dr Nicolas Straube (b.1980) is an ichthyologist and biologist at the Zoologische Staatssammlung München, Germany, where he gained his doctorate (2011). He was a post-doctoral research fellow at the College of Charleston (2011–2013) and is now doing post-doctoral research at Jena University, Germany. Among his publications are the co-wrtten *Molecular Phylogeny and Node Time Estimation of Bioluminescent Lantern Sharks (Elasmobranchii: Etmopteridae)* (2010) and *Capturing Protein-coding Genes across Highly Divergent Species* (2013).

Strickrott

Strickrott's Hagfish *Eptatretus strickrotti* Møller & Jones, 2007

W. Bruce Strickrott was the pilot of the Alvin submarine that was used to take the holotype. Describer W. Joe Jones told the Woods Hole Oceanographic Institution that without Alvin pilots oceanographers would not get their jobs done. 'We saw this little thing swimming like a worm and I told Bruce, "There is no way you are going to catch it"', said Jones. However, Strickrott managed to get the submarine behind the hagfish and vacuum it up using a device known as a slurp gun. 'I was like, man, this guy has skills and deserves recognition. The naming was a way to express our gratitude.' Strickrott said, 'It's a feather in my cap. It's recognition from researchers for my contributions to the advancement of science.'

Struhsaker

Megamouth Shark Family *Megachasmidae* Taylor, Compagno & Struhsaker, 1983

Megamouth Shark genus *Megachasma* Taylor, Compagno & Struhsaker, 1983

Megamouth Shark *Megachasma pelagios* Taylor, Compagno & Struhsaker, 1983

Dr Paul James Struhsaker (b.1935) was a fishery biologist and ichthyologist. The University of Hawaii, Honolulu awarded his master's (1967) and doctorate in zoology (1973). He has participated in fisheries investigations in the western North Atlantic, Gulf of Mexico, western Caribbean, Alaska, US west coast and the Hawaiian Islands. He worked at the National Marine Fisheries Service, Southeast Fisheries Center (1957, 1959–1965, 1981–82) and at the Honolulu Laboratory (1969–1977). He was associated with the Seattle, Juneau and Woods Hole Laboratories for brief periods. He was a research associate at the Bishop Museum, Hawaii. He co-wrote *Observations on the Biology and Distribution of the Thorny Stingray*, Dasyatis centroura *(Pisces: Dasyatidae)* (1969) and *Megamouth: A New Species, Genus, and Family of Lamnoid Shark* (Megachasma pelagios, *Family Megachasmidae) from the Hawaiian Islands* (1983). He has contributed to much research including initiating modern bottom trawl surveys in the Hawaiian Islands and the first studies of daily growth rings in otoliths of tropical marine fishes. Three other fish species are named after him.

Suckley

Spotted Spiny Dogfish *Squalus suckleyi* Girard, 1854

Dr George Suckley (1830–1869) was an American Army surgeon and naturalist. He was appointed as Assistant Surgeon and Naturalist of the Pacific Railway Survey between Minnesota and the Puget Sound (1853). Later, he explored the Oregon and Washington territories, which had not yet been admitted as States of the Union. He resigned from the Army (1856) to concentrate on natural history, then rejoined the Union Army and served as a surgeon throughout the Civil War. He co-wrote *Natural History of Washington Territory* (1859). He collected the holotype. Three birds are named after him.

Suess

Torpedo sp. *Torpedo suessii* Steindachner, 1898

The original description contains no etymology but we think that this species is named after Eduard Suess (1831–1914), who was an Austrian palaeontologist, geologist and malacologist. He is famous for his hypothesis of a mega-continent 'Gondwana' (1861) and the Tethys Ocean (1893). He wrote the three-volume *Das Antlitz der Erde* (1885–1901) in which he was the first to posit the idea of the biosphere. Craters on both Mars and the Moon are named after him.

Suvorov

Okhotsk Skate *Bathyraja violacea* Suvorov, 1935

Evgenii Konstantinovich Suvorov (1880–1953) was a Russian zoologist and ichthyologist. He graduated from the University of St Petersburg (1903). He took part in and, later, led expeditions to study fish biology and fishing methods and to discover new fishing regions in many places from the Baltic to the Pacific and from the White Sea to the Caspian (1904–1920). He was Director of the Polytechnic Institute of Fish Culture, Leningrad (St Petersburg) (1921–1931). He was a professor at Leningrad University (1931–1952) and head of a sub-department (1949). Among his achievements was to introduce the artificial breeding of Atlantic salmon (1920).

Swart

Legskates *Anacanthobatidae* von Bonde & Swart, 1924

Legskate genus *Anacanthobatis* von Bode & Swart, 1923

Spotted Legskate *Anacanthobatis marmoratus* von Bonde & Swart, 1923
Smoothnose Legskate *Cruriraja durbanensis* von Bonde & Swart, 1923
Roughnose Legskate *Cruriraja parcomaculata* von Bonde & Swart, 1923
Munchkin Skate *Rajella caudaspinosa* von Bonde & Swart, 1923
Leopard Skate *Rajella leopardus* von Bonde & Swart, 1923

D. B. Swart was an ichthyologist at the Zoology Department, University of Cape Town, which he joined as a demonstrator (1921), and became an assistant lecturer (1924). With Cecil von Bonde he published the paper, *The Platosomia (Skates and Rays) Collected by 'SS Pickle'*, a report for the Fisheries Marine Biological Survey (1923).

T

Taaf

Whiteleg Skate *Amblyraja taaf* Meissner, 1987

This species is named after a location, Territoire des Terres australes et antarctiques françaises (TAAF).

Tachikawa

Longnose Sawtail Catshark *Galeus longirostris* Tachikawa & Taniuchi, 1987
Blurred Lantern Shark *Etmopterus bigelowi* Shirai & Tachikawa, 1993

Dr Hiroyuki Tachikawa is a Japanese ichthyologist and herpetologist, specialising in sea turtles, at Ogasawara Marine Centre, Tokyo (1993) and at the coastal branch of the Natural History Museum and Institute, Chiba (2004).

Takagi

Sleeper Torpedo *Crassinarke brasiliensis* Takagi, 1951

Dr Kazunori Takagi (fl.1989) was a Japanese ichthyologist at the Tokyo University of Fisheries. He wrote *A Study on the Scale of the Gobiid Fishes of Japan* (1951).

Takahashi

Ocellate Japanese Topeshark *Hemitriakis complicofasciata* Takahashi & Nakaya, 2004

Dr Tetsumi Takahashi is a Japanese marine biologist at the Laboratory of Animal Ecology, Department of Zoology, Graduate School of Science, Kyoto University. He specialises in cichlid fishes of Lake Tangyanika.

Talwar

Indian Swellshark *Cephaloscyllium silasi* Talwar, 1974
Quilon Electric Ray *Heteronarce prabhui* Talwar, 1981

Dr Purnesh Kumar Talwar (b.1934) is an Indian ichthyologist whose master's degree and doctorate were awarded by the University of Rajasthan. He was joint director, Northern Regional Station, Zoological Survey of India, Dehra Dun. He co-wrote *Inland Fishes of India and Adjacent Countries* (1991).

Tanaka I

Squaliform Shark genus *Cirrhigaleus* Tanaka I, 1912

Jordan's Chimaera *Chimaera jordani* Tanaka I, 1905
Owston's Chimaera *Chimaera owstoni* Tanaka I, 1905
Trapezoid Torpedo *Torpedo tokionis* Tanaka I, 1908
Cloudy Catshark *Scyliorhinus torazame* Tanaka I, 1908
Flathead Catshark *Apristurus macrorhynchus* Tanaka I, 1909
Borneo Catshark *Apristurus platyrhynchus* Tanaka I, 1909
Mandarin Dogfish *Cirrhigaleus barbifer* Tanaka I, 1912
Finback Catshark sp. *Proscyllium venustum* Tanaka I, 1912
Frog Shark *Somniosus longus* Tanaka I, 1912
Dapple-bellied Softnose Skate *Rhinoraja kujiensis* Tanaka I, 1916
Japanese Shortnose Spurdog *Squalus brevirostris* Tanaka I, 1917

Shigeho Tanaka I (1878–1974) was a Japanese ichthyologist who was Professor of Zoology at the Imperial University, Tokyo. He wrote a great many papers on fish including sharks and rays and co-wrote a book on Japanese fishes with Jordan (q.v.), *A Catalogue of the Fishes of Japan* (1913).

Tanaka II

Japanese Velvet Dogfish *Zameus ichiharai* Yano & Tanaka II, 1984

Dr Shigeho 'Sho' Tanaka II is an ichthyologist who is a professor at the Department of Fisheries, Tokai University, Shimizu City, Japan.

Tang

Yellow-spotted Fanray *Platyrhina tangi* Iwatsuki, Zhang & Nakaya, 2011

Amoy Fanray *Platyrhina limboonkengi* Tang, 1933
[Syn. *Platyrhina sinensis*]

D-S. Tang was a Chinese ichthyologist at the University of Amoy. Iwatsuki *et al.* synonymised Tang's *Platyrhina limboonkengi* with *Platyrhina sinensis*. One supposes that they felt guilty at removing his name from the canon of eponyms so re-established it with this ray name. He wrote *The Elasmobranchiate Fishes of Amoy* (1934).

Taniuchi

Whitetail Dogfish *Scymnodalatias albicauda* Taniuchi & Garrick, 1986

Longnose Sawtail Catshark *Galeus longirostris*
Tachikawa & Taniuchi, 1987

Dr Toru Taniuchi is an ichthyologist who was a professor at the Department of Fisheries, Faculty of Agriculture, University of Tokyo (1996–2000) and since 2000 has been a professor at the College of Bioresource Sciences, Nihon University. His bachelor's degree (1964) and doctorate (1969) were awarded by the Faculty of Agriculture, University of Tokyo. He wrote *Variation in the Teeth of the Sand Shark*, Odontaspis taurus *(Rafinesque) Taken from the East China Sea* (1970).

Tao

Hagfish sp. *Eptatretus cheni* S-C Shen & Tao, 1975
Hagfish sp. *Eptatretus taiwanae* S-C Shen & Tao, 1975

Hsi-Jen Tao (b.1946) is Associate Professor in the Department of Life Science, National Taiwan University. Her bachelor's degree was awarded by the National Taiwan Normal University. Her areas of special interest are comparative anatomy of the vertebrates and vertebrate paleontology. The species were described in her co-written paper, *Systematic Studies on the Hagfish (Eptatretidae) in the Adjacent Waters around Taiwan with Description of Two New Species.* (1975). She continues to publish widely, including *Fossil Chondrichthyes Fishes of Chia-hsien, Kaoshung County, Taiwan* (2008).

Taranetz

Mud Skate *Bathyraja taranetzi* Dolganov, 1983

Anatoly Yakovlevich Taranetz (1910–1941) was a Russian marine biologist and ichthyologist who was an eminent expert on Far East fishes. He graduated from the Vladivostock Industrial College (1929), becoming an 'observer' in the raw materials sector of the Pacific Fisheries Research Centre, part of the Russian Academy of Science. He then became (1932) a Marine Researcher at TIRH Complex Pacific Expedition of the State Hydrological Institute before working (1933) at the Leningrad Zoological Institute, where he presented his thesis, *Freshwater Fish of the North-Western Basin in the Sea of Japan*. He conducted a number of expeditions on the Amur River (1930s) including to Sahalin (1934). He was a 'group leader' studying salmon (1939) and edited a guide to the fishing industry of the Far East, *Guide to the Fishes of the Soviet Far East and Adjacent Waters* (1937) one of around 30 papers he wrote. He was drafted into the army (November 1941) and was killed the following

month when his echelon were destroyed by enemy aircraft. Several other fishes are named after him.

Tatiana

River Stingray sp. *Potamotrygon tatianae* JP da Silva & Carvalho, 2011

Tatiana Raso de Moraes Possato (1978–2006) was a biologist whose bachelor's degree was awarded by the Universidade de São Paulo, where she was studying for a master's degree when she died, tragically young. She had been an enthusiastic researcher of chondrichthyans, in particular potamotrygonids.

Taylor, FH

Australian Sharpnose Shark *Rhizoprionodon taylori* Ogilby, 1915

Frank Henry Taylor (1886–1945) was an entomologist at the Institute of Tropical Medicine, Townsville, Queensland, Australia. He collected the holotype. He wrote a number of entomological papers including *Medical Entomology in Australia* (1934) following his discovery of a mite that might have been a vector for Mossman fever.

Taylor, LR

Megamouth Shark Family *Megachasmidae* LR Taylor, Compagno & Struhsaker, 1983

Megamouth Shark genus *Megachasma* LR Taylor, Compagno & Struhsaker, 1983

Longnose Catshark *Apristurus kampae* LR Taylor, 1972
Mexican Horn Shark *Heterodontus mexicanus* LR Taylor & Castro-Aguirre, 1972
Megamouth Shark *Megachasma pelagios* LR Taylor, Compagno & Struhsaker, 1983

Dr Leighton R. Taylor Jr (b.1940) is a marine biologist who was Curator of Fishes, Waikiki Aquarium, Hawaii and later Director (1975–1986). He was also a professor of biology at the University of Hawaii. For a time he combined his old calling with being a vintner in the Napa Valley, California producing Cabernet Sauvignon and Merlot, but he and his wife, Linda, appear to have sold their winery (2008) and returned to live in Hawaii, where he became a member of the board of directors of a Hawaiian community group (2013). He co-wrote Holacanthus griffisi, *A New Species of Angelfish from the Central Pacific Ocean* (1981).

Teague

Freshwater Stingray sp. *Potamotrygon brumi*
Devincenzi & Teague, 1942
[Junior Syn. *Potamotrygon brachyura*]

Gerald Warren Teague (1885–1974) was a British businessman, diplomat and naturalist who was a British Vice-Consul in Uruguay and Director of the Midland Uruguay Railway Company. He was an honorary collaborator with the Natural History Museum in Montevideo. A reptile is also named after him.

Tee-Van

Prickly Brown Ray *Dipturus teevani* Bigelow & Schroeder, 1951

Spiny Skate *Bathyraja spinosissima* Beebe & Tee-Van, 1941
Ecuador Skate *Dipturus ecuadoriensis* Beebe & Tee-Van, 1941
Pacific Chupare *Himantura pacifica* Beebe & Tee-Van, 1941

John Tee-Van (1897–1967) joined the staff of the Bronx Zoo (1911) as an assistant keeper in the birdhouse. He became Beebe's assistant in the tropical research department and accompanied him on 24 expeditions. He retired having been Director of both the Bronx Zoo and the Coney Island Aquarium (1952–1962). He was honoured 'in appreciation of his helpful assistance' as editor-in-chief of the monograph series, *Fishes of the Western North Atlantic*.

Temminck

Broadfin Shark *Lamiopsis temminckii* Müller & Henle, 1839

Japanese Butterfly Ray *Gymnura japonica* Temminck & Schlegel, 1850
Japanese Sleeper Ray *Narke japonica* Temminck & Schlegel, 1850

Coenraad Jacob Temminck (1778–1858) was a Dutch ornithologist, illustrator and collector. He was the first director of the Rijksmuseum van Natuurlijke Historie in Leiden (1820–1858). He was a wealthy man who had a very large collection of specimens and live birds. His first ornithological task was cataloguing his father's (Jacob) extensive collection, after whom he may have named some birds. Le Vaillant (q.v.) collected specimens for Jacob. He issued his *Manuel d'ornithologie, ou Tableau Systém-*

atique des Oiseaux qui se Trouvent en Europe (1815) and wrote *Nouveau Recueil de Planches Coloriées d'Oiseaux* (1820). Twenty-six birds, 13 mammals, 2 reptiles, and 7 or more fishes are named after him. He 'shared the museum's treasures' with Henle and Müller during a visit (1837).

Teng

Straight-tooth Weasel Shark *Paragaleus tengi* JTF Chen, 1963

Hagfish sp. *Eptatretus yangi* Teng, 1958
Taiwan Gulper Shark *Centrophorus niaukang* Teng, 1959
Taiwan Saddled Carpetshark *Cirrhoscyllium formosanum* Teng, 1959
Smalleyed Guitarfish *Rhinobatos microphthalmus* Teng, 1959
Smalleye Pygmy Shark *Squaliolus aliae* Teng, 1959
Bigeyed Sixgill Shark *Hexanchus nakamurai* Teng, 1962

Dr Teng Huo-Tu (1911–1978) was an ichthyologist at the Taiwan Fisheries Research Institute. His doctorate was awarded by Kyoto University, Japan (1962). He was involved in the classification of chondrichthyes, especially sharks.

Thetis

Thorntail Stingray *Dasyatis thetidis* Ogilby, 1899
[Alt. Black or Longtail Stingray]

HMCS 'Thetis' was on a trawling expedition off the coast of New South Wales (1898) during which the holotype was collected. The Thetis was named after a sea nymph in Greek mythology.

Thiel

Shark genus *Planonasus* Weigmann, Stehmann & Thiel, 2013
Leg Skate genus *Indobatis* Weigmann, Stehmann & Thiel, 2014

Dwarf False Catshark *Planonasus parini* Weigmann, Stehmann & Thiel, 2013
Sparsely-thorned Skate *Rajella paucispinosa* Weigmann, Stehmann & Thiel, 2014

Dr Ralf Thiel (b. 1960) is a German marine biologist and ichthyologist whose diploma in biology (1986) and doctorate (1991) were awarded by the University of Rostock and who became (2006) Head of the Ichthyology Section and Professor (2014) at Biocenter Grindel and Zoological Museum, University of Hamburg, Germany. Previously he worked at the University of Hamburg (1995–2001), as Curator of Ichthyology at the German Oceanographic Museum, Stralsund (2001–2003) and as head of the department of marine biology there (2003–2006), a post he combined with lecturing on fish ecology at the University of Rostock (2002–2006).

Thompson

Backwater Butterfly Ray *Gymnura natalensis* Gilchrist & Thompson, 1911

William Wardlaw Thompson was a South African ichthyologist who was Gilchrist's assistant at the South African Museum. He co-wrote *The Blennidae of South Africa* (1908).

Thomsen

Jespersen's Hagfish *Myxine jespersenae* Møller, Feld, Poulsen, Thomsen & Thormar, 2005

Dr Philip Francis Thomsen (b.1979) is doing post-doctoral research at the Center for GeoGenetics, Natural History Museum of Denmark, University of Copenhagen, which had awarded his bachelor's degree (2005), master's (2008) and doctorate (2013). His research interests are mainly focused on the applications of DNA from environmental samples, in both freshwater and marine environments as well as several other sources, such as terrestrial sediments, human coprolites and leech gut contents. He also works in areas such as animal evolution, ecology and conservation. He has published a number of papers including as main author of *Monitoring Endangered Freshwater Biodiversity using Environmental DNA* (2012), *Detection of a Diverse Marine Fish Fauna Using Environmental DNA from Seawater Samples* (2012) and *Non-destructive Sampling of Ancient Insect DNA* (2009).

Thormar

Jespersen's Hagfish *Myxine jespersenae* Møller, Feld, Poulsen, Thomsen & Thormar, 2005

Jonas Thormar (b.1980) is a PhD student on seagrass ecosystems at the Department of Biosciences, University of Oslo (2010–2014). He obtained his master's degree from the Zoological Museum, University of Copenhagen (2010) on the taxonomy of the Phylum *Kinorhyncha*. He is a keen diver and underwater photographer, guiding divers and conducting scientific research in the seas from Antarctica to the high Arctic. Among his most recent papers is *Mer-*

istoderes gen. nov., A New Kinorhynch Genus, with the Description of Two New Species and their Implications for Echinoderid Phylogeny (2012).

Thorson

Freshwater Stingray genus *Plesiotrygon* Rosa, Castello & Thorson, 1987

Long-tailed River Stingray *Plesiotrygon iwamae* Rosa, Castello & Thorson, 1987

Dr Thomas B. Thorson (1917–1999) was an ichthyologist at the University of Nebraska-Lincoln from which he retired (1982) as Professor Emeritus of Zoology. St Olaf College, Minnesota awarded his bachelor's degree (1938) and the University of Washington his master's (1941). He was a captain in the US Air Force in the Second World War in Europe (1942–1945), after which he returned to the University of Nebraska (1948–1950) and completed his doctorate at the University of Washington (1952). He taught at San Francisco State University (1952–1954) and South Dakota State University (1954–1956). He returned to Nebraska as an assistant professor (1956), becoming full professor (1961). He was Chairman of the Department of Zoology (1967–1971) and Vice-Director of the School of Life Sciences (1975–1977). He wrote *Ichthyology of the Lakes of Nicaragua: Historical Perspective* (1976), in which he explained how bull sharks had adapted to life in fresh water. The book was based on his experience as consultant to Jacques Cousteau's team when they filmed in Lake Nicaragua (1974).

Thouin

Shaw's Shovelnose Guitarfish *Rhinobatos thouiniana* Shaw, 1804

André Thouin (1746–1824) was a botanist and a colleague of Lacépède's at the Jardin des Plantes (Muséum Nationale d'Histoire Naturelle), Paris. Shaw's original description has no etymology but states that it was described following a description (probably by Lacépède in 1798) of a fish called *Raja thouin*, an example of which was in Paris. Thouin was honoured as he had helped secure a specimen in Holland and transport it to France.

Thurston

Smoothtail Mobula *Mobula thurstoni* Lloyd, 1908

Dr Edgar Thurston (1855–1935) was an ethnographer, natural historian and museologist who qualified as a physician in England (1877). He was Superintendent of the Government Museum, Madras (Chennai), establishing the natural history and anthropology sections (1885–1910), during which time he gave access to Lloyd to study specimens. He returned to England (1910) and eventually settled in Cornwall, where he was a noted plant collector (1915–1926). He mainly published on ethnography but wrote a book on the amphibians of southern India. A reptile is named after him.

Tilesius

Arctic Lamprey *Lethenteron camtschaticum* Tilesius, 1811

Wilhelm Gottlieb Tilesius von Tilenau (1769–1857) was a German naturalist, explorer, physician, draughtsman and engraver. He studied medicine at the University of Leipzig, graduating with a master's (1795) and doctorates in philosophy (1797) and in medicine (1801). Studying marine animals en route, he took ship to Portugal (1796–1796), where he practised medicine. He became a professor at Moscow University (1803), during which time he also took passage as ship's doctor on the 'Nadezhda' on the Russian circumnavigation (1803–1806). Interestingly he would often play his violin during the voyage. They visited, among other places, Japan, Kamchatka, Cape Horn, the Canary Islands, Brazil and China. His illustrated account of the voyage was published (1814); his description of its natural history finds became *Fruits of the Natural History of the First Russian Circumnavigation of the Earth* (1813).

Tilston

Tilston's Whaler Shark *Carcharhinus tilstoni* Whitley, 1950
[Alt. Whitley's Blacktip Shark, Australian Blacktip Shark]

Richard Tilston was a naturalist who trained as a physician at Guy's Hospital, London (1841) and became a naval surgeon (1842). He was a Royal Navy assistant surgeon at Port Essington, New South Wales (now Northern Territory), Australia (1849), the year in which the settlement was abandoned. His will (1850) is in the Public Records Office, Kew. The type was collected at Port Essington during his time there.

Timle

Blackspotted Numbfish *Narcine timlei* Bloch & JGT Schneider, 1801

The original description has no etymology but there is a hint that 'timle' may derive from a Malay name for this fish. Another possibility contained in the description is that it refers to Tamil Nadu, where the type locality, Tranquebar, is located: '*Habitat ad Tranquebariam Timlei Malais dicta*'.

Ting

Taiwan Angelshark *Squatina formosa* S-C Shen & Ting, 1972

Wai-Hwa Ting was a Taiwanese ichthyologist who co-wrote *Ecological and Morphological Study on Fish-fauna from the Waters around Taiwan and its Adjacent Islands. 2. Notes on Some Rare Continental Shelf Fishes and Description of Two New Species* (1972).

Tobije

Japanese Eagle Ray *Myliobatis tobijei* Bleeker, 1854

The original description has no etymology but a cryptic note gives the impression that Tobijei might be a local Japanese name for this fish. In fact *tobi* is the Japanese name for a bird, the kite (referring to the ray's bird-like wingspan formed by its fused pectoral fins) and *jei* is Japanese for ray, so the name translates as 'kite ray'.

Tobituka

Leadhued Skate *Notoraja tobitukai* Hiyama, 1940

T. Tobituka directed the trawling fishery survey that collected the type specimen.

Tokarev

Multispine Giant Stingray *Dasyatis multispinosa* Tokarev, 1959

Aleksey K. Tokarev (1915–1957) was a Russian biologist who was a member of the Soviet Antarctic expeditions (1955–1956) on board the research vessel 'Ob'. He died while returning from the Antarctic. He co-wrote *Ichthyofauna, The Study of Ichthyofauna and the Goals of Investigations* (1958), which was published after his death as was the description of this species. Cape Tokarev and Tokarev Island, both in the Antarctic, are named after him.

Tokubee

Izu Catshark *Scyliorhinus tokubee* Shirai, Hagiwara & Nakaya, 1992

This species is named after the fishing boat and

private lodge of Toshiyuki Iida, who captured type specimens and is familiarly known as 'Tokubee-san', 'Tokubee' being an old-fashioned male name in Japan.

Torre

Dwarf Catshark *Scyliorhinus torrei* Howell-Rivero, 1936

Professor Carlos de la Torre y la Huerta (1858–1950) of Havana University was a malacologist and was regarded as the foremost Cuban naturalist of his generation. He was closely associated with the Smithsonian in Washington, DC before Castro took power. He was a leading figure in the Academia de Ciencias Médicas, Físicas y Naturales de la Habana. It was he who first recognised this species as new and granted Howell Rivero permission to study and describe it. An extinct Cuban mammal, a bird and a reptile are named after him.

Tortonése

Tortonése's Stingray *Dasyatis tortonesei* Capapé, 1975

Dr Enrico Tortonése (1911–1987) was an Italian ichthyologist and specialist in echinoderms. He was a professor at the Genoa Museum of Natural History. Capapé frequently cites Tortonése's work on Mediterranean sharks and rays.

Tosh

Brown Whipray *Himantura toshi* Whitley, 1939

Dr James Ramsey Tosh (1872–1917) was a marine biologist and inland waterway engineer who was employed by the government of Queensland as a fisheries expert (1900–1903). The University of St Andrews, where he was employed as an assistant to the Professor of Natural History, awarded his master's degree (1894). He was working for the British Red Cross Society in Mesopotamia (Iraq) when he succumbed and died of heat stroke. He mentions the whipray in a report he wrote (1902–1903).

Townsend

Sandpaper Skate *Bathyraja interrupta* Gill & Townsend, 1897
Blunt Skate *Rhinoraja obtuse* Gill & Townsend, 1897

Dr Charles Haskins Townsend (1859–1944) was an American zoologist who was Chief of the Fisheries Division of the US Fish Commission (1897–1902), then became Director of the New York Aquarium (1902–1937). He explored northern California

(1883–1884) and the Kobuk River, Alaska (1885). He wrote extensively on fisheries, whaling, fur sealing and deep-sea exploration, including *Guide to the New York Aquarium* (1919). A mammal, two reptiles and ten birds are named after him.

Trautman

Southern Brook Lamprey *Ichthyomyzon gagei* CL Hubbs & Trautman, 1937
Mountain Brook Lamprey *Ichthyomyzon greeleyi* CL Hubbs & Trautman, 1937
Silver Lamprey *Ichthyomyzon unicuspis* CL Hubbs & Trautman, 1937

Milton Bernhard Trautman (1899–1991) was a self-taught ornithologist and ichthyologist from Ohio. He wrote *The Fishes of Ohio* (1957). He worked for the State of Ohio Department of Fish and Game (1926–1934) and was Assistant Curator of fishes at the Museum of Zoology, University of Michigan and Assistant Director and research biologist for the Michigan Department of Conservation (1934–1939). He was a research biologist at the Franz Theodore Stone Laboratory of Ohio State University (1939–1955), Curator of vertebrates at the Ohio State Museum of the Ohio Historical Society (1955–1970) and Curator of birds at the Ohio State University Museum of Zoology (1970–1991). Lacking formal education he received an honorary PhD from the College of Wooster (1951).

Troll

Pointy-nosed Blue Chimaera *Hydrolagus trolli* Didier & Séret, 2002

Dr Raymond 'Ray' Michael Troll (b.1954) is an American artist (he calls himself a 'fin artist'), musician, humourist and illustrator. He studied art at Bethany College, Lindsborg where he was awarded a BA in printmaking (1977). He did various day jobs to finance his artwork, then (1981) went to Washington State University to do his master's (1981), where he focused on his drawing skills. When he left he took a teaching post back at Bethany College. He moved to Alaska (1983), worked in his sister's retail fish market and also taught an art class and rented a studio. He began angling and drew what he caught, including a ratfish, which became his talisman. He started printing T-shirts (1984) with fishy themes and earns a good living from it. He put together a travelling show, 'Dancing to the Fossil Record', blending science and art (including the infamous 'Evolvo' art car). It ran from San Francisco (1995), ending in

Denver, Colorado (1999). He has since been involved with other shows, including acting as art director for the Miami Museum of Science's *Amazon Voyage* traveling exhibit. He has a music radio show in Alaska where he lives and is a rock-and-roll fanatic. He has had various bands, including The Squawking Fish, Zulu and the Robot Slave Boys, and the Rapping Ratfish Brothers and now plays in The Ratfish Wranglers. His nom de soirée is 'Ratfish Ray' and his personalised number plate 'RATFSH'. His pictures inspired Didier to name the chimaera after him saying, 'It's kind of nice to name a species for someone, and I thought: here's a chance to name one for someone who's really interested. It kind of looks like him, but with facial hair.' Troll was awarded the Alaska Governor's award for the arts (2006), a gold medal for 'distinction in the natural history arts' by the Academy of Natural Sciences, Philadelphia (2007) and an honorary doctorate in fine arts by the University of Alaska Southeast (2008). Among his publications is the children's book, *Sharkabet, a Sea of Sharks from A to Z* (2002).

Tschudi

Apron Ray *Discopyge tschudii* Heckel, 1846

Baron Dr Johann Jakob von Tschudi (1818–1889) was a Swiss explorer, physician, diplomat, naturalist, hunter, anthropologist, cultural historian, language researcher and statesman. He travelled to Peru (1838) where he spent five years exploring and collecting. He was appointed Swiss ambassador to Brazil (1860–1868). He wrote *Untersuchungen über die Fauna Peruana Ornithologie* (1844–1846). He also collected the type of the ray. Ten birds, six reptiles, five amphibians and five mammals are named after him.

Tuna

Tuna's Skate *Bathyraja tunae* Stehmann, 2005

Dr María Cristina Oddone Franco, who was a post-doctoral researcher at the Instituto de Ciências Biológicas, Universidade Federal do Rio Grande, RS, Brazil, was nicknamed 'Tuna'. Her bachelor's degree was awarded by Universidad de la República (Udelar) (2000), her master's by the Universidade Federal do Rio Grande (2003) and her doctorate by Universidade Estadual Paulista (2007). She is one of the scientific editors of the *Pan-American Journal of Aquatic Sciences*. She has already written a number of papers, including *Size at Maturity of the Smallnose Fanskate* Sympterygia bonapartii *(Müller & Henle, 1841) (Pisces, Elasmobranchii, Rajidae) in the SW Atlantic* (2004).

Tuniyev

Nina'a Lamprey *Lethenteron ninae* Naseka, Tuniyev & Renaud, 2009

Dr Sako B. Tuniyev (b.1983) is a Russian biologist, ichthyologist and herpetologist who works as the leading researcher in the Sochi National Park, Russia. He graduated in biology from Kuban State University (2005), which awarded his doctorate (2008). He has written more than 40 scientific papers, including co-writing *On Distribution and Taxonomic Status of Rock Lizard* Darevskia brauneri Szczerbaki (2012).

Tzinovsky

Creamback Skate *Bathyraja tzinovskii* Dolganov, 1985

Dr Vladimir Diodorovich Tzinovskiy (Tzinovsky) (b.1946), a researcher at the P.P. Shirshov Institute of Oceanology, Moscow (part of the Russian Academy of Science), is an Arctic oceanographer who collected the holotype. During his career he undertook many oceanic expeditions including to the Arctic basin. He co-wrote *Hydrobiological Studies in the Arctic Ocean at NP-23 (May–October 1977)* (1978).

U

Ulrich

Carolina Hammerhead *Sphyrna gilberti* Quattro, Driggers, Grady, Ulrich & MA Roberts, 2013

Dr Glenn F. Ulrich (b.1944) is a fisheries biologist who was with the South Carolina Department of Natural Resources, Charleston from 1983 until he retired. In retirement he led a study for the Port Royal Sound Foundation into the red drum fish (2007–2011). He wrote *Incidental Catch of Loggerhead Turtles by South Carolina Commercial Fisheries* (1978).

Ushie

Cow Stingray *Dasyatis ushiei* Jordan & CL Hubbs, 1925

'Ushiei' is the Japanese word for 'Cow-Ray'.

V

Vaillant

Lesser Guinean Devil Ray *Mobula rochebrunei* Vaillant, 1879

Thorny River Stingray *Potamotrygon constellate* Vaillant, 1880

Atlantic Sawtail Catshark *Galeus atlanticus* Vaillant, 1888

Smallmouth Knifetooth Dogfish *Scymnodon obscurus* Vaillant, 1888

Professor Léon Louis Vaillant (1834–1914) was a French herpetologist, ichthyologist and malacologist at the Muséum National d'Histoire Naturelle, Paris. He was on four French naval expeditions aboard the 'Travailleur' (1880, 1881 and 1882) and the 'Talisman' (1883). Three reptiles and two amphibians are named after him.

Valenciennes

Smalleye Hammerhead *Sphyrna tudes* Valenciennes, 1822

Achille Valenciennes (1794–1865) was a zoologist, ichthyologist and conchologist. In his early career he classified much of Humboldt's neo-tropical collection and they became friends. He also made important contributions to parasitology. He worked as an assistant at the Muséum National d'Histoire Naturelle, Paris. He co-wrote with Cuvier (q.v.), under whom he had studied, the 22-volume *Histoire Naturelle des Poissons* (1828–1849), which he completed after Cuvier's death. Two reptiles are named after him.

Van Hasselt

Butterfly Ray genus *Gymnura* Van Hasselt, 1823
Cownose Ray genus *Rhinoptera* Van Hasselt, 1824

Dr Johan Coenraad van Hasselt (1797–1823) was a Dutch physician, zoologist, botanist and mycologist. He studied medicine at the University of Groningen. He undertook an expedition for the Netherlands Commission for Natural Sciences to Java (1820) with his friend Heinrich Kuhl, to study the fauna and flora of the island. He was the first person to climb Mount Pangrango. They left from Texel stopping at Madeira, the Cape of Good Hope and Cocos Island en route. Kuhl died after eight months; van Hasselt continued the work for another two years before he too died of disease and exhaustion. However, during their time there they sent the Museum of Leiden 200 skeletons, 200 skins of mammals from 65 species, 2,000 bird skins, 1,400 fishes, 300 reptiles and amphibians, and many insects and crustaceans. A bird and a mammal are named after him.

Vélez

Velez Ray *Raja velezi* Chirichigno, 1973

Juan José Vélez Diéguez is an ichthyologist at the Universidad Nacional del Callao, Peru where he is a professor. He was at the Instituto del Mar del Perú. He wrote *Peces Marinos Clave Artificial para Identificar los Peces Marinos Comunes en la Costa Central del Perú* (1980). He and the describer have often published together. He was honoured for his dedication to ichthyology in general and for collaborations with the author.

Vespertilio

Ornate Eagle Ray *Aetomylaeus vespertilio* Bleeker, 1852

Linnaeus named a genus of bats *Vespertilio* (1758) and used it also as a binomial *Ogcocephalus vespertilio* (1758) when describing a batfish. We can only assume that Bleeker used this term because of the association between the flight in air of a bat and in water of a ray, or for its bat-like fused pectoral wings.

Vidthayanon

Roughback Whipray *Himantura kittipongi* Vidthayanon & TR Roberts, 2005

Dr Chavalit Vidthayanon (b.1959) is an ichthyologist in Thailand, where he works for the WWF as a senior freshwater specialist. Tokyo University of Fisheries awarded his doctorate. He co-wrote *Systematic Revision of the Asian Catfish Family Pangasiidae, with Biological Observations and Descriptions of Three New Species* (1991).

Vincent

Western Shovelnose Ray *Aptychotrema vincentiana* Haacke, 1885

St. Vincent's Gulf Dogfish *Asymbolus vincenti* Zietz, 1908
[Alt Gulf catshark]

These species are named after the location where the holotypes were caught: the Gulf of St Vincent, near Kangaroo Island.

Vladykov

Po Brook Lamprey *Lethenteron zanandreai* Vladykov, 1955

Vladykov's Lamprey *Eudontomyzon vladykovi* Oliva & Zanandrea, 1959 [Alt. Danubian Brook Lamprey]

Western Brook Lamprey *Lampetra richardsoni* Vladykov & Follett, 1965

Modoc Brook Lamprey *Entosphenus folletti* Vladykov & Kott, 1976

Kern Brook Lamprey *Entosphenus hubbsi* Vladykov & Kott, 1976

Alaskan Brook Lamprey *Lethenteron alaskense* Vladykov & Kott, 1978

Lamprey sp. *Lethenteron matsubarai* Vladykov & Kott, 1978

Klamath River Lamprey *Entosphenus similis* Vladykov & Kott, 1979

Macedonia Brook Lamprey *Eudontomyzon hellenicus* Vladykov, Renaud, Kott & Economidis, 1982

Dr Vadim Dimitrievitch Vladykov (1898–1986) was a Ukrainian-born Canadian zoologist. Charles University, Prague awarded his degree in zoology and anthropology. He took a post as Assistant Scientist at the Fisheries Research Board of Canada (1930) and became a Canadian citizen (1936). He then worked at the Laboratoire de Zoologie, Université de Montréal as Professor of Ichthyology. He then became a professor at Ottawa University (1958–1973) until retirement and thereafter Professor Emeritus. He was also Research Associate at the National Museum of Natural Sciences. A prolific writer, he published more than 290 papers and contributed to longer works. He is also honoured in the name of a ship, 'Canadian Coast Guard Ship Vladykov'.

Von Bonde

Legskates *Anacanthobatidae* von Bonde & Swart, 1924

Legskate genus *Anacanthobatis* von Bonde & Swart, 1923

Spotted Legskate *Anacanthobatis marmoratus* von Bonde & Swart, 1923

Smoothnose Legskate *Cruriraja durbanensis* von Bonde & Swart, 1923

Roughnose Legskate *Cruriraja parcomaculata* von Bonde & Swart, 1923

Munchkin Skate *Rajella caudaspinosa* von Bonde & Swart, 1923

Leopard Skate *Rajella leopardus* von Bonde & Swart, 1923

Dr Cecil von Bonde (1895–1983) was a South African zoologist, ichthyologist and oceanographer. The University of Cape Town, where he became a Senior Lecturer (1918–1923), awarded his doctorate. He studied oceanography and lectured in zoology at the University of Liverpool (1924–1925). He returned to South Africa and became acting Head of the Zoology Department (1926) upon Gilchist's death. He was Director of fisheries and Government Marine Biologist (1928–1952) and was Managing Director of the Fisheries Development Corporation of South Africa (1952–1960). He and Swart together published the paper, *The Platosomia (Skates and Rays) Collected by 'SS Pickle'*, a report for the Fisheries Marine Biological Survey (1923).

Von Martens

Java Stingaree *Urolophus javanicus* von Martens, 1864

Dr Karl Eduard von Martens (1831–1904) was a German conchologist and zoologist. He was in the East Indies (Indonesia) (1860–1863). His doctorate was awarded by the University of Tübingen (1855), after which he moved to Berlin where he worked at the university and the Berlin Zoological Museum for the rest of his life. He was Professor and Director of the Berlin Zoological Museum (1883–1887) and was Curator of malacology and of other invertebrates (1864–1904). He was a member of the 'Thetis' expedition to the Far East (1860–1862) and stayed on in southeast Asia (1863–1864) before returning to Berlin. A bird is named after him.

Vongpanich

Magnificent Catshark *Proscyllium magnificum* Last & Vongpanich, 2004

Vararin Vongpanich (b.1971) is a Thai malacologist and marine biologist at the Phuket Marine Biological Center, whose master's degree in marine science was awarded by the University of Aarhus (1999). Her area of study is marine bivalves, particularly *Arcidae, Mactridae*. She wrote *Revision of Shell Morphology of Indo-Pacific Mactridae* (1999).

Vooren

Hidden Angelshark *Squatina occulta* Vooren & KG da Silva, 1992

Brazilian Blind Electric Ray *Benthobatis kreffti* Rincón, Stehmann & Vooren, 2001

Striped Rabbitfish *Hydrolagus matallanasi* Soto & Vooren, 2004

Dr Carolus Maria Vooren (b.1941) is a Dutch-born Brazilian ichthyologist and ornithologist at the Departamento de Oceanografia, Universidade Federal do Rio Grande, where he is a professor. He co-wrote *Guia de Albatrozes e Petréis do sul do Brasil* (1989) and *Distribution and Abundance of the Lesser Electric Ray* Narcine brasiliensis *(Olfers, 1831)* (Elasmobranchii: Narcinidae) *in Southern Brazil in Relation to Environmental Factors* (2009).

W

Wadd

Blind Shark *Brachaelurus waddi* Bloch & JGT
Schneider, 1801

The original description is little help. The shark was described on the basis of Latham's drawing of it (now lost); the description mentions that Dr John Latham called it 'Waddi'. It could be based on an aboriginal word 'waddi' or 'waddy' meaning a war club or type of tree, which in turn may refer to the shark's shape.

Wagner, MH

Peppered Catshark *Galeus piperatus* S Springer & MH
Wagner, 1966

Mary Hayes Wagner (fl.1967) was a biologist at the Bureau of Commercial Fisheries Ichthyological Field Station, Stanford, California (1966). She wrote *Shark Fishing Gear: A Historical Review* (1966), co-wrote *Field Guide to Eastern Pacific and Hawaiian Sharks* (1967) and was an illustrator of other people's work.

Wagner, NP

Caspian Lamprey *Caspiomyzon wagneri* Kessler, 1870

Professor Dr Nicolai Petrovitch Wagner (1829–1907) was a Russian zoologist particularly interested in entomology. He was a professor (1871–1894) at St Petersburg University where Karl Fedorovich Kessler (q.v.) was a colleague. The University of Kazan awarded his first degree (1851) and doctorate (1854) and appointed him Professor of Zoology (1860). He is most famous for discovering paedogenesis (larval reproduction) in insects (1862). During his time at St Petersburg he mainly worked on zoological research at the White Sea culminating in the publication *The White Sea Invertebrates* (1855). Later in life he became President of the Russian Society of Experimental Psychology (1891). He was also a novelist and writer of children's stories such as *Fairy Tales of Tomcat Murlyki* (1881).

Waite

Ratfish sp. *Hydrolagus waitei* Fowler, 1907
Southern Round Skate *Irolita waitii* McCulloch, 1911

Skate genus *Arhynchobatis* Waite, 1909
Sleeper Ray genus *Typhlonarke* Waite, 1909

Ratfish sp. *Hydrolagus ogilbyi* Waite, 1898
Australian Blackspot Catshark *Aulohalaelurus labiosus*
Waite, 1905
Longtail skate *Arhynchobatis asperrimus* Waite, 1909
Plunket Shark *Proscymnodon plunketi* Waite, 1910

Edgar Ravenswood Waite (1866–1928) was an all-round zoologist and museum director. After schooling he worked in an accounts office but his interest in natural history led him to read biology at Owens College, Manchester. He became Assistant Curator (1888) of the museum of the Leeds Philosophical and Literary Society (later Leeds City Museum), later (1891) becoming Curator. After marrying (1892), he went to Australia (1893) to become Assistant Curator of vertebrates at the Australian Museum (1893–1905). During this time he undertook a scientific expedition aboard the trawler 'HMCS Thetis' (1899). He was in New Zealand (1906–1914) as Curator of the Canterbury Museum, Christchurch. He was General Director of the South Australian Public Library, Museum and Art Gallery (1914–1928). He wrote *The Fishes of South Australia* (1923). He contracted malaria in New Guinea, which compromised his health; he died of enteric fever while attending a meeting in Hobart, Tasmania. There is no mention of which Waite the southern round skate was named after in his etymology but McCulloch often cited Edgar Waite's works on fishes in his writings. However, he was honoured in the name of the ratfish for contributions to Australian ichthyology. A reptile and a bird are named after him.

Walbaum

Manta Ray genus *Manta* Walbaum, 1792

Giant Oceanic Manta Ray *Manta birostris* Walbaum, 1792
Chala Guitarfish *Rhinobatos percellens* Walbaum, 1792
Dicus Ray *Paratrygon ajereba* Walbaum, 1792

Johann Julius Walbaum (1724–1799) was a physician, taxonomist and naturalist. He described a number of new taxa, particularly fishes. His extensive collection formed the basis of the Naturhistorische Museum, Lübeck, which was destroyed in the Second World War.

Walker, HJ

Hagfish sp. *Eptatretus walkeri* CB McMillan & Wisner,
2004

H. J. Walker Jr is a senior museum scientist at the Scripps Institution of Oceanography, University of California, San Diego. His research interests include the taxonomy, systematics and zoogeography of marine fishes. He has described numerous new species and has written a number of papers such as *The World's Smallest Vertebrate*, Schindleria brevipinguis, *a New Paedomorphic Species in the Family Schindleriidae* (2004).

Walker, T

Eastern Spotted Gummy Shark *Mustelus walkeri*
WT White & Last, 2008

Dr Terence I. 'Terry' Walker is an ichthyologist and expert on sharks at the Department of Primary Industries, Queenscliff, Victoria, Australia, where he was Acting Director (2008). White honoured him for 'dedicating a lifetime to the ecology and fisheries management of Australian chondrichthyans'.

Wallace

Yellow-spotted Skate *Leucoraja wallacei* Hulley, 1970

Deepwater Stingray genus *Plesiobatis* Wallace, 1967

Black Legskate *Anacanthobatis ori* Wallace, 1967
Blackspot Skate *Dipturus campbelli* Wallace, 1967
Rattail Skate *Dipturus lanceorostratus* Wallace, 1967
Roughbelly Skate *Dipturus springeri* Wallace, 1967
Prownose Skate *Dipturus stenorhynchus* Wallace, 1967
Davie's Stingaree *Plesiobatis daviesi* Wallace, 1967

Dr John H. Wallace (fl.1981) was head of research at the Oceanographic Research Institute, Durban, South Africa and later Director of the Port Elizabeth Museum. His (1970) study covering the west and south coasts is a companion to Hulley's (1967) study of east coast South African *rajiform* fishes.

Walsh

Angelshark sp. *Squatina caillieti* Walsh, Ebert &
Compagno, 2011

Jonathan H. Walsh is an American ichthyologist who was at the Pacific Shark Research Center, Moss Landing Maine Laboratories, California when this fish was described. He now works for Pacific Gas and Electricity. He wrote *A Review of the Systematics of Western North Pacific Angel Sharks, Genus Squatina, with Redescriptions of* Squatina formosa, S. japonica, *and* S. nebulosa *(Chondrichthyes: Squatiniformes, Squatinidae)* (2007).

Wanderley

Freshwater Stingray sp. *Potamotrygon boesemani*
Rosa, Carvalho & Wanderley, 2008

Cristiane Wanderley Almeida Leal (b.1985) is a Brazilian ichthyologist at the Department of Systematics and Ecology, Universidade Federal da Paraíba from where she graduated with a master's degree in biological sciences – zoology (2010). Her emphasis has been on the taxonomy of recent groups, particularly the systematics and taxonomy of neo-tropical freshwater stingrays of the Potamotrygonidae family, and the skeletal morphology of this family.

Ward

Northern Wobbegong *Orectolobus wardi* Whitley, 1939

Charles Melbourne Ward (1903–1966) was an actor, naturalist and marine collector. As his parents were entertainers his childhood was peripatetic and his education divided between Sydney and New York. He became an actor and musician (1919) and played jazz saxophone and clarinet. He became a marine zoologist after a crab was named *Cleistostoma wardi* after him (1926). He collected all over the world from Australia to the Atlantic coast of the USA. He became an honorary zoologist at the Australian Museum, Sydney (1929) where Whitley became a friend. He visited Papua New Guinea (1932) and collected ethnographic artefacts there. During the Second World War he both entertained the troops and instructed them on tropical hygiene. He opened a gallery of natural history and native art at Medlow Bath in the Blue Mountains (1943). It was described as combining 'old curiosity shop and scientific exhibits'. He suffered from diabetes and died of a coronary occlusion, leaving his scientific collections and library to the Australian Museum. He collected the type specimen.

Warren

Sixgill Sawshark *Pliotrema warreni* Regan, 1906

Professor Dr Ernest Warren (1871–1945) was an English zoologist and first Director of the Natal Museum, Pietermaritzburg (1903–1935) and was the first to send specimens of the sawshark to the BMNH. He was also involved in the establishment (1910) of Natal University College in Pietermaritzburg (now part of the University of KwaZulu-Natal). He championed the establishment of national parks in Natal (1920s–1930s). A reptile is named after him.

Wayuu

Hagfish sp. *Eptatretus wayuu* Mok, Saavedra-Diaz & Acero, 2001

This species is named after the Wayuu, the aborigines who live on the coastal region of the Guajira Peninsular, near to where the holotype was caught.

Weber

Pale Catshark *Apristurus sibogae* Weber, 1913

Siboga Skate *Fenestraja sibogae* Weber, 1913

Annandale's Skate *Rajella annandalei* Weber, 1913

Max Carl Wilhelm Weber van Bosse (1852–1937) was a German-born Dutch physician and zoologist, Director of the Zoological Museum in Amsterdam (1883), when he became a naturalised Dutch citizen. He was educated in Germany at Bonn and Berlin. He did military service in the German army, half the time as a doctor and half as a hussar. He made a voyage in the small schooner 'Willem Barents' (1881), appropriately to the Barents Sea. He combined the roles of watch-keeping officer, ship's doctor and naturalist. His wife was a skilled and learned botanist and after their marriage the Webers spent three summers in Norway, where he could dissect whales and she could collect algae, her specialty. They made a number of other voyages to Sumatra, Java, Sulawesi and Flores (1888) and to South Africa (1894). He was co-author with De Beaufort of the authoritative *The Fishes of the Indo-Australian Archipelago*. He also wrote of 'Weber's Line', an important zoogeographical line between Sulawesi and the Moluccas, which is sometimes preferred over Wallace's line (between Sulawesi and Borneo) as the dividing line between the Oriental and Australasian faunas. He dedicated his great work on fishes to his wife Anna van Bosse 'who has been always a joyful and helpful travelling-companion to me, in the extreme North, in South Africa, in the Indo-Australian Archipelago and also during the Siboga Expedition'. The 'Siboga' expedition, to Indonesian waters, was carried out under Weber's personal leadership (1899–1900). Two mammals, three reptiles, two amphibians and a bird are named after him.

Weed

Skate genus *Dactylobatus* BA Bean & Weed, 1909

Blind Torpedo *Benthobatis marcida* BA Bean & Weed, 1909

Skilletskate *Dactylobatus armatus* BA Bean & Weed, 1909

Alfred Cleveland Weed (1881–1953) was an ichthyologist whose bachelor's degree was awarded by Cornell University. He was at the Field Museum, Chicago (1920–1942) and became Assistant Curator, ichthyology and herpetology (1920–1922) and Assistant Curator for ichthyology only (1922). He was on an expedition to Greenland (1927 or 1928) and was later employed at the Smithsonian as an ichthyologist. He was a member of the Washington Biologists' Field Club (1912–1945). He wrote *The Alligator Gar* (1923).

Weigmann

Shark genus *Planonasus* Weigmann, Stehmann & Thiel, 2013

Leg Skate genus *Indobatis* Weigmann, Stehmann & Thiel, 2014

Dwarf False Catshark *Planonasus parini* Weigmann, Stehmann & Thiel, 2013

Human's Whaler Shark *Carcharhinus humani* WT White & Weigmann, 2014

Sparsely-thorned Skate *Rajella paucispinosa* Weigmann, Stehmann & Thiel, 2014

Simon Weigmann (b.1985) is a German biologist and ichthyologist at the Biocenter Grindel and Zoological Museum, University of Hamburg, Germany, where he was awarded his bachelor's (2008) and master's (2010) degrees and where he is now a PhD student, having previously been a student assistant for the Senckenberg Museum, Frankfurt-am-Main. He wrote *Contribution to the Taxonomy and Distribution of Eight Ray Species* (Chondrichthyes, Batoidea) *from Coastal Waters of Thailand* (2011) and is lead author of *Contribution to the Taxonomy and Distribution of Pristiophorus nancyae* (Elasmobranchii: Pristiophoriformes) *from the Deep Western Indian Ocean* (2014). He has just (August 2014) published Rajella paucispinosa *n. sp., a New Deep-water Skate* (Elasmobranchii, Rajidae) *from the Western Indian Ocean off South Mozambique, and a Revised Generic Diagnosis* (2014).

Weng

Weng's Skate *Dipturus wengi* Séret & Last, 2008

Dr Herman Ting-Chen Weng is a Queensland fisheries biologist. The University of Adelaide, South Australia awarded his doctorate (1971). He wrote *The Black Bream*, Acanthopagrus butcheri (*Munro): Its Life History and Fishery in South Australia* (1971). He established his own company in Brisbane, Fishing

Weng Publications, through which he published and marketed a number of his own works including *Wonderful Fishes* (1982), a book for children. He was honoured as he 'showed an enthusiastic interest in skates and collected the first validated Australian specimens of this species in 1983'.

Werner

Chupare Whipray *Himantura schmardae* Werner, 1904

Professor Dr Franz J. M. Werner (1867–1939) was an Austrian explorer, zoologist and herpetologist who taught at the Natural History Museum, Vienna. Here the Director, Steindachner (q.v.) disliked him and forbade him access to the herpetological collection. He collected in North and East Africa, being in Egypt (1904) and the Sudan (1905), and made regular visits south to Uganda and west to Morocco until the outbreak of the First World War. His publications include *Amphibien und Reptilien* (1910). He described dozens of amphibians, reptiles and a few fishes. Twenty-eight reptiles, six amphibians and several arachnids and fishes are named after him.

White, EI

Sawfish genus *Anoxypristis* EI White & Moy-Thomas, 1941

Dr Errol Ivor White CBE (1901–1985) was a British geologist. He graduated with his major being geology (1921). He went on to achieve a PhD (1927) and DSc (1936). He was at the BMNH for over forty years (1922–1966), rising to Deputy Keeper (1938) and Keeper of Palaeontology (1955) with a particular interest in fossil fishes. He was on a number of trips for the Museum starting (1929) with one to Madagascar. He was elected to the Royal Society (1956) and was President of the Linnaean Society. He died of cancer.

White, WT

Bali Catshark *Atelomycterus baliensis* WT White, Last & Dharmadi, 2005

Australian Weasel Shark *Hemigaleus australiensis* WT White, Last & Compagno, 2005

Australian Grey Smooth-Hound *Mustelus ravidus* WT White & Last, 2006

White-fin Smooth-Hound *Mustelus widodoi* WT White & Last, 2006

Indonesian Shovelnose Ray *Rhinobatos penggali* Last, WT White & Fahmi, 2006

Southern Mandarin Dogfish *Cirrhigaleus australis* WT White, Last & Stevens, 2007

Indonesian Speckled Catshark *Halaelurus maculosus* WT White, Last & Stevens, 2007

Rusty Catshark *Halaelurus sellus* WT White, Last & Stevens, 2007

Eastern Highfin Spurdog *Squalus albifrons* Last, WT White & Stevens, 2007

Western Highfin Spurdog *Squalus altipinnis* Last, WT White & Stevens, 2007

Greeneye Spurdog *Squalus chloroculus* Last, WT White & Motomura, 2007

Edmund's Spurdog *Squalus edmundsi* Last, WT White & Stevens, 2007

Eastern Longnose Spurdog *Squalus grahami* Last, WT White & Stevens, 2007

Indonesian Shortsnout Spurdog *Squalus hemipinnis* WT White, Last & Yearsley, 2007

Western Longnose Spurdog *Squalus nasutus* Last, Marshall & WT White, 2007

Bartail Spurdog *Squalus notocaudatus* Last, WT White & Stevens, 2007

Bighead Catshark *Apristurus bucephalus* WT White, Last & Pogonoski, 2008

Whitefin Swellshark *Cephaloscyllium albipinnum* Last, Motomura & WT White, 2008

Cook's Swellshark *Cephaloscyllium cooki* Last, Séret & WT White, 2008

Australian Reticulate Swellshark *Cephaloscyllium hiscosellum* WT White & Ebert, 2008

Painted Swellshark *Cephaloscyllium pictum* Last, Séret & WT White, 2008

Flagtail Swellshark *Cephaloscyllium signourum* Last, Séret & WT White, 2008

Speckled Swellshark *Cephaloscyllium speccum* Last, Séret & WT White, 2008

Saddled Swellshark *Cephaloscyllium variegatum* Last & WT White, 2008

Narrowbar Swellshark *Cephaloscyllium zebrum* Last & WT White, 2008

Western Gulper Shark *Centrophorus westraliensis* WT White, Ebert & Compagno, 2008

Southern Dogfish *Centrophorus zeehaani* WT White, Ebert & Compagno, 2008

Eastern Australian Sawshark *Pristiophorus peroniensis* Yearsley, Last & WT White, 2008

Whitefin Chimaera *Chimaera argiloba* Last, WT White & Pogonoski, 2008

Southern Chimaera *Chimaera fulva* Didier, Last & WT White, 2008

Longspine Chimaera *Chimaera macrospina* Didier, Last & WT White, 2008

Shortspine Chimaera *Chimaera obscura* Didier, Last & WT White, 2008

Dwarf Black Stingray *Dasyatis parvonigra* Last & WT White, 2008

Deepwater Skate *Dipturus acrobelus* Last, WT White &
Pogonoski, 2008

Pale Tropical Skate *Dipturus apricus* Last, WT White &
Pogonoski, 2008

Heald's Skate *Dipturus healdi* Last, WT White &
Pogonoski, 2008

Blacktip Skate *Dipturus melanospilus* Last, WT White &
Pogonoski, 2008

Queensland Deepwater Skate *Dipturus
queenslandicus* Last, WT White & Pogonoski, 2008

Northern River Shark *Glyphis garricki* Compagno,
WT White & Last, 2008

White Spotted Gummy Shark *Mustelus stevensi*
WT White & Last, 2008

Eastern Spotted Gummy Shark *Mustelus walkeri*
WT White & Last, 2008

Peppered Maskray *Neotrygon picta* Last & WT White,
2008

Network Wobbegong *Orectolobus reticulatus* Last,
Pogonoski & WT White, 2008

Tropical Sawshark *Pristiophorus delicatus* Yearsley,
Last & WT White, 2008

Eastern Angelshark *Squatina albipunctata* Last &
WT White, 2008

Indonesian Angelshark *Squatina legnota* Last &
WT White, 2008

Western Angelshark *Squatina pseudocellata* Last &
WT White, 2008

Indonesian Houndshark *Hemitriakis indroyonoi*
WT White, Compagno & Dharmadi, 2009

Borneo River Shark *Glyphis fowlerae* Compagno,
WT White & Cavanagh, 2010

Ningaloo Maskray *Neotrygon ningalooensis* Last,
WT White & Puckridge, 2010

Indonesian Wobbegong *Orectolobus leptolineatus*
Last, Pogonoski & WT White, 2010

Dogfish Shark sp. *Squalus formosus* WT White &
Iglésias, 2011

Merauke Stingray *Dasyatis longicauda* Last &
WT White, 2013

Stingray sp. *Himantura javaensis* Last & WT White, 2013

Naru Eagle Ray *Aetobatus narutobiei* WT White,
Furumitsu & Yamaguchi, 2013

Human's Whaler Shark *Carcharhinus humani* White
WT & Weigmann, 2014

Dr William Toby White (b.1977) is an ichthyologist
at the CSIRO fish collection (Marine & Atmospheric
Research), Hobart, Tasmania. He has worked on
the ecology, taxonomy and biogeography of sharks
and rays for over 12 years but in recent years he has
focused on the shark and ray resources of Australia,
Indonesia and Papua New Guinea. He has published numerous scientific articles, books and book
chapters on these topics, including co-writing *Sharks
and Rays of Borneo* (2010).

Whitley

Wedgenose Skate *Dipturus whitleyi* Iredale, 1938

Whitley's Blacktip Shark *Carcharhinus tilstoni* Whitley,
1950

[Alt. Tilston's Whaler Shark, Australian Blacktip Shark]

Sawtail Catshark genus *Figaro* Whitley, 1928

Requiem Shark genus *Rhizoprionodon* Whitley, 1929

Softnose Skate genus *Irolita* Whitley, 1931

Catshark genus *Asymbolus* Whitley, 1939

Peacock Skate genus *Pavoraja* Whitley, 1939

Cobbler Wobbegong genus *Sutorectus* Whitley, 1939

Skate genus *Zearaja* Whitley, 1939

Lemon Shark genus *Negaprion* Whitley, 1940

Whiskery Shark genus *Furgaleus* Whitley, 1951

Australian Sawtail Catshark *Figaro boardmani*
Whitley, 1928

[Alt. Banded Shark]

Yellowback Stingaree *Urolophus sufflavus* Whitley,
1929

Indonesian Greeneye Spurdog *Squalus montalbani*
Whitley, 1931

Shorttail Torpedo *Torpedo macneilli* Whitley, 1932

Graceful Shark *Carcharhinus amblyrhynchoides*
Whitley, 1934

Australian Marbled Catshark *Atelomycterus macleayi*
Whitley, 1939

Coate's Shark *Carcharhinus coatesi* Whitley, 1939

White-spotted Skate *Dipturus cerva* Whitley, 1939

Moller's Slendertail Lantern Shark *Etmopterus molleri*
Whitley, 1939

Brown Whipray *Himantura toshi* Whitley, 1939

Blackfin Ghost Shark *Hydrolagus lemurs* Whitley, 1939

Freckled Skate *Leucoraja garmani* Whitley, 1939

Northern Wobbegong *Orectolobus wardi* Whitley,
1939

White-spotted Wedgefish *Rhynchobatus australiae*
Whitley, 1939

Southern Sleeper Shark *Somniosus antarcticus*
Whitley, 1939

Western Shovelnose Stingaree *Trygonoptera mucosa*
Whitley, 1939

Greenback Skate *Dipturus gudgeri* Whitley, 1940

Hale's Wobbegong *Orectolobus halei* Whitley, 1940

Creek Whaler *Carcharhinus fitzroyensis* Whitley, 1943

Whiskery Shark *Furgaleus macki* Whitley, 1943

Nervous Shark *Carcharhinus cautus* Whitley, 1945

Papual Epaulette Carpetshark *Hemiscyllium hallstromi*
  Whitley, 1967
Hooded Carpetshark *Hemiscyllium strahani* Whitley,
  1967

Gilbert Percy Whitley (1903–1975) was a British-born Australian ichthyologist and malacologist. Born and educated in England, he migrated to Australia with his family (1921) and joined the staff of the Australian Museum, Sydney (1922), while studying zoology at Sydney Technical College and the University of Sydney. He was appointed ichthyologist (1925) (later Curator of Fishes) at the Museum, a position he held until retirement (1964). During his term of office he doubled the size of the ichthyological collection to 37,000 specimens through many collecting expeditions. He was also a major force in the Royal Zoological Society of New South Wales, of which he was made a fellow (1934) and which he served as President (1940–1941, 1959–1960 and 1973–1974). He also edited its publications (1947–1971). He is commemorated by the Royal Zoological Society of New South Wales Whitley Awards for excellence in zoological publications relating to Australasian fauna. The name of the skate was named after him as a replacement name (which was preoccupied) for the *Raja scabra*, which Whitley had used in a recent article: 'Mr. Whitley's oversight', wrote Iredale, 'is the more remarkable as he and I pride ourselves that we carefully check all of our references many times, yet even with our meticulousness errors may slip through.' Later, Whitley wrote that Iredale 'supplied a barrage of vigorous criticism'.

Whitney

Humpack Smooth-Hound *Mustelus whitneyi*
  Chirichigno, 1973

Dr Richard R. Whitney (1927–2011) was a fisheries biologist. The University of Utah awarded his bachelor's degree (1949) and his master's (1951) and Iowa State University his doctorate (1955). He was at the University of California, Los Angeles as a research biologist (1954–1957), at the Chesapeake Biological Laboratory (1957–1960), at the Bureau of Commercial Fisheries (1960–1967), unit leader at the Washington Cooperative Fishery Research Unit (1967–1983) and Professor, School of Fisheries, University of Washington, Seattle (1983–1993), where he is now Professor Emeritus. He co-wrote *Inland Fishes of Washington* (2003). He was honoured for 'teachings and guidance in the study of sharks'.

Widodo

White-fin Smooth-Hound *Mustelus widodoi* WT White
  & Last, 2006

Dr Johannes A. O. Widodo (b.1944) is a biologist at the Research Institute of Marine Fisheries, Jakarta, Indonesia and has done much work on the shark and ray fisheries of Indonesia. He wrote: *A Check-list of Fishes Collected by Multiara 4 from November 1974 to November 1975* (1976). White said that Widodo's 'research on the shark and ray fisheries of Indonesia has provided important baseline data for this important faunal region'.

Wilms

Lana's Sawshark *Pristiophorus lanae* Ebert & Wilms, 2013

Hana A. Wilms was a fisheries technician at the Marine Biology Department, University of California, Santa Cruz, California, the university that awarded her marine biology bachelor's degree (2014). She has worked as an intern at the Oceans Research Institute (2011), The National Parks Service (2011) and the Ballona Institute (2010), and is now with the Department of Fish and Wildlife.

Wisner

Hagfish sp. *Paramyxine wisneri* Kuo, Huang & Mok, 1994
Hagfish sp. *Eptatretus wisneri* CB McMillan, 1999

Hagfish sp. *Nemamyxine kreffti* CB McMillan & Wisner,
  1982
Giant Hagfish *Eptatretus carlhubbsi* CB McMillan &
  Wisner, 1984
Hagfish sp. *Eptatretus laurahubbsae* CB McMillan &
  Wisner, 1984
Hagfish sp. *Eptatretus strahani* CB McMillan & Wisner,
  1984
Hagfish sp. *Eptatretus nanii* Wisner & CB McMillan,
  1988
Guadaloupe Hagfish *Eptatretus fritzi* Wisner & CB
  McMillan, 1990
Shorthead Hagfish *Eptatretus mcconnaugheyi* Wisner
  & CB McMillan, 1990
Cortez Hagfish *Eptatretus sinus* Wisner & CB McMillan,
  1990
Hagfish sp. *Myxine debueni* Wisner & CB McMillan,
  1995
Hagfish sp. *Myxine dorsum* Wisner & CB McMillan,
  1995
Hagfish sp. *Myxine fernholmi* Wisner & CB McMillan,
  1995

Hagfish sp. *Myxine hubbsi* Wisner & CB McMillan, 1995

Hagfish sp. *Myxine hubbsoides* Wisner & CB McMillan, 1995

Hagfish sp. *Myxine knappi* Wisner & CB McMillan, 1995

Hagfish sp. *Myxine mccoskeri* Wisner & CB McMillan, 1995

Hagfish sp. *Myxine pequenoi* Wisner & CB McMillan, 1995

Hagfish sp. *Myxine robinsorum* Wisner & CB McMillan, 1995

Hagfish sp. *Eptatretus fernholmi* CB McMillan & Wisner, 2004

Hagfish sp. *Eptatretus moki* CB McMillan & Wisner, 2004

Hagfish sp. *Eptatretus walkeri* CB McMillan & Wisner, 2004

Robert Lester Wisner (1912–2005) was an American ichthyologist. After military service San Diego State University awarded his bachelor's degree (1947). He worked at the Scripps Institution of Oceanography (1947) for his entire professional life. While there he took part in numerous collecting voyages including observing a nuclear detonation. He has written many papers and longer works, including *The Taxonomy and Distribution of Lanternfishes (Family Myctophidae) of the Eastern Pacific Ocean* (2011). He was McMillan's colleague and she honoured him in the name of the hagfish for his invaluable assistance with her hagfish research and other contributions to ichthyology. He was also a keen fly fisherman.

Wongratana

Hagfish sp. *Eptatretus indrambaryai* Wongratana, 1983

Dr Thosaporn Wongratana is a Thai ichthyologist who is Professor Emeritus at the Department of Biology, Faculty of Science, Chulalongkorn University and Curator at the University's Museum of Natural History. Kasetsart University awarded his first degree and the University of London his doctorate. He also studied at both the Smithsonian and the Australian Museum, Sydney and has carried out studies for the Food and Agriculture Organisation of the United Nations. He wrote many papers describing new fish species (1977–1988) such as *Diagnoses of 24 New Species and Proposal of a New Name for a Species of Indo-Pacific Clupeoid Fishes* (1983). He has undertaken a number of pelagic research trips, particularly to the Indian Ocean.

Wu

Wu's Skate *Dipturus wuhanlingi* Jeong & Nakabo, 2008

Skate sp. *Okamejei mengae* Jeong, Nakabo & Wu, 2007

Dr Wu Han-Lin is Professor of Ichthyology at Shanghai Ocean University and Director of the Fisheries Laboratory (2012), having previously been at Shanghai Fisheries University (2000). He co-wrote *Fauna Sinica, Chinese Fishes of the Suborder Gobioidei* (2010). He was honoured in the skate name for his 'great contributions' to Chinese ichthyology.

X

Xiong

East China Legskate *Anacanthobatis donghaiensis* Deng, Xiong & Zhan, 1983

Fat Catshark *Aprocturus pinguis* Deng, Xiong & Zhan, 1984

Gulper Shark sp. *Centrophorus robustus* Deng, Xiong & Zhan, 1985

Shortnose Demon Catshark *Apristurus internatus* Deng, Xiong & Zhan, 1988

Xiong Guo-Qiang was (fl.1981–1983) an ichthyologist at the East China Sea Fisheries Institute, National Bureau of Aquatic Products of China, which is mainly devoted to fishery research activities for the East China Sea and Yangtze River estuary. He co-wrote *Two New Species of Deep Water Sharks from the East China Sea* (1985).

Y

Yagolkowski

Otorongo Ray *Potamotrygon castexi* Castello &
Yagolkowski, 1969

Daniel Ricardo Yagolkowski (b.1947) is an Argentinian marine biologist, educationalist and public translator for English; he also translates from Portuguese and French. He has translated into Spanish a number of Arthur C. Clarke's science fiction novels and Michael Crichton's *Jurassic Park*. He studied zoology at the Universidad de Buenos Aires and held a scholarship at the Universidad del Salvador where he studied sharks and stingrays under Castex (q.v.). He started courses in audiovisual translation; these became very popular and others have since copied him by starting to lecture on the subject. He has also conducted research into and published on artificial intelligence. When most of his papers got lost, he decided to switch to teaching and translating, but before abandoning research in biology he had written a number of papers, including *Redescripción de* P. schroederi, *Yépez 1957 Designación de nos Paratipos de* P. pauckei. *Castex 1963 y Descripción de los Mimos* (1970).

Yamaguchi

Naru Eagle Ray *Aetobatus narutobiei* WT White,
Furumitsu & Yamaguchi, 2013

Dr Atsuko Yamaguchi is an ichthyologist at the Faculty of Fisheries, Nagasaki University, Nagasaki, Japan, where she is Professor of Fisheries. She co-wrote *Age, Growth and Age at Sexual Maturity of Fan Ray* Platyrhina sinensis *(Batoidea: Platyrhinidae) in Ariake Bay, Japan* (2008).

Yang

Hagfish sp. *Eptatretus yangi* Teng, 1958
Taiwanese Blind Electric Ray *Benthobatis yangi*
Carvalho, Compagno & Ebert, 2003

Hung-Chia (Hung-Jia) Yang was an ichthyologist and highly talented illustrator working at the Taiwanese Fisheries Research Institute, Kaohsiung. He carried out extensive research into Taiwanese cartilaginous fishes and was honoured for this and his 'superb fish illustrations'.

Yano

Japanese Velvet Dogfish *Zameus ichiharai* Yano &
Tanaka, 1984
Japanese Roughshark *Oxynotus japonicus* Yano &
Murofushi, 1985
Splendid Lantern Shark *Etmopterus splendidus* Yano,
1988
Circle-blotch Pygmy Swellshark *Cephaloscyllium
circulopullum* Yano, Ahmad & Gambang, 2005
Sarawak Pygmy Swellshark *Cephaloscyllium
sarawakensis* Yano, Ahmad & Gambang, 2005

Dr Kazunari Yano (1956–2006) was a leading marine biologist and ichthyological researcher. Tokai University, Japan awarded his doctorate (1986). He wrote *Biology of the Megamouth Shark* (1997). He died from a brain tumour.

Yearsley

Skate genus *Insentiraja* Yearsley & Last, 1992

Eastern Looseskin Skate *Insentiraja laxipella* Yearsley
& Last, 1992
Roughnose Stingray *Pastinachus solocirostris* Last,
Manjaji & Yearsley, 2005
Kapala Stingaree *Urolophus kapalensis* Yearsley &
Last, 2006
Fatspine Spurdog *Squalus crassispinus* Last, Edmunds
& Yearsley, 2007
Indonesian Shortsnout Spurdog *Squalus hemipinnis*
WT White, Last & Yearsley, 2007
Sandy Skate *Pavoraja arenaria* Last, Mallick & Yearsley,
2008
Mosaic Skate *Pavoraja mosaic* Last, Mallick & Yearsley,
2008
False Peacock Skate *Pavoraja pseudonitida* Last,
Mallick & Yearsley, 2008
Dusky Skate *Pavoraja umbrosa* Last, Mallick &
Yearsley, 2008
Tropical Sawshark *Pristiophorus delicatus* Yearsley,
Last & WT White, 2008
Eastern Australian Sawshark *Pristiophorus peroniensis*
Yearsley, Last & WT White, 2008
Yellow Shovelnose Stingaree *Trygonoptera galba* Last
& Yearsley, 2008

Gordon Kenneth Yearsley (b.1966) was an ichthyologist and fisheries taxonomist at CSIRO Marine Research, Hobart, Tasmania, which he joined in 1986. He is now following various systematic research interests as an honorary fellow. The University of Tasmania awarded his bachelor's degree (1988). He has written or co-written many papers and several books including contributing to *Sharks and Rays of Borneo* (2010).

Yépez

Maracaibo River Stingray *Potamotrygon yepezi* Castex & Castello, 1970

(See **Fernández-Yépez**)

Yorozu

Pinocchio Catshark *Apristurus australis* Sato, Nakaya & Yorozu, 2008

Michikazu Yorozu worked at Yokohama Hakkeijima Sea Paradise, a leisure park also housing one of the top aquariums in Japan.

Z

Zanandrea

Po Brook Lamprey *Lethenteron zanandreai* Vladykov,
1955

Vladykov's Lamprey *Eudontomyzon vladykovi* Oliva
& Zanandrea, 1959 [Alt. Danubian Brook Lamprey]

Dr Giuseppe Zanandrea was a Jesuit priest at the
Instituto di Anatomia Comparata della Università di
Bologna. He was honoured as he had made several
interesting biometrical and biological studies of
lampreys from northern Italy. He wrote *Speciation
Among Lampreys* (1959).

Zantedeschi

Taillight Shark *Euprotomicroides zantedeschia* Hulley
& Penrith, 1966

The South African arum lily is called *Zantedeschia
aethiopica* after the Italian physician. The botanical genus *Zantedeschia* was named by the German
botanist Kurt Sprengel after Giovanni Zantedeschi
(1773–1846), an Italian physician and botanist who
discovered several new species. The botanical
museum in Molina (*Museo Botanico della Lessinia di
Molina*) is dedicated to him. The trawler that caught
the first shark specimen was called 'Arum' after
the lily and the shark was, in turn, named after the
trawler! The shark derives its name via a convoluted
route!

Zeehaan

Southern Dogfish *Centrophorus zeehaani* WT White,
Ebert & Compagno, 2008

Zeehaan is the name of an Australian trawler that
caught a giant squid *Architeuthis dux* near Portland,
Victoria (2008) and from which the first specimens
of the dogfish were collected from Tasmanian waters
(1979). The unusual catch weighed about 250 kilo-
grams.

Zhan

East China Leg Skate *Anacanthobatis donghaiensis* Deng,
Xiong & Zhan, 1983

Fat Catshark *Apristurus pinguis* Deng, Xiong & Zhan, 1983

Shortnose Demon Catshark *Apristurus internatus*
Deng, Xiong & Zhan, 1988

Gulper Shark sp. *Centrophorus robustus* Deng, Xiong
& Zhan, 1985

Zhan Hong-Xi is a Chinese ichthyologist at the East
China Sea Fisheries Institute, National Bureau of
Aquatic Products of China. He co-wrote *On Three
New Species of Sharks of the Genus Carcharhinus from
China* (1981).

Zhang

Yellow-spotted Fanray *Platyrhina tangi* Iwatsuki,
Zhang & Nakaya, 2011

Dr Zhang Jie is a Chinese ichthyologist at the Institute of Zoology, Chinese Academy of Sciences,
Beijing where she was a post-doctoral fellow
(2003–2006) and subsequently Assistant Professor
(2006–2009) and Associate Professor (2010). Her
bachelor's degree was awarded by Huazhong Agricultural University (1990), after which she was an
assistant engineer, Beijing Fishery Science Institute
(1990–1995). Her master's (1998) and doctorate
(2001) were both awarded by Nagasaki University,
Japan. She wrote *Biodiversity and Conservation of
Salangidae Species* (2008).

Zhu

South China Cookiecutter Shark *Isistius labialis* Meng,
Zhu & Li, 1985

(See **Chu**)

Zietz

Gulf Catshark *Asymbolus vincenti* Zietz, 1908

Amandus Heinrich Christian Zietz (1840–1921) was
a Danish-born Australian ichthyologist. He started
working life as a teacher but did not enjoy it and
turned to his passion – natural history. After getting
a job at the Godefroi Museum as a collector, he
moved to the Kiel Museum where he worked for a
number of years, particularly on their fish collection,
and was appointed Curator. He was invited to join
the staff of the Adelaide Museum (1884), eventually
rising to Assistant Director before retiring (1910). He
took part in a number of collecting expeditions in
Australia notably in New South Wales with his son
(1906). He was always interested in birds and fishes
and published a number of papers. Two birds are
named after him.

Zuge

Pale-edged Stingray *Dasyatis zugei* Müller & Henle, 1841

This species binomial *zugei* is after *zugu-ei*, the Japanese name for this species.

Zugmayer

Baluchistan Torpedo *Torpedo zugmayeri* Engelhardt, 1912

Professor Dr Erich Johann Georg Zugmayer (1879–1938) was an Austrian explorer, zoologist, ichthyologist and herpetologist at the Bavarian State Zoological Collection, Munich. He visited Iceland (1902). He explored the area around Lake Urmia, Persia (Iran), and collected in Tibet, Ladakh and Baluchistan (then part of India, now of Pakistan) where he collected the holotype. He wrote, or co-wrote, a number of papers and longer works including *Die Fische von Balutschistan* (1913). Two reptiles, an amphibian and two birds are named after him.

APPENDIX

Scientific name	Common name	Describer
Aculeola	Dogfish genus	de Buen
Aculeola nigra	Hooktooth Dogfish	de Buen
Aetobatus	Shark genus	Blainville
Aetobatus flagellum	Longheaded Eagle Ray	Bloch
Aetobatus guttatus	Sharpwing Eagle Ray	Shaw
Aetobatus narinari	Spotted Eagle Ray	Euphrasen
Aetobatus narutobiei	Naru Eagle Ray	White WT
Aetobatus ocellatus	Ocellated Eagle Ray	Kuhl
Aetomylaeus	Eagle Ray genus	Garman
Aetomylaeus maculatus	Mottled Eagle Ray	Gray
Aetomylaeus milvus	Eagle Ray	Müller
Aetomylaeus nichofii	**Nieuhof's Eagle Ray**	Bloch
Aetomylaeus vespertilio	Ornate Eagle Ray	Bleeker
Alopias	Thresher Shark genus	Rafinesque
Alopias pelagicus	Pelagic Thresher	Nakamura
Alopias vulpinus	Common Thresher	Bonaterre
Alopiidae	Thresher Shark family	Bonaparte
Amblyraja	Skate genus	Malm
Amblyraja badia	Broad Skate	Garman
Amblyraja doellojuradoi	Southern Thorny Skate	Pozzi
Amblyraja frerichsi	Thickbody Skate	Krefft G
Amblyraja georgiana	Antarctic Starry Skate	Norman
Amblyraja hyperborean	Arctic Skate	Collett
Amblyraja jenseni	Shorttail Skate	Bigelow
Amblyraja radiata	Starry Skate	Donovan
Amblyraja reversa	Reversed Skate	Lloyd
Amblyraja robertsi	**Roberts Bigmouth Skate**	Hulley
Amblyraja taaf [Syn. *Raja taaf*]	Whiteleg Skate	Meisner
Anacanthobatidae	Leg Skate family	von Bonde
Anacanthobatis	Leg Skate genus	von Bonde
Anacanthobatis americanus	American Leg Skate	Bigelow
Anacanthobatis donghaiensis	East China Leg Skate	Deng
Anacanthobatis folirostris	Leaf-nose Leg Skate	Bigelow
Anacanthobatis longirostris	Longnose Leg Skate	Bigelow
Anacanthobatis marmoratus	Spotted Leg Skate	von Bonde
Anacanthobatis nanhaiensis	South China Leg Skate	Meng
Anacanthobatis stenosoma	Narrow Leg Skate	Li
Anoxypristis	Sawfish genus	White EI

Scientific name	Common name	Describer
Anoxypristis cuspidata	Knifetooth Sawfish	Latham
Apristurus	Catshark genus	Garman
Apristurus albisoma	Catshark sp.	Nakaya
Apristurus ampliceps	Roughskin Catshark	Sasahara
Apristurus aphyodes	White Ghost Catshark	Nakaya
Apristurus australis	**Pinocchio Catshark**	Sato
Apristurus brunneus	Brown Catshark	Gilbert
Apristurus bucephalus	Bighead Catshark	White WT
Apristurus canutus	Hoary Catshark	Springer S
Apristurus exsanguis	Flaccid Catshark	Sato
Apristurus fedorovi	**Fedorov's Catshark**	Dolganov
Apristurus garricki	**Garrick's Catshark**	Sato
Apristurus gibbosus	Humpback Catshark	Chu
Apristurus herklotsi	Longfin Catshark	Fowler HW
Apristurus indicus	Smallbelly Catshark	Brauer
Apristurus internatus	Shortnose Demon Catshark	Deng
Apristurus investigatoris	Broadnose Catshark	Misra
Apristurus japonicus	Japanese Catshark	Nakaya
Apristurus kampae	Longnose Catshark	Taylor LR
Apristurus laurussonii	Iceland Catshark	Sæmundsson
Apristurus longicephalus	Longhead Catshark	Nakaya
Apristurus macrorhynchus	Flathead Catshark	Tanaka I
Apristurus macrostomus	Broadmouth Catshark	Meng
Apristurus manis	Ghost Catshark	Springer S
Apristurus melanoasper	Black Roughscale Catshark	Iglesias
Apristurus microps	Smalleye Catshark	Gilchrist
Apristurus micropterygeus	Smalldorsal Catshark	Meng
Apristurus nakayai	Milk-eye Catshark	Iglesias
Apristurus nasutus	Largenose Catshark	de Buen
Apristurus parvipinnis	Smallfin Catshark	Springer S
Apristurus pinguis	Fat Catshark	Deng
Apristurus platyrhynchus	Borneo Catshark	Tanaka I
Apristurus profundorum	Deep-water Catshark	Goode
Apristurus riveri	**Rivero's Catshark**	Bigelow
Apristurus saldanha	Saldanha Catshark	Barnard
Apristurus sibogae	Pale Catshark	Weber
Apristurus sinensis	South China Catshark	Chu
Apristurus sp. X	**Galbraith's Catshark**	Undescribed
Apristurus spongiceps	Spongehead Catshark	Gilbert
Apristurus stenseni	Panama Ghost Catshark	Springer S
Aptychotrema	Guitarfish genus	Norman
Aptychotrema bougainvillii	Short-snouted Shovelnose Ray	Müller

Scientific name	Common name	Describer
Aptychotrema rostrata	Eastern Shovelnose Ray	Shaw
Aptychotrema timorensis	Spotted Shovelnose Ray	Last
Aptychotrema vincentiana	Western Shovelnose Ray	Haacke
Arhynchobatis	Skate genus	Waite
Arhynchobatis asperrimus	Longtail skate	Waite
Asymbolus	Catshark genus	Whitley
Asymbolus analis	Australian Spotted Catshark	Ogilby
Asymbolus funebris	Blotched Catshark	Compagno
Asymbolus galacticus	Starry Catshark	Séret
Asymbolus occiduus	Western Spotted Catshark	Last
Asymbolus pallidus	Pale Spotted Catshark	Last
Asymbolus parvus	Dwarf Catshark	Compagno
Asymbolus rubiginosus	Orange Spotted Catshark	Last
Asymbolus submaculatus	Variagated Catshark	Compagno
Asymbolus vincenti	**St Vincent's Gulf Dogfish**	Zietz
Atelomycterus	Catshark genus	Garman
Atelomycterus baliensis	Bali Catshark	White WT
Atelomycterus fasciatus	Banded Sand Catshark	Compagno
Atelomycterus macleayi	Australian Marbled Catshark	Whitley
Atelomycterus marnkalha	Eastern Banded Catshark	Jacobsen
Atlantoraja	Skate genus	Menni
Atlantoraja castelnaui	Spotback Skate	Miranda-Ribeiro
Atlantoraja cyclophora	Eyespot Skate	Regan
Atlantoraja platana	La Plata Skate	Günther
Aulohalaelurus	Catshark genus	Fowler HW
Aulohalaelurus kanakorum	New Caledonian Catshark	Séret
Aulohalaelurus labiosus	Australian Blackspot Catshark	Waite
Bathyraja	Skate genus	Ishiyama
Bathyraja abyssicola	Deep-sea Skate	Gilbert
Bathyraja aguja	Aguja Skate	Kendall
Bathyraja albomaculata	White-dotted Skate	Norman
Bathyraja aleutica	Aleutian Skate	Gilbert
Bathyraja andriashevi	Little-eyed Skate	Dolganov
Bathyraja bergi	Bottom Skate	Dolganov
Bathyraja brachyurops	Broadnose Skate	Fowler HW
Bathyraja caeluronigricans	Purple-black Skate	Ishiyama
Bathyraja cousseauae	**Cousseau's Skate**	Díaz de Astarloa
Bathyraja diplotaenia	Dusky-pink Skate	Ishiyama
Bathyraja eatonii	**Eaton's Skate**	Günther
Bathyraja fedorovi	**Fedorov's Skate**	Dolganov
Bathyraja griseocauda	Graytail Skate	Norman
Bathyraja hesperafricana	West African Skate	Stehmann

Scientific name	Common name	Describer
Bathyraja interrupta	Sandpaper Skate	Gill
Bathyraja irrasa	Kerguelen Sandpaper Skate	Hureau
Bathyraja ishiharai	Abyssal Skate	Stehmann
Bathyraja isotrachys	Raspback Skate	Günther
Bathyraja leucomelanos	Domino Skate	Iglesias
Bathyraja lindbergi	Commander Skate	Ishiyama
Bathyraja longicauda	Slimtail Skate	de Buen
Bathyraja maccaini	**McCain's Skate**	Springer S
Bathyraja macloviana	Patagonian Skate	Norman
Bathyraja maculata	White-blotched Skate	Ishiyama
Bathyraja magellanica	Magellan Skate	Philippi
Bathyraja mariposa	Butterfly Skate	Stevenson
Bathyraja meridionalis	Dark-belly Skate	Stehmann
Bathyraja minispinosa	Smallthorn Skate	Ishiyama
Bathyraja multispinis	Multispine Skate	Norman
Bathyraja murrayi	**Murray's Skate**	Günther
Bathyraja notoroensis	Notoro Skate	Ishiyama
Bathyraja pallida	Pale Ray	Forster GR
Bathyraja panthera	Leopard Skate	Orr
Bathyraja papilionifera	Butterfly skate	Stehmann
Bathyraja parmifera	Alaska Skate	Bean TH
Bathyraja pseudoisotrachys	Bottom Skate	Ishiyama
Bathyraja richardsoni	**Richardson's Ray**	Garrick
Bathyraja scaphiops	Cuphead Skate	Norman
Bathyraja schroederi	Whitemouth Skate	Krefft G
Bathyraja shuntovi	Longnose Deep-sea Skate	Dolganov
Bathyraja simoterus	Hokkaido Skate	Ishiyama
Bathyraja smirnovi	Golden Skate	Soldatov
Bathyraja smithii	African Softnose Skate	Müller
Bathyraja spinicauda	Spinytail Skate	Jensen
Bathyraja spinosissima	Spiny Skate	Beebe
Bathyraja taranetzi	Mud Skate	Dolganov
Bathyraja trachouros	Eremo Skate	Ishiyama
Bathyraja trachura	Roughtail Skate	Gilbert
Bathyraja tunae	**Tuna's Skate**	Stehmann
Bathyraja tzinovskii	Creamback Skate	Dolganov
Bathyraja violacea	Okhotsk Skate	Suvorov
Benthobatis	Blind Ray genus	Alcock
Benthobatis kreffti	Brazilian Blind Electric Ray	Rincón
Benthobatis marcida	Blind Torpedo	Bean BA
Benthobatis moresbyi	Dark Blind Ray	Alcock
Benthobatis yangi	Taiwanese Blind Electric Ray	Carvalho

Scientific name	Common name	Describer
Brachaeluridae	Blind Shark family	Applegate
Brachaelurus	Carpetshark genus	Ogilby
Brachaelurus waddi	Blind Shark	Bloch
Breiiraja	Ray genus	Bigelow
Breviraja abasiriensis	Skate sp.	Ishiyama
Breviraja claramaculata	Brightspot Skate	McEachran
Breviraja colesi	Lightnose Skate	Bigelow
Breviraja marklei	Nova Scotia Skate	McEachran
Breviraja mouldi	Blacknose Skate	McEachran
Breviraja nigriventralis	Blackbelly Skate	McEachran
Breviraja spinosa	Spinose Skate	Bigelow
Brochiraja	Deepsea Skate genus	Last
Brochiraja aenigma	Enigma Skate	Last
Brochiraja albilabiata	White-lipped Skate	Last
Brochiraja asperula	Smooth Deep-sea Skate	Garrick
Brochiraja heuresa	New Zealand Eureka Skate	Last
Brochiraja leviveneta	Smooth Blue Skate	Last
Brochiraja microspinifera	Small Prickly Skate	Last
Brochiraja spinifera	Prickly Deep-sea Skate	Garrick
Brochiraja vittacauda	Deep-sea Skate sp.	Séret
Bythaelurus	Catshark genus	Compagno
Bythaelurus alcockii	Arabian Catshark	Garman
Bythaelurus canescens	Dusky Catshark	Günther
Bythaelurus clevai	Broadheaded Catshark	Séret
Bythaelurus dawsoni	New Zealand Catshark	Springer S
Bythaelurus giddingsi	Jaguar Catshark	McCosker
Bythaelurus hispidus	Bristly Catshark	Alcock
Bythaelurus immaculatus	Spotless Catshark	Chu
Bythaelurus incanus	Sombre Catshark	Last
Callorhinchus	Chimaera Genus	Lacépède
Callorhinchus callorynchus	Ploughnose Chimaera	Linnaeus
Callorhinchus capensis	Cape Elephantfish	Duméril AHA
Callorhinchus milii	Australian Ghost Shark	Bory de Saint-Vincent
Carcharhinidae	Requiem Shark family	Jordan
Carcharhiniformes	Ground Shark order	Compagno
Carcharhinus	Requiem Shark genus	Blainville
Carcharhinus acronotus	Blacknose Shark	Poey
Carcharhinus albimarginatus	Silvertip Shark	Ruppell
Carcharhinus altimus	Bignose Shark	Springer S
Carcharhinus amblyrhynchoides	Graceful Shark	Whitley
Carcharhinus amblyrhynchos	Gray Reef Shark	Bleeker
Carcharhinus amboinensis	Pigeye Shark	Müller

Scientific name	Common name	Describer
Carcharhinus borneensis	Borneo Shark	Bleeker
Carcharhinus brachyurus	Copper Shark	Günther
Carcharhinus brevipinna	Spinner Shark	Müller
Carcharhinus cautus	Nervous Shark	Whitley
Carcharhinus cerdale	Pacific Smalltail Shark	Gilbert
Carcharhinus coatesi	**Coates' Shark**	Whitley
Carcharhinus dussumieri	Whitecheek Shark	Müller
Carcharhinus falciformis	Silky Shark	Müller
Carcharhinus fitzroyensis	Creek Whaler	Whitley
Carcharhinus galapagensis	Galapagos Shark	Snodgrass
Carcharhinus hemiodon	Pondicherry Shark	Müller
Carcharhinus humani	**Human's Whaler Shark**	White WT
Carcharhinus isodon	Finetooth Shark	Müller
Carcharhinus leiodon	Smooth Tooth Blacktip Shark	Garrick
Carcharhinus leucas	Bull Shark	Müller
Carcharhinus limbatus	Blacktip Shark	Müller
Carcharhinus longimanus	Oceanic Whitetip Shark	Poey
Carcharhinus macloti	Hardnose Shark	Müller
Carcharhinus macrops	Requiem Shark sp.	Liu JX
Carcharhinus melanopterus	Blacktip Reef Shark	Quoy
Carcharhinus obscurus	Dusky Shark	Lesueur
Carcharhinus perezii	Caribbean Reef Shark	Poey
Carcharhinus plumbeus	Sandbar Shark	Nardo
Carcharhinus porosus	Smalltail Shark	Ranzani
Carcharhinus sealei	Blackspot Shark	Pietschmann
Carcharhinus signatus	Night Shark	Poey
Carcharhinus sorrah	Spot-tail Shark	Müller
Carcharhinus tilstoni	**Tilston's Whaler Shark**	Whitley
Carcharhinus tilstoni	**Whitley's Blacktip Shark**	Whitley
Carcharhinus tjutjot	White-cheeked Shark	Bleeker
Carcharias	Great White Shark genus	Rafinesque
Carcharias macrurus	Common Whaler	Ramsay
Carcharias taurus	Sand Tiger Shark	Rafinesque
Carcharias tricuspidatus	Indian Sand Tiger	Day
Carcharodon	Megalodon genus	Smith A
Carcharodon carcharias	Great White Shark	Linnaeus
Carcharodon megalodon	Megatooth Shark	Agassiz
Caspiomyzon wagneri	Caspian Lamprey	Kessler
Centrophoridae	Gulper Shark family	Bleeker
Centrophorus	Gulper Shark genus	Müller
Centrophorus acus	Needle Dogfish	Garman
Centrophorus atromarginatus	Dwarf Gulper Shark	Garman

Scientific name	Common name	Describer
Centrophorus granulosus	Gulper Shark	Bloch
Centrophorus harrissoni	Dumb Gulper Shark	McCulloch
Centrophorus isodon	Blackfin Gulper Shark	Chu
Centrophorus lusitanicus	Lowfin Gulper Shark	Bocage
Centrophorus moluccensis	Smallfin Gulper Shark	Bleeker
Centrophorus niaukang	Taiwan Gulper Shark	Teng
Centrophorus robustus	Gulper Shark sp.	Deng
Centrophorus seychellorum	Seychelles Gulper Shark	Baranes
Centrophorus squamosus	Leafscale Gulper Shark	Bonnaterre
Centrophorus tessellatus	Mosaic Gulper Shark	Garman
Centrophorus westraliensis	Western Gulper Shark	White WT
Centrophorus zeehaani	Southern Dogfish	White WT
Centroscyllium	Deepwater Dogfish genus	Müller
Centroscyllium excelsum	Highfin Dogfish	Shirai
Centroscyllium fabricii	Black Dogfish	Reinhardt
Centroscyllium granulatum	Granular Dogfish	Günther
Centroscyllium kamoharai	Bareskin Dogfish	Abe
Centroscyllium nigrum	Combtooth Dogfish	Garman
Centroscyllium ornatum	Ornate Dogfish	Alcock
Centroscyllium ritteri	Whitefin Dogfish	Jordan
Centroscymnus	Portuguese Dogfish genus	Bocage
Centroscymnus coelolepis	Portuguese Dogfish	Bocage
Centroscymnus cryptacanthus	Shortnose Velvet Dogfish	Regan
Centroscymnus owstoni	Roughskin Dogfish	Garman
Centroselachus	Velvet Dogfish genus	Garman
Centroselachus crepidater	Longnose Velvet Dogfish	Bocage
Cephaloscyllium	Catshark genus	Gill
Cephaloscyllium albipinnum	Whitefin Swellshark	Last
Cephaloscyllium circulopullum	Circle-blotch Pygmy Swellshark	Yano
Cephaloscyllium cooki	**Cook's Swellshark**	Last
Cephaloscyllium fasciatum	Reticulated Swellshark	Chan
Cephaloscyllium hiscosellum	Australian Reticulate Swellshark	White WT
Cephaloscyllium isabellum	Draughtsboard Swellshark	Bonaterre
Cephaloscyllium laticeps	Australian Swellshark	Duméril AHA
Cephaloscyllium maculatum	Spotted Swellshark	Schaaf-Da Silva
Cephaloscyllium pardelotum	Leopard-spotted Swellshark	Schaaf-Da Silva
Cephaloscyllium pictum	Painted Swellshark	Last
Cephaloscyllium sarawakensis	Sarawak Pygmy Swellshark	Yano
Cephaloscyllium signourum	Flagtail Swellshark	Last
Cephaloscyllium silasi	Indian Swellshark	Talwar
Cephaloscyllium speccum	Speckled Swellshark	Last
Cephaloscyllium stevensi	**Steven's Swellshark**	Clark

Scientific name	Common name	Describer
Cephaloscyllium sufflans	Balloon Swellshark	Regan
Cephaloscyllium variegatum	Saddled Swellshark	Last
Cephaloscyllium ventriosum	Swellshark	Garman
Cephaloscyllium zebrum	Narrowbar Swellshark	Last
Cephalurus	Catshark genus	Bigelow
Cephalurus cephalus	Lollipop Catshark	Gilbert
Cetorhinidae	Basking Shark family	Gill
Cetorhinus	Basking Shark genus	Blainville
Cetorhinus maximus	Basking Shark	Gunnerus
Chaenogaleus	Weasel Shark genus	Gill
Chaenogaleus macrostoma	Hooktooth Shark	Bleeker
Chiloscyllium	Bamboo Shark genus	Müller
Chiloscyllium arabicum	Arabian Carpetshark	Gubanov
Chiloscyllium burmensis	Burmese Bamboo Shark	Dingerkus
Chiloscyllium caerulopunctatum	Bluespotted Bamboo Shark	Pellegrin
Chiloscyllium griseum	Gray Bamboo Shark	Müller
Chiloscyllium hasseltii	**Hasselt's Bamboo Shark**	Bleeker
Chiloscyllium indicum	Slender Bamboo Shark	Gmelin
Chiloscyllium punctatum	Brownbanded Bamboo Shark	Müller
Chimaera argiloba	Whitefin Chimaera	Last
Chimaera bahamaensis	Bahamas Ghostshark	Kemper
Chimaera cubana	Chimaera sp.	Howell-Rivero
Chimaera fulva	Southern Chimaera	Didier
Chimaera jordani	**Jordan's Chimaera**	Tanaka I
Chimaera lignaria	**Carpenter's Chimaera**	Didier
Chimaera macrospina	Longspine Chimaera	Didier
Chimaera monstrosa	Rabbitfish	Linnaeus
Chimaera notafricana	Cape Chimaera	Kemper
Chimaera obscura	Shortspine Chimaera	Didier
Chimaera opalescens	Chimaera sp.	Luchetti
Chimaera owstoni	**Owston's Chimaera**	Tanaka I
Chimaera panther	Leopard Chimaera	Didier
Chimaera phantasma	Silver Chimaera	Jordan
Chlamydoselachidae	Frilled Shark family	Garman
Chlamydoselachus	Frilled Shark genus	Garman
Chlamydoselachus africana	Southern African Frilled Shark	Ebert
Chlamydoselachus anguineus	Frilled Shark	Garman
Cirrhigaleus	Squaliform Shark genus	Tanaka I
Cirrhigaleus asper	Roughskin Spurdog	Merrett
Cirrhigaleus australis	Southern Mandarin Dogfish	White WT
Cirrhigaleus barbifer	Mandarin Dogfish	Tanaka I
Cirrhoscyllium	Carpetshark genus	Smith HM

Scientific name	Common name	Describer
Cirrhoscyllium expolitum	Barbelthroat Carpetshark	Smith HM
Cirrhoscyllium formosanum	Taiwan Saddled Carpetshark	Teng
Cirrhoscyllium japonicum	Saddle Carpetshark	Kamohara
Crassinarke brasiliensis	Sleeper Torpedo	Takagi
Cruriraja	Skate genus	Bigelow
Cruriraja andamanica	Andaman Leg Skate	Lloyd
Cruriraja atlantis	Atlantic Leg Skate	Bigelow
Cruriraja cadenati	Broadfoot Leg Skate	Bigelow
Cruriraja durbanensis	Smoothnose Leg Skate	von Bonde
Cruriraja hulleyi	Roughnose Leg Skate	Aschliman
Cruriraja parcomaculata	Roughnose Leg Skate	von Bonde
Cruriraja poeyi	Cuban Leg Skate	Bigelow
Cruriraja rugosa	Rough Leg Skate	Bigelow
Ctenacis	Harlequin Shark genus	Compagno
Ctenacis fehlmanni	Harlequin Catshark	Springer S
Dactylobatus	Skate genus	Bean BA
Dactylobatus armatus	Skilletskate	Bean BA
Dactylobatus clarkia	Hookskate	Bigelow
Dalatias	Kitefin Shark genus	Rafinesque
Dalatias licha	Kitefin Shark	Bonnaterre
Dalatiidae	Kitefin Shark family	Gray
Dasyatidae	Whiptail Stingray family	Jordan
Dasyatis	Skate genus	Rafinesque
Dasyatis acutirostra	Sharpnose Stingray	Nishida
Dasyatis akajei	Whip Stingray	Müller
Dasyatis americana	Southern Stingray	Hildebrand
Dasyatis bennetti	**Bennett's Stingray**	Müller
Dasyatis brevicaudata	Short-tail Stingray	Hutton
Dasyatis centroura	Roughtail Stingray	Mitchill
Dasyatis chrysonota	Blue Stingray	Smith A
Dasyatis colarensis	Colares Stingray	Santos
Dasyatis dipterura	Diamond Stingray	Jordan
Dasyatis fluviorum	Estuary Stingray	Ogilby
Dasyatis garouaensis	Smooth Freshwater Stingray	Stauch
Dasyatis geijskesi	Sharpsnout Stingray	Boesemann
Dasyatis gigantean	Giant Stumptail Stingray	Lindberg
Dasyatis guttata	Longnose	Bloch
Dasyatis hastate	Stingray sp.	DeKay
Dasyatis hypostigma	Groovebelly Stingray	Santos
Dasyatis izuensis	Izu Stingray	Nakaya
Dasyatis laevigata	Yantai Stingray	Chu
Dasyatis laosensis	Mekong Stingray	Roberts TR

Scientific name	Common name	Describer
Dasyatis lata	Brown Stingray	Garman
Dasyatis longa	Longtail Stingray	Garman
Dasyatis longicauda	Merauke Stingray	Last
Dasyatis margarita	Daisy Stingray	Günther
Dasyatis margaritella	Pearl Stingray	Compagno
Dasyatis marianae	Brazilian Large-eyed Stingray	Gomes
Dasyatis marmorata	Marbled Stingray	Steindachner
Dasyatis matsubarai	Pitted Stingray	Miyosi
Dasyatis microps	Smalleye Stingray	Annandale
Dasyatis multispinosa	Multispine Giant Stingray	Tokarev
Dasyatis navarrae	Blackish Stingray	Steindachner
Dasyatis parvonigra	Dwarf Black Stingray	Last
Dasyatis pastinaca	Common Stingray	Linnaeus
Dasyatis rudis	Smalltooth Stingray	Lesueur
Dasyatis sabina	Atlantic Stingray	Lesueur
Dasyatis say	Bluntnose Stingray	Lesueur
Dasyatis sinensis	Chinese Stingray	Steindachner
Dasyatis thetidis	Thorntail Stingray	Ogilby
Dasyatis tortonesei	**Tortonese's Stingray**	Capape
Dasyatis ukpam	Thorny Freshwater Stingray	Smith JA
Dasyatis ushiei	Cow Stingray	Jordan
Dasyatis zugei	Pale-edged Stingray	Müller
Deania	Shark genus	Jordan
Deania calcea	Birdbeak Dogfish	Lowe RT
Deania hystricosa	Rough Longnose Dogfish	Garman
Deania profundorum	Arrowhead Dogfish	Smith HM
Deania quadrispinosum	Longsnout Dogfish	McCulloch
Dentiraja	Skate genus	Richardson J
Dentiraja lemprieri	Thornback skate	Richardson J
Diplobatis	Electric Ray genus	Bigelow
Diplobatis colombiensis	Colombian Electric Ray	Fechhelm
Diplobatis guamachensis	Brownband Numbfish	Martín Salazar
Diplobatis ommata	Ocellate Electric Ray	Jordan
Diplobatis pictus	Painted Electric Ray	Palmer
Dipturus	Skate genus	Rafinesque
Dipturus acrobelus	Deepwater Skate	Last
Dipturus amphispinus	Ridgeback Skate	Last
Dipturus apricus	Pale Tropical Skate	Last
Dipturus argentinensis	Argentine Skate	Díaz de Astarloa
Dipturus australis	Sydney Skate	Macleay
Dipturus batis	Blue Skate	Linnaeus
Dipturus bullisi	**Bullis Skate**	Bigelow

Scientific name	Common name	Describer
Dipturus campbelli	Blackspot Skate	Wallace
Dipturus canutus	Gray Skate	Last
Dipturus cerva	White-spotted Skate	Whitley
Dipturus confuses	Longnose Skate	Last
Dipturus crosnieri	Madagascar Skate	Séret
Dipturus diehli	Thorny-tail Skate	Soto
Dipturus doutrei	Violet Skate	Cadenat
Dipturus ecuadoriensis	Ecuador Skate	Beebe
Dipturus endeavouri	**The Endeavour Skate**	Last
Dipturus falloargus	False Argus Skate	Last
Dipturus flavirostris	Skate sp.	Philippi
Dipturus flindersi	Pygmy Thornback Skate	Last
Dipturus garricki	San Blas Skate	Bigelow
Dipturus gigas	Giant Skate	Ishiyama
Dipturus grahami	**Grahams' Skate**	Last
Dipturus gudgeri	Greenback Skate	Whitley
Dipturus healdi	**Heald's Skate** [Alt. **Leyland's Skate**]	Last
Dipturus innominatus	New Zealand Smooth Skate	Garrick
Dipturus johannisdavisi	Travancore Skate	Alcock
Dipturus kwangtungensis	Kwangtung Skate	Chu
Dipturus laevis	Barndoor Skate	Mitchill
Dipturus lanceorostratus	Rattail Skate	Wallace
Dipturus leptocaudus	Thintail Skate	Krefft G
Dipturus linteus	Sailray	Fries
Dipturus macrocauda	Bigtail Skate	Ishiyama
Dipturus melanospilus	Blacktip Skate	Last
Dipturus mennii	South Brazilian Skate	Gomes
Dipturus nidarosiensis	Norwegian Skate	Storm
Dipturus oculus	Ocellate Skate	Last
Dipturus olseni	Spreadfin Skate	Bigelow
Dipturus oregoni	Hooktail Skate	Bigelow
Dipturus oxyrinchus	Longnosed Skate	Linnaeus
Dipturus polyommata	Argus Skate	Ogilby
Dipturus pullopunctatus	Slime Skate	Smith JLB
Dipturus queenslandicus	Queensland Deepwater Skate	Last
Dipturus springeri	Roughbelly Skate	Wallace
Dipturus stenorhynchus	Prownose Skate	Wallace
Dipturus teevani	Prickly Brown Ray	Bigelow
Dipturus tengu	Acutenose Skate	Jordan
Dipturus trachydermus	Roughskin Skate	Krefft G
Dipturus wengi	**Weng's Skate**	Last
Dipturus whitleyi	Wedgenose Skate	Iredale

Scientific name	Common name	Describer
Dipturus wuhanlingi	**Wu's Skate**	Jeong
Discopyge	Apron Ray genus	Heckel
Discopyge castelloi	Electric Ray sp.	Menni
Discopyge tschudii	Apron Ray	Heckel
Echinorhinidae	Squaliform Shark family	Gill
Echinorhinus	Squaliform Shark genus	Blainville
Echinorhinus brucus	Bramble Shark	Bonnaterre
Echinorhinus cookei	Prickly Shark	Pietschmann
Electrolux addisoni	Ornate Sleeper Ray	Compagno
Entosphenus folletti	Modoc Brook Lamprey	Vladykov
Entosphenus hubbsi	Kern Brook Lamprey	Vladykov
Entosphenus lethophagus	Pit-Klamath Brook Lamprey	Hubbs CL
Entosphenus macrostomus	Vancouver Lamprey	Beamish
Entosphenus minimus	Miller Lake Lamprey	Bond
Entosphenus similis	Klamath River Lamprey	Vladykov
Eptatretus alastairi	Australian Hagfish sp.	Mincarone
Eptatretus ancon	Hagfish sp.	Mok
Eptatretus astrolabium	Hagfish sp.	Fernholm
Eptatretus atami	Brown Hagfish	Dean
Eptatretus bischoffii	Hagfish sp.	Schneider AF
Eptatretus burgeri	Inshore Hagfish	Girard
Eptatretus caribbeaus	Hagfish sp.	Fernholm
Eptatretus carlhubbsi	Giant Hagfish	McMillan CB
Eptatretus cheni	Hagfish sp.	Shen S-C
Eptatretus chinensis	Hagfish sp.	Kuo
Eptatretus cirrhatus	Broadgilled Hagfish	Forster JR
Eptatretus deani	Black Hagfish	Evermann
Eptatretus eos	Hagfish sp.	Fernholm
Eptatretus fernholmi	Hagfish sp.	McMillan CB
Eptatretus fritzi	Guadaloupe Hagfish	Wisner
Eptatretus goliath	Goliath Hagfish	Mincarone
Eptatretus gomoni	Hagfish sp.	Mincarone
Eptatretus grouseri	Hagfish sp.	McMillan CB
Eptatretus indrambaryai	Hagfish sp.	Wongratana
Eptatretus lakeside	Hagfish sp.	Mincarone
Eptatretus laurahubbsae	Hagfish sp.	McMillan CB
Eptatretus longipinnis	Longfinned Hagfish	Strahan
Eptatretus lopheliae	Hagfish sp.	Fernholm
Eptatretus mcconnaugheyi	Shorthead Hagfish	Wisner
Eptatretus mccoskeri	Hagfish sp.	McMillan CB
Eptatretus mendozai	Medoza's Hagfish	Hensley
Eptatretus menezesi	Hagfish sp.	Mincarone

Scientific name	Common name	Describer
Eptatretus minor	Hagfish sp.	Fernholm
Eptatretus moki	Hagfish sp.	McMillan CB
Eptatretus multidens	Hagfish sp.	Fernholm
Eptatretus nanii	Hagfish sp.	Wisner
Eptatretus nelsoni	Hagfish sp.	Kuo
Eptatretus octatrema	Eightgill Hagfish	Barnard
Eptatretus okinoseanus	Hagfish sp.	Dean
Eptatretus profundus	Fivegill Hagfish	Barnard
Eptatretus rubicundus	Hagfish sp.	Kuo
Eptatretus sheni	Hagfish sp.	Kuo
Eptatretus sinus	Cortez Hagfish	Wisner
Eptatretus springeri	Gulf Hagfish	Bigelow
Eptatretus stoutii	Pacific Hagfish	Lockington
Eptatretus strahani	Hagfish sp.	McMillan CB
Eptatretus strickrotti	**Strickrott's Hagfish**	Møller
Eptatretus taiwanae	Hagfish sp.	Shen S-C
Eptatretus walkeri	Hagfish sp.	McMillan CB
Eptatretus wayuu	Hagfish sp.	Mok
Eptatretus wisneri	Hagfish sp.	McMillan CB
Eridacnis	Ribbontail Catshark genus	Smith HM
Eridacnis barbouri	Cuban Ribbontail Catshark	Bigelow
Eridacnis radcliffei	Pygmy Ribbontail Catshark	Smith HM
Eridacnis sinuans	African Ribbontail Catshark	Smith JLB
Etmopteridae	Lantern Shark family	Fowler HW
Etmopterus	Lantern Shark genus	Rafinesque
Etmopterus abernethyi	Blackbelly Lanternshark	Garrick
Etmopterus baxteri	New Zealand Lantern Shark	Garrick
Etmopterus bigelowi	Blurred Lantern Shark	Shirai
Etmopterus brachyurus	Short-tail Lantern Shark	Smith HM
Etmopterus bullisi	Lined Lantern Shark	Bigelow
Etmopterus burgessi	Broad-snouted Lantern Shark	Schaaf-Da Silva
Etmopterus carteri	**Carter Gilbert's Lantern Shark**	Springer S
Etmopterus caudistigmus	Tailspot Lantern Shark	Last
Etmopterus compagnoi	Lantern Shark sp.	Fricke
Etmopterus decacuspidatus	Comtooth Lantern Shark	Chan
Etmopterus dianthus	Pink Lantern Shark	Last
Etmopterus dislineatus	Lined Lantern Shark	Last
Etmopterus evansi	Blackmouth Lantern Shark	Last
Etmopterus fusus	Pygmy Lantern Shark	Last
Etmopterus gracilispinis	Broadbanded Lantern Shark	Krefft G
Etmopterus granulosus	Southern Lantern Shark	Günther
Etmopterus hillianus	Caribbean Lantern Shark	Poey

Scientific name	Common name	Describer
Etmopterus joungi	Shortfin Smooth Lanternshark	Knuckey
Etmopterus litvinovi	Smalleye Lantern Shark	Dolganov
Etmopterus lucifer	Blackbelly Lantern Shark	Jordan
Etmopterus molleri	**Moller's Slendertail Lantern Shark**	Whitley
Etmopterus perryi	Dwarf Lantern Shark	Springer S
Etmopterus polli	African Lantern Shark	Bigelow
Etmopterus princeps	Great Lantern Shark	Collett
Etmopterus pseudosqualiolus	False Lantern Shark	Last
Etmopterus pycnolepis	Dense-scale Lantern Shark	Kotlyar
Etmopterus robinsi	West Indian Lantern Shark	Schofield
Etmopterus schmidti	Lanternshark sp.	Dolganov
Etmopterus schultzi	Fringefin Lantern Shark	Bigelow
Etmopterus sculptus	Sculpted Lantern Shark	Ebert
Etmopterus sentosus	Thorny Lantern Shark	Bass
Etmopterus spinax	Velvet Belly Lantern Shark	Linnaeus
Etmopterus splendidus	Splendid Lantern Shark	Yano
Etmopterus tasmaniensis	Tasmanian Lantern Shark	Myagkov
Etmopterus unicolor	Brown Lantern Shark	Engelhardt
Etmopterus viator	Lantern Shark sp.	Straube
Etmopterus villosus	Hawaiian Lantern Shark	Gilbert
Etmopterus virens	Green Lantern Shark	Bigelow
Eucrossorhinus	Wobbegong genus	Regan
Eucrossorhinus dasypogon	Tasselled Wobbegong	Bleeker
Eudontomyzon danfordi	Carpathian Lamprey	Regan
Eudontomyzon graecus	Epirus Brook Lamprey	Renaud
Eudontomyzon hellenicus	Macedonia Brook Lamprey	Vladykov
Eudontomyzon mariae	Ukranian Brook Lamprey	Berg LS
Eudontomyzon morii	Korean Lamprey	Berg LS
Eudontomyzon stankokaramani	Drin Brook Lamprey	Karaman MS
Eudontomyzon vladykovi	Vladykov's Lamprey	Oliva
Euprotomicroides	Taillight Shark genus	Hulley
Euprotomicroides zantedeschia	Taillight Shark	Hulley
Euprotomicrus	Pygmy Shark genus	Gill
Euprotomicrus bispinatus	Pygmy Shark	Quoy
Eusphyra	Winghead Shark genus	Gill
Eusphyra blochii	Winghead Shark	Cuvier
Fenestraja	Pygmy Skate genus	McEachran
Fenestraja atripinna	Blackfin Pygmy Skate	Bigelow
Fenestraja cubensis	Cuban Pygmy Skate	Bigelow
Fenestraja ishiyamai	Plain Pygmy Skate	Bigelow
Fenestraja maceachrani	Madagascar Pygmy Skate	Séret
Fenestraja mamillidens	Prickly Skate	Alcock

Scientific name	Common name	Describer
Fenestraja plutonia	Pluto Skate	Garman
Fenestraja sibogae	**Siboga Skate**	Weber
Fenestraja sinusmexicanus	Gulf of Mexico Pygmy Skate	Bigelow
Figaro	Sawtail Catshark genus	Whitley
Figaro boardmani	Australian Sawtail Catshark	Whitley
Figaro striatus	Northern Sawtail Catshark	Gledhill
Furgaleus	Whiskery Shark genus	Whitley
Furgaleus macki	Whiskery Shark	Whitley
Galeocerdo	Requiem Shark genus	Müller
Galeocerdo cuvier	Tiger Shark	Péron
Galeorhinus	Tope Shark genus	Blainville
Galeorhinus galeus	School Shark	Linnaeus
Galeus	Catshark genus	Rafinesque
Galeus antillensis	Antilles Catshark	Springer S
Galeus arae	Roughtail Catshark	Nichols
Galeus atlanticus	Atlantic Sawtail Catshark	Vaillant
Galeus cadenati	Longfin Sawtail Catshark	Springer S
Galeus eastmani	Gecko Catshark	Jordan
Galeus gracilis	Slender Sawtail Catshark	Compagno
Galeus longirostris	Longnose Sawtail Catshark	Tachikawa
Galeus melastomus	Blackmouth Catshark	Rafinesque
Galeus mincaronei	Southern Sawtail Catshark	Soto
Galeus murinus	Mouse Catshark	Collett
Galeus nipponensis	Broadfin Sawtail Catshark	Nakaya
Galeus piperatus	Peppered Catshark	Springer S
Galeus polli	African Sawtail Catshark	Cadenat
Galeus priapus	Phallic Catshark	Séret
Galeus sauteri	Blacktip Sawtail Catshark	Jordan
Galeus schultzi	Dwarf Sawtail Catshark	Springer S
Galeus springeri	**Springer's Sawtail Catshark**	Konstantinou
Ginglymostoma	Nurse Shark genus	Müller
Ginglymostoma cirratum	Nurse Shark	Bonnaterre
Ginglymostomatidae	Nurse Shark familly	Gill
Glaucostegus	Ray genus	Bonaparte
Glaucostegus granulatus	Granulated Guitarfish	Cuvier
Glaucostegus halavi	Halavi Ray	Forsskål
Glaucostegus typus	**Bennett's Shovelnose Guitarfish**	Bennett ET
Glyphis	River Shark genus	Agassiz
Glyphis fowlerae	Borneo River Shark	Compagno
Glyphis gangeticus	Ganges Shark	Müller
Glyphis garricki	Northern River Shark	Compagno
Glyphis glyphis	Speartoothed Shark	Müller

Scientific name	Common name	Describer
Glyphis siamensis	Irrawaddy River Shark	Steindachner
Gogolia	Houndshark genus	Compagno
Gogolia filewoodi	Sailback Houndshark	Compagno
Gollum	Smoothhound Shark genus	Compagno
Gollum attenuates	Slender Smooth-hound	Garrick
Gollum sp.	Whitemarked Gollumshark	Undescribed
Gollum suluensis	Sulu Gollumshark	Last
Gurgesiella	Skate genus	de Buen
Gurgesiella atlantica	Atlantic Pygmy Skate	Bigelow
Gurgesiella dorsalifera	Onefin Skate	Compagno
Gurgesiella furvescens	Dusky Finless Skate	de Buen
Gymnura	Butterfly Ray genus	Van Hasselt
Gymnura afuerae	Peruvian Butterfly Ray	Hildebrand
Gymnura altavela	Spiny Butterfly Ray	Linnaeus
Gymnura australis	Australian Butterfly Ray	Ramsey
Gymnura bimaculata	Twin-spot Butterfly Ray	Norman
Gymnura crebripunctata	Longsnout Butterfly Ray	Peters
Gymnura crooki	Butterfly Ray sp.	Fowler HW
Gymnura japonica	Japanese Butterfly Ray	Temminck
Gymnura marmorata	California Butterfly Ray	Cooper
Gymnura micrura	Smooth Butterfly Ray	Bloch
Gymnura natalensis	Backwater Butterfly Ray	Gilchrist
Gymnura poecilura	Longtail Butterfly Ray	Shaw
Gymnura tentaculata	Tentacled Butterfly Ray	Müller
Gymnura zonura	Zonetail Butterfly Ray	Bleeker
Gymnuridae	Butterfly Ray family	Fowler HW
Halaelurus	Catshark genus	Gill
Halaelurus boesemani	Speckled Catshark	Springer S
Halaelurus buergeri	Blackspotted Catshark	Müller
Halaelurus lineatus	Lined Catshark	Bass
Halaelurus maculosus	Indonesian Speckled Catshark	White WT
Halaelurus natalensis	Tiger Catshark	Regan
Halaelurus quagga	Quagga Catshark	Alcock
Halaelurus sellus	Rusty Catshark	White WT
Haploblepharus	Catshark genus	Garman
Haploblepharus edwardsii	Puffadder Shyshark	Schinz
Haploblepharus fuscus	Brown Shyshark	Smith JLB
Haploblepharus kistnasamyi	Natal Shyshark	Human
Haploblepharus pictus	Dark Shyshark	Müller
Harriotta	Chimaera Genus	Goode
Harriotta haeckeli	Smallspine Spookfish	Karrer
Harriotta raleighana	Pacific Longnose Chimaera	Goode

Scientific name	Common name	Describer
Heliotrygon	Round Ray genus	Carvalho
Heliotrygon gomesi	**Gomes' Round Ray**	Carvalho
Heliotrygon rosai	**Rosa's Round Ray**	Carvalho
Hemigaleidae	Weasel Shark family	Compagno
Hemigaleus	Sicklefin Weasel Shark genus	Bleeker
Hemigaleus australiensis	Australian Weasel Shark	White WT
Hemigaleus machlani	Weasle Shark sp.	Herre
Hemigaleus microstoma	Sicklefin Weasel Shark	Bleeker
Hemipristis	Weasel Shark genus	Agassiz
Hemipristis elongatus	Snaggletooth Shark	Klunzinger
Hemiscylliidae	Bamboo Shark family	Gill
Hemiscyllium	Bamboo Shark genus	Müller
Hemiscyllium freycineti	Indian Speckled Carpetshark	Quoy
Hemiscyllium galei	Walking Shark	Allen
Hemiscyllium hallstromi	Papual Epaulette Carpetshark	Whitley
Hemiscyllium halmahera	Bamboo Shark sp.	Allen
Hemiscyllium henryi	Walking Shark	Allen
Hemiscyllium michaeli	**Michael's Epaulette Shark**	Allen
Hemiscyllium ocellatum	Epaulette Carpetshark	Bonnaterre
Hemiscyllium strahani	Hooded Carpetshark	Whitley
Hemiscyllium trispeculare	Speckled Carpetshark	Richardson J
Hemitriakis	Houndshark genus	Herre
Hemitriakis abdita	Deepwater Sicklefin Houndshark	Compagno
Hemitriakis complicofasciata	Ocellate Japanese Topeshark	Takahashi
Hemitriakis falcate	Sicklefin Houndshark	Compagno
Hemitriakis indroyonoi	Indonesian Houndshark	White WT
Hemitriakis japonicus	Japanese Topeshark	Müller
Hemitriakis leucoperiptera	Whitefin Japanese Topeshark	Herre
Heptranchias	Sharpnose Shark genus	Rafinesque
Heptranchias perlo	Sharpnose Sevengill Shark	Bonnaterre
Heterodontidae	Bullhead Shark family	Gray
Heterodontiformes	Bullhead Shark order	Berg LS
Heterodontus	Bull Shark genus	Blainville
Heterodontus francisci	Horn Shark	Girard
Heterodontus galeatus	Crested Bullhead Shark	Günther
Heterodontus japonicus	Japanese Bullhead Shark	Maclay
Heterodontus mexicanus	Mexican Horn Shark	Taylor LR
Heterodontus omanensis	Oman Bullhead Shark	Baldwin ZH
Heterodontus portusjacksoni	Port Jackson Shark	Meyer
Heterodontus quoyi	Galapagos Bullhead Shark	Freminville
Heterodontus ramalheira	Whitespotted Bullhead Shark	Smith JLB
Heterodontus sp X	Cryptic Horn Shark	Castro

Scientific name	Common name	Describer
Heterodontus zebra	Zebra Bullhead Shark	Gray
Heteronarce	Sleeper Ray genus	Regan
Heteronarce bentuviai	Elat Electric Ray	Baranes
Heteronarce garmani	Natal Electric Ray	Regan
Heteronarce mollis	**Lloyd's Electric Ray**	Lloyd
Heteronarce prabhui	Quilon Electric Ray	Talwar
Heteroscyllium	Carpet Shark genus	Regan
Heteroscyllium colcloughi	**Colclough's Shark**	Ogilby
Heteroscymnoides	Pygmy Shark family	Fowler HW
Heteroscymnoides marleyi	Longnose Pygmy Shark	Fowler HW
Hexanchidae	Cow Sharks family	Gray
Hexanchiformes	Primitive Shark order	de Buen
Hexanchus	Sixgill Shark genus	Rafinesque
Hexanchus griseus	Bluntnose Sixgill Shark	Bonnaterre
Hexanchus nakamurai	**Bigeyed Sixgill Shark**	Teng
Hexatrygon	Stingray genus	Heemstra
Hexatrygon bickelli	Sixgill Stingray	Heemstra
Hexatrygonidae	Stingray family	Heemstra
Himantura	Stingray genus	Müller
Himantura alcockii	Pale-spot Whipray	Annandale
Himantura astra	Black-spotted Whipray	Last
Himantura bleekeri	**Bleeker's Whipray**	Blyth
Himantura chaophraya	Giant Freshwater Stingray	Monkolprasit
Himantura dalyensis	Freshwater Whipray	Last
Himantura draco	Dragon Stingray	Compagno
Himantura fai	Pink Whipray	Jordan
Himantura fava	Honeycomb Whipray	Annandale
Himantura fluviatilis	Ganges Stingray	Hamilton F
Himantura gerrardi	**Gerrard's Stingray**	Gray
Himantura granulate	Mangrove Whipray	Macleay
Himantura hortlei	**Hortle's Whipray**	Last
Himantura imbricate	Scaly Whipray	Bloch
Himantura javaensis	Stingray sp.	Last
Himantura jenkinsii	**Jenkins' Whipray**	Annandale
Himantura kittipongi	**Kittipong's Stingray** [Alt. Roughback Whipray]	Vidthayanon
Himantura krempfi	Marbled Freshwater Whipray	Chabanaud
Himantura leoparda	Leopard Whipray	Manjaji-Matsumoto
Himantura lobistoma	Tubemouth Whipray	Manjaji-Matsumoto
Himantura marginata	Blackedge Whipray	Blyth
Himantura microphthalma	Smalleye Whipray	Chen JTF
Himantura oxyrhyncha	Marbled Whipray	Sauvage

Scientific name	Common name	Describer
Himantura pacifica	Pacific Chupare	Beebe
Himantura pareh	Stingray sp.	Bleeker
Himantura pastinacoides	Round Whipray	Bleeker
Himantura randalli	Arabian Whipray	Last
Himantura schmardae	Chupare Whipray	Werner
Himantura signifier	White-rimmed Whipray	Compagno
Himantura toshi	Brown Whipray	Whitley
Himantura tutul	Fine-spotted Leopard Whipray	Borsa
Himantura uarnacoides	Whitenose Whipray	Bleeker
Himantura uarnak	Reticulate Whipray	Gmelin
Himantura undulata	Leopard Whipray	Bleeker
Himantura walga	Dwarf Whipray	Müller
Holohalaelurus	Catshark genus	Fowler HW
Holohalaelurus favus	Honeycomb Izak	Human
Holohalaelurus grennian	Grinning Izak	Human
Holohalaelurus melanostigma	Crying Izak	Norman
Holohalaelurus punctatus	White-spotted Izak	Gilchrist
Holohalaelurus regani	Izak Catshark	Gilchrist
Hongeo	Korean Skate genus	Jeong
Hydrolagus affinis	Smalleyed Rabbitfish	Capello
Hydrolagus africanus	African Chimaera	Gllchrist
Hydrolagus alberti	Ratfish sp.	Bigelow
Hydrolagus alphus	White Spot Chimaera	Quaranta
Hydrolagus barbouri	Ratfish sp.	Garman
Hydrolagus bemisi	Pale Ghost Shark	Didier
Hydrolagus colliei	Spotted Ratfish	Lay
Hydrolagus deani	Philippine Chimaera	Smith HM
Hydrolagus eidolon	Ratfish sp.	Jordan
Hydrolagus homonycteris	Black Ghostshark	Didier
Hydrolagus lusitanicus	Ghost Shark sp.	Moura
Hydrolagus macrophthalmus	Ghost Shark sp.	de Buen
Hydrolagus marmoratus	Marbled Ghost Shark	Didier
Hydrolagus matallanasi	Striped Rabbitfish	Soto
Hydrolagus mccoskeri	Galápagos Ghostshark	Barnett
Hydrolagus melanophasma	Eastern Pacific Black Ghostshark	Didier
Hydrolagus mirabilis	Large-eyed Rabbitfish	Collett
Hydrolagus mitsukurii	Spookfish	Jordan
Hydrolagus novaezealandiae	Dark Ghost Shark	Fowler HW
Hydrolagus ogilbyi	Ratfish sp.	Waite
Hydrolagus pallidus	Ratfish sp.	Hardy
Hydrolagus purpurescens	Purple Chimaera	Gilbert CH
Hydrolagus trolli	Pointy-nosed Blue Chimaera	Didier

Scientific name	Common name	Describer
Hydrolagus waitei	Ratfish sp.	Fowler HW
Hypnos	Electric Ray genus	Duméril AHA
Hypnos monopterygius	Coffin Ray	Duméril AHA
Hypogaleus	Houndshark genus	Smith JLB
Hypogaleus hyugaensis	Blacktip Tope	Miyosi
Iago	Houndshark genus	Compagno
Iago garricki	Longnose Houndshark	Fourmanoir
Iago omanensis	Bigeye Houndshark	Norman
Ichthyomyzon bdellium	Ohio Lamprey	Jordan
Ichthyomyzon fossor	Northern Brook Lamprey	Reighard
Ichthyomyzon gagei	Southern Brook Lamprey	Hubbs CL
Ichthyomyzon greeleyi	Mountain Brook Lamprey	Hubbs CL
Ichthyomyzon unicuspis	Silver Lamprey	Hubbs CL
Indobatis	Leg Skate genus	Weigmann
Indobatis ori	Black Leg Skate	Wallace
Insentiraja	Skate genus	Yearsley
Insentiraja laxipella	Eastern Looseskin Skate	Yearsley
Insentiraja subtilispinosa	Velvet Skate	Stehmann
Irolita	Softnose Skate genus	Whitley
Irolita waitii	Southern Round Skate	McCulloch
Irolita westraliensis	Western Round skate	Last
Isistius	Dogfish Shark genus	Gill
Isistius brasiliensis	Cookiecutter Shark	Quoy
Isistius labialis	South China Cookiecutter Shark	Meng
Isistius plutodus	South China Cookiecutter Shark	Garrick
Isogomphodon	Requiem Shark genus	Gill
Isogomphodon oxyrhynchus	Daggernose Shark	Müller
Isurus	Mackerel Shark genus	Rafinesque
Isurus oxyrinchus	Shortfin Mako	Rafinesque
Isurus paucus	Longfin Mako	Guitart-Manday
Lamiopsis	Broadfin Shark genus	Gill
Lamiopsis temminckii	Broadfin Shark	Müller
Lamna	Mackerel Shark genus	Cuvier
Lamna ditropis	Salmon Shark	Hubbs CL
Lamna nasus	Porbeagle	Bonnaterre
Lamnidae	Mackerel Shark family	Müller
Lamniformes	Mackerel Shark order	Berg LS
Lampetra aepyptera	Least Brook Lamprey	Abbott
Lampetra lamottei	Brook Lamprey	Lesueur
Lampetra lanceolata	Turkish Brook Lamprey	Kux
Lampetra planeri	European Brook Lamprey	Bloch
Leptocharias	Houndshark genus	Smith A

Scientific name	Common name	Describer
Leptocharias smithii	Barbeled Houndshark	Müller
Leptochariidae	Barbeled Houndshark family	Gray
Lethenteron	Brook Lamprey sub-genus	Creaser
Lethenteron alaskense	Alaskan Brook Lamprey	Vladykov
Lethenteron camtschaticum	Arctic Lamprey	Tilesius
Lethenteron kessleri	Siberian Brook Lamprey	Anikin
Lethenteron matsubarai	Lamprey sp.	Vladykov
Lethenteron ninae	**Nina's Lamprey**	Naseka
Lethenteron reissneri	Far Eastern Brook Lamprey	Dybowski
Lethenteron zanandreai	Po Brook Lamprey	Vladykov
Leucoraja	Hardnose Skate genus	Malm
Leucoraja caribbaea	Maya Skate	McEachran
Leucoraja circularis	Sandy Skate	Couch
Leucoraja compagnoi	Tigertail Skate	Stehmann
Leucoraja erinacea	Little Skate	Mitchill
Leucoraja fullonica	Shagreen Skate	Linnaeus
Leucoraja garmani	Freckled Skate	Whitley
Leucoraja lentiginosa	Speckled Skate	Bigelow
Leucoraja leucosticta	Whitedappled Skate	Stehmann
Leucoraja melitensis	Maltese Skate	Clark RS
Leucoraja naevus	Cuckoo Skate	Müller
Leucoraja ocellata	Winter Skate	Mitchill
Leucoraja pristispina	Sawback Skate	Last
Leucoraja virginica	Virginia Skate	McEachran
Leucoraja wallacei	Yellow-spotted Skate	Hulley
Leucoraja yucatanensis	Yucatán Skate	Bigelow
Loxodon	Sliteye Shark genus	Müller
Loxodon macrorhinus	Sliteye Shark	Müller
Makararaja	Freshwater Ray genus	Roberts TR
Makararaja chindwinensis	Freshwater Ray sp.	Roberts TR
Malacoraja	Soft Skate genus	Stehmann
Malacoraja kreffti	**Krefft's Skate**	Stehmann
Malacoraja obscura	Brazilian Soft Skate	Carvalho
Malacoraja senta	Smooth Skate	Garman
Malacoraja spinacidermis	Soft Skate	Barnard
Manta	Manta Ray genus	Walbaum
Manta alfredi	**Prince Alfred's Ray**	Krefft JLG
Manta birostris	Giant Oceanic Manta Ray	Walbaum
Megachasma	Megamouth Shark genus	Taylor LR
Megachasma pelagios	Megamouth Shark	Taylor LR
Megachasmidae	Megamouth Shark Family	Taylor LR
Miroscyllium	Rasptooth Dogfish genus	Shirai

Scientific name	Common name	Describer
Miroscyllium sheikoi	Rasptooth Dogfish	Dolganov
Mitsukurina	Goblin Shark genus	Jordan
Mitsukurina owstoni	Goblin Shark	Jordan
Mitsukurinidae	Goblin Shark family	Jordan
Mobula	Smoothhound Sharke genus	Linck
Mobula	Eagle Ray genus	Rafinesque
Mobula coilloti	Sicklefin Devil Ray	Cadenat
Mobula eregoodootenkee	Pygmy Devil Ray	Bleeker
Mobula hypostoma	Lesser Devil Ray	Bancroft
Mobula japonica	Spinetail Mobula	Müller
Mobula kuhlii	Shortfin Devil Ray	Müller
Mobula mobular	Devil Fish	Bonnaterre
Mobula munkiana	**Munk's Devil Ray**	Notarbartolo di Sciara
Mobula rancureli	Devil Ray	Cadenat
Mobula rochebrunei	Lesser Guinean Devil Ray	Vaillant
Mobula tarapacana	Chilean Devil Ray	Philippi
Mobula thurstoni	Smoothtail Mobula	Lloyd
Mollisquama	Pocket Shark genus	Dolganov
Mollisquama parini	Pocket Shark	Dolganov
Mordacia praecox	Non-parasitic Lamprey	Potter
Mustelus albipinnis	White-margin Fin Houndshark	Castro-Aguirre
Mustelus antarcticus	Gummy Shark	Günther
Mustelus asterias	Starry Smooth-Hound	Cloquet
Mustelus californicus	Gray Smooth-Hound	Gill
Mustelus canis	Dusky Smooth-Hound	Mitchill
Mustelus dorsalis	Sharptooth Smooth-Hound	Gill
Mustelus fasciatus	Striped Smooth-Hound	Garman
Mustelus griseus	Spotless Smooth-Hound	Pietschmann
Mustelus henlei	Brown Smooth-Hound	Gill
Mustelus higmani	Smalleye Smooth-Hound	Springer S
Mustelus lenticulatus	Spotted Estuary Smooth-Hound	Phillipps
Mustelus lunulatus	Sicklefin Smooth-Hound	Jordan
Mustelus manazo	Starspotted Smooth-Hound	Bleeker
Mustelus mangalorensis	Smoothhound sp.	Cubelio
Mustelus mento	Speckled Smooth-Hound	Cope
Mustelus minicanis	Dwarf Smooth-Hound	Heemstra
Mustelus mosis	Arabian Smooth-Hound	Hemprich
Mustelus mustelus	Common Smooth-Hound	Linnaeus
Mustelus norrisi	Narrowfin Smooth-Hound	Springer S
Mustelus palumbes	Whitespotted Smooth-Hound	Smith JLB
Mustelus punctulatus	Blackspotted Smooth-Hound	Risso
Mustelus ravidus	Australian gray Smooth-Hound	White WT

Scientific name	Common name	Describer
Mustelus schmitti	Narrownose Smooth-Hound	Springer S
Mustelus sinusmexicanus	Gulf Smooth-Hound	Heemstra
Mustelus stevensi	White Spotted Gummy Shark	White WT
Mustelus walkeri	Eastern Spotted Gummy Shark	White WT
Mustelus whitneyi	Humpback Smooth-Hound	Chirichigno
Mustelus widodoi	White-fin Smooth-Hound	White WT
Myliobatidae	Tropical Stingray family	Bonaparte
Myliobatiformes	Stingrays & Relatives order	Compagno
Myliobatis	Eagle Ray genus	Cuvier
Myliobatis aquila	Common Eagle Ray	Linnaeus
Myliobatis australis	Australian Bull Ray	Macleay
Myliobatis californica	Bat Eagle Ray	Gill
Myliobatis chilensis	Chilean Eagle Ray	Philippi
Myliobatis freminvillii	Bullnose Eagle Ray	Lesueur
Myliobatis goodei	Southern Eagle Ray	Garman
Myliobatis hamlyni	Purple Eagle Ray	Ogilby
Myliobatis longirostris	Snouted Eagle Ray	Applegate
Myliobatis peruvianus	Peruvian Eagle Ray	Garman
Myliobatis ridens	Shortnose Eagle Ray	Ruocco
Myliobatis tenuicaudatus	New Zealand Eagle Ray	Hector
Myliobatis tobijei	Japanese Eagle Ray	Bleeker
Myxine australis	Southern Hagfish	Jenyns
Myxine debueni	Hagfish sp.	Wisner
Myxine dorsum	Hagfish sp.	Wisner
Myxine fernholmi	Hagfish sp.	Wisner
Myxine formosana	Hagfish sp.	Mok
Myxine garmani	Hagfish sp.	Jordan
Myxine glutinosa	Atlantic Hagfish	Linnaeus
Myxine hubbsi	Hagfish sp.	Wisner
Myxine hubbsoides	Hagfish sp.	Wisner
Myxine ios	White-headed Hagfish	Fernholm
Myxine jespersenae	**Jespersen's Hagfish**	Møller
Myxine knappi	Hagfish sp.	Wisner
Myxine kuoi	Hagfish sp.	Mok
Myxine mcmillanae	Hagfish sp.	Hensley
Myxine pequenoi	Hagfish sp.	Wisner
Myxine robinsorum	Hagfish sp.	Wisner
Myxine sotoi	Hagfish sp.	Mincarone
Narcine	Electric Ray genus	Henle
Narcine atzi	**Atz's Numbfish**	Carvalho
Narcine bancroftii	**Bancroft's Numbfish**	Griffith
Narcine brasiliensis	Brazilian Electric Ray	Olfers

Scientific name	Common name	Describer
Narcine brevilabiata	Shortlip Electric Ray	Besednov
Narcine brunnea	Brown Numbfish	Annandale
Narcine entemedor	Giant Electric Ray	Jordan
Narcine insolita	Madagascar Numbfish	Carvalho
Narcine lasti	**Last's Numbfish**	Carvalho
Narcine leoparda	Electric Ray sp.	Carvalho
Narcine lingual	Chinese Numbfish	Richardson J
Narcine maculate	Darkfinned Numbfish	Shaw
Narcine nelsoni	Eastern Numbfish	Carvalho
Narcine oculifera	Bigeye Electric Ray	Carvalho
Narcine ornata	Ornate Numbfish	Carvalho
Narcine prodorsalis	Tonkin Numbfish	Besednov
Narcine rierai	Slender Electric Ray	Lloris
Narcine tasmaniensis	Tasmanian Numbfish	Richardson J
Narcine timlei	Blackspotted Numbfish	Bloch
Narcine vermiculatus	Vermiculate Electric Ray	Breder
Narcine westraliensis	**McCain's Skate**	McKay
Narcinidae	Numbfishes family	Gill
Narke	Numbfish genus	Kaup
Narke capensis	Onefin Electric Ray	Gmelin
Narke dipterygia	Numbray	Bloch
Narke japonica	Japanese Sleeper Ray	Temminck
Nasolamia	White-nose Shark genus	Compagno
Nasolamia velox	Whitenose Shark	Gilbert
Nebrius	Carpet Shark genus	Ruppell
Nebrius ferrugineus	Tawny Nurse Shark	Lesson
Negaprion	Lemon Shark genus	Whitley
Negaprion acutidens	Sicklefin Lemon Shark	Ruppell
Negaprion brevirostris	Lemon Shark	Poey
Nemamyxine kreffti	Hagfish sp.	McMillan CB
Neoharriotta	Chimaera Genus	Bigelow
Neoharriotta carri	Dwarf Sicklefin Chimaera	Bullis
Neoharriotta pinnata	Sicklefin Chimaera	Schnakenbeck
Neoharriotta pumila	Dwarf Chimaera	Didier
Neomyxine biniplicata	Hagfish sp.	Richardson LR
Neoraja	Pygmy Skate genus	McEachran
Neoraja africana	West African Pygmy Skate	Stehmann
Neoraja caerulea	Blue Ray	Stehmann
Neoraja carolinensis	Carolina Pygmy Skate	McEachran
Neoraja iberica	Iberian Pygmy Skate	Stehmann
Neoraja stehmanni	African Pygmy Skate	Hulley
Neotrygon	Maskray genus	Castelnau

Scientific name	Common name	Describer
Neotrygon annotata	Plain Maskray	Last
Neotrygon kuhlii	**Kuhl's Stingray**	Müller
Neotrygon leylandi	Painted Maskray	Last
Neotrygon ningalooensis	Ningaloo Maskray	Last
Neotrygon picta	Peppered Maskray	Last
Neotrygon trigonoides	New Caledonian Maskray	Castelnau
Notomyxine tridentiger	Hagfish sp.	Garman
Notomyxine	Hagfish genus	Nani
Notoraja	Skate genus	Ishiyama
Notoraja alisae	Deepwater Skate sp.	Séret
Notoraja azurea	Blue Skate	McEachran
Notoraja fijiensis	Deepwater Skate sp.	Séret
Notoraja hirticauda	Ghost Skate	Last
Notoraja inusitata	Deepwater Skate sp.	Séret
Notoraja lira	Broken Ridge Skate	McEachran
Notoraja longiventralis	Deepwater Skate sp.	Séret
Notoraja ochroderma	Pale Skate	McEachran
Notoraja sapphire	Sapphire Skate	Séret
Notoraja sticta	Blotched Skate	McEachran
Notoraja tobitukai	Leadhued Skate	Hiyama
Notorynchus cepedianus	Broadnose Sevengill Shark	Peron
Notorynchus maculatus	Seven Gill Shark sp.	Ayres
Odontaspididae	Sand Shark family	Müller
Odontaspis	Sand Shark genus	Agassiz
Odontaspis ferox	Smalltooth Sand Tiger	Risso
Odontaspis noronhai	Bigeye Sand Tiger	Maul
Okamejei	Skate genus	Ishiyama
Okamejei acutispina	Sharpspine Skate	Ishiyama
Okamejei arafurensis	Arafura Skate	Last
Okamejei boesemani	**Boeseman's Skate**	Ishiyama
Okamejei cairae	Borneo Sand Skate	Last
Okamejei heemstrai	East African Skate	McEachran
Okamejei hollandi	Yellow-spotted Skate	Jordan
Okamejei jensenae	Sulu Sea Skate	Last
Okamejei kenojei	Ocellate Spot Skate	Müller
Okamejei leptoura	Thintail Skate	Last
Okamejei meerdervoortii	Bigeye Skate	Bleeker
Okamejei mengae	Skate sp.	Jeong
Okamejei philipi	Skate sp.	Lloyd
Okamejei pita	Pita Skate	Fricke
Okamejei powelli	Indian Ringed Skate	Alcock
Okamejei schmidti	Browneye Skate	Ishiyama

Scientific name	Common name	Describer
Orectolobidae	Wobbegong Shark family	Gill
Orectolobiformes	Carpet Shark order	Applegate
Orectolobus	Carpet Shark genus	Bonaparte
Orectolobus floridus	Floral Banded Wobbegong	Last
Orectolobus halei	**Hale's Wobbegong**	Whitley
Orectolobus hutchinsi	Western Wobbegong	Last
Orectolobus japonicus	Japanese Wobbegong	Regan
Orectolobus leptolineatus	Indonesian Wobbegong	Last
Orectolobus maculatus	Spotted Wobbegong	Bonnaterre
Orectolobus ornatus	Ornate Wobbegong	De Vis
Orectolobus parvimaculatus	Dwarf Spotted Wobbegong	Last
Orectolobus reticulatus	Network Wobbegong	Last
Orectolobus wardi	Northern Wobbegong	Whitley
Oxynotidae	Rough Shark family	Gill
Oxynotus	Rough Shark genus	Rafinesque
Oxynotus bruniensis	Prickly Dogfish	Ogilby
Oxynotus caribbaeus	Caribbean Roughshark	Cervigon
Oxynotus centrina	Angular Roughshark	Linnaeus
Oxynotus japonicus	Japanese Roughshark	Yano
Oxynotus paradoxus	Sailfin Roughshark	Frade
Paragaleus	Atlantic Weasel Shark genus	Budker
Paragaleus leucolomatus	Whitetip Weasel Shark	Compagno
Paragaleus pectoralis	Atlantic Weasel Shark	Garman
Paragaleus randalli	Slender Weasel Shark	Compagno
Paragaleus tengi	Straight-tooth Weasel Shark	Chen JTF
Paramyxine fernholmi	Hagfish sp.	Kuo
Paramyxine wisneri	Hagfish sp.	Kuo
Parascylliidae	Collared Carpet Shark family	Gill
Parascyllium	Carpet Shark genus	Gill
Parascyllium collare	Collared Carpetshark	Ramsey
Parascyllium elongatum	Elongate Carpetshark	Last
Parascyllium ferrugineum	Rusty Carpetshark	McCulloch
Parascyllium sparsimaculatum	Ginger Carpetshark	Schaaf-Da Silva
Parascyllium variolatum	Necklace Carpetshark	Duméril AHA
Paratrygon	Ray genus	Duméril AHA
Paratrygon ajereba	Dicus Ray	Walbaum
Parmaturus	Catshark genus	Garman
Parmaturus albimarginatus	White-tip Catshark	Séret
Parmaturus albipenis	White-clasper Catshark	Séret
Parmaturus bigus	Beige Catshark	Séret
Parmaturus campechiensis	Campeche Catshark	Springer S
Parmaturus lanatus	Velvet Catshark	Séret

Scientific name	Common name	Describer
Parmaturus macmillani	**McMillan's Catshark**	Hardy
Parmaturus melanobranchus	Blackgill Catshark	Chan
Parmaturus pilosus	Salamander Shark	Garman
Parmaturus xaniurus	Filetail Catshark	Gilbert
Pastinachus atrus	Cowtail Stingray	Macleay
Pastinachus gracilicaudus	Narrowtail Stingray	Last
Pastinachus sephen	Feathertail Stingray	Forsskål
Pastinachus solocirostris	Roughnose Stingray	Last
Pastinachus stellurostris	Starrynose Stingray	Last
Pavoraja	Peacock Skate genus	Whitley
Pavoraja alleni	**Allen's Skate**	McEachran
Pavoraja arenaria	Sandy Skate	Last
Pavoraja mosaic	Mosaic Skate	Last
Pavoraja nitida	Peacock Skate	Günther
Pavoraja pseudonitida	False Peacock Skate	Last
Pavoraja umbrosa	Dusky Skate	Last
Pentanchus	Catshark genus	Smith HM
Pentanchus profundicolus	Onefin Catshark	Smith HM
Petromyzon marinus	Sea Lamprey	Linnaeus
Planonasus	Shark genus	Weigmann
Platyrhina	Fanray genus	Müller
Platyrhina hyugaensis	Hyuga Fanray	Iwatsuki
Platyrhina limboonkengi	Amoy Fanray	Tang
Platyrhina sinensis	Fanray	Bloch
Platyrhina tangi	Yellow-spotted Fanray	Iwatsuki
Platyrhinidae	Thornback Ray family	Jordan
Platyrhinoidis	Guitarfish genus	Garman
Platyrhinoidis triseriata	Thornback Guitarfish	Jordan
Plesiobatidae	Deepwater Stingray	Nishida
Plesiobatis	Deepwater Stingray genus	Wallace
Plesiobatis daviesi	**Davies' Stingaree**	Wallace
Plesiotrygon	Freshwater Stingray genus	Rosa
Plesiotrygon iwamae	**Long-tailed River Stingray**	Rosa
Plesiotrygon nana	Long-tailed River Stingray	Carvalho
Pliotrema warreni	Sixgill Sawshark	Regan
Poroderma	Catshark genus	Smith A
Poroderma africanum	Striped Catshark	Gmelin
Poroderma pantherinum	Leopard Catshark	Müller
Potamotrygon	Freshwater Stingray genus	Garman
Potamotrygon amandae	Freshwater Stingray sp.	Loboda
Potamotrygon boesemani	Freshwater Ray sp.	Rosa
Potamotrygon brachyura	Short-tailed River Stingray	Günther

Scientific name	Common name	Describer
Potamotrygon brumi	Freshwater Stingray sp.	Devicenzi
Potamotrygon castexi	Otorongo Ray	Castello
Potamotrygon constellate	Thorny River Stingray	Vaillant
Potamotrygon falkneri	Large-spot River Stingray	Castex
Potamotrygon henlei	Bigtooth River Stingray	Castelnau
Potamotrygon humerosa	Freshwater Stingray sp	Garman
Potamotrygon hystrix	Porcupine River Stingray	Müller
Potamotrygon labradori	Freshwater Stingray sp.	Castex
Potamotrygon leopoldi	White-blotched River Stingray	Castello
Potamotrygon limai	Freshwater Stingray sp	Fontelle
Potamotrygon magdalenae	**Magdalena River Stingray**	Duméril AHA
Potamotrygon marinae	Freshwater Stingray sp.	Deynat
Potamotrygon menchacai	Freshwater Stingray sp.	Achenbach
Potamotrygon motoro	Ocellate River Stingray	Müller
Potamotrygon ocellata	Red-blotched River Stingray	Engelhardt
Potamotrygon orbignyi	Smooth Back River Stingray	Castelnau
Potamotrygon pantanensis	Freshwater Stingray sp.	Loboda
Potamotrygon pauckei	Freshwater Stingray sp.	Castex
Potamotrygon schroederi	Rosette River Stingray	Fernández-Yépez
Potamotrygon schuhmacheri	**Schuhmacher's stingray**	Castex
Potamotrygon scobina	Raspy River Stingray	Garman
Potamotrygon signata	Parnaiba River Stingray	Garman
Potamotrygon tatianae	River Stingray sp.	Da Silva
Potamotrygon tigrina	Tiger Ray sp.	Carvalho
Potamotrygon yepezi	Maracaibo River Stingray	Castex
Potamotrygonidae	River Stingray family	Garman
Prionace	Requiem Shark genus	Cantor
Prionace glauca	Blue Shark	Linnaeus
Pristidae	Sawfish family	Bonaparte
Pristiophoridae	Saw Shark family	Müller
Pristiophoriformes	Saw Shark order	Berg LS
Pristiophorus	Sawshark genus	Müller
Pristiophorus cirratus	Longnose Sawshark	Latham
Pristiophorus delicatus	Tropical Sawshark	Yearsley
Pristiophorus japonicus	Japanese Sawshark	Günther
Pristiophorus lanae	**Lana's Sawshark**	Ebert
Pristiophorus nancyae	African Dwarf Sawshark	Ebert
Pristiophorus nudipinnis	Shortnose Sawshark	Günther
Pristiophorus peroniensis	Eastern Australian Sawshark	Yearsley
Pristiophorus schroederi	Bahamas Sawshark	Springer S
Pristis	Sawfish genus	Linck
Pristis clavata	Dwarf Sawfish	Garman

Scientific name	Common name	Describer
Pristis microdon	**Leichhardt's Sawfish**	Latham
Pristis pectinata	Smalltooth Sawfish	Latham
Pristis perotteti	Large-tooth Sawfish	Müller
Pristis pristis	Common Sawfish	Linnaeus
Pristis zijsron	Longcomb Sawfish	Bleeker
Proscylliidae	Finback Catshark genus	Compagno
Proscyllium	Catshark genus	Hilgendorf
Proscyllium habereri	Graceful Catshark	Hilgendorf
Proscyllium magnificum	Magnificent Catshark	Last
Proscyllium venustum	Finback Catshark sp.	Tanaka I
Proscymnodon	Dogfish genus	Fowler HW
Proscymnodon macracanthus	Largespine Velvet Dogfish	Regan
Proscymnodon plunketi	**Lord Plunket's Shark**	Waite
Psammobatis	Ray genus	Günther
Psammobatis bergi	Blotched Sand Skate	Marini
Psammobatis extenta	Zipper Sand Skate	Garman
Psammobatis lentiginosa	Freckled Sand Skate	McEachran
Psammobatis normani	Shortfin Sand Skate	McEachran
Psammobatis parvacauda	Smalltail Sand Skate	McEachran
Psammobatis rudis	Smallthorn Sand Skate	Günther
Psammobatis rutrum	Spade Sand Skate	Jordan
Psammobatis scobina	Raspthorn Sand Skate	Philippi
Pseudocarcharias	Crocodile Shark genus	Cadenat
Pseudocarcharias kamoharai	**Kamohara's Sand-shark**	Matsubara
Pseudocarchariidae	Crocodile Shark family	Compagno
Pseudoginglymostoma	Shark genus	Dingerkus
Pseudoginglymostoma brevicaudatum	Short-tail Nurse Shark	Günther
Pseudoraja	Skate genus	Bigelow
Pseudoraja fischeri	**Fanfin skate**	Bigelow
Pseudotriakidae	Ground Shark family	Gill
Pseudotriakis	False Catshark genus	Brito Capello
Pseudotriakis microdon	False Catshark	Brito Capello
Pteromylaeus	Bull Ray genus	Garman
Pteromylaeus asperrimus	Rough Eagle Ray	Gilbert
Pteromylaeus bovinus	Bull Ray	Saint-Hilaire
Pteroplatytrygon	Stingray genus	Fowler HW
Pteroplatytrygon violacea	Pelagic Stingray	Bonaparte
Raja	Skate genus	Linnaeus
Raja ackleyi	**Ackley's Ocellate Ray**	Garman
Raja africana	African Skate	Capape
Raja asterias	Mediterranean Starry Ray	Delaroche
Raja bahamensis	Bahama Skate	Bigelow

Scientific name	Common name	Describer
Raja binoculata	Big Skate	Girard
Raja brachyura	Blonde Ray	Lafont
Raja cervigoni	Finspot Ray	Bigelow
Raja chinensis	Chinese Ray	Basilewsky
Raja clavata	Thornback Ray	Linnaeus
Raja cortezensis	**Cortez' Ray**	McEachran
Raja eglanteria	Clearnose Skate	Bosc
Raja equatorialis	Equatorial Ray	Jordan
Raja herwigi	Cape Verde Skate	Krefft G
Raja inornata	California Ray	Jordan
Raja koreana	Korea Skate	Jeong
Raja madarensis	Madeiran Ray	Lowe RT
Raja microocellata	Small-eyed Ray	Montagu
Raja miraletus	Brown Ray	Linnaeus
Raja montagui	Spotted Ray	Fowler HW
Raja polystigma	Speckled Ray	Regan
Raja pulchra	Mottled Skate	Liu FH
Raja radula	Rough Ray	Delaroche
Raja rhina	Longnose Skate	Jordan
Raja rondeleti	**Rondelet's Ray**	Bougis
Raja rouxi	Ray sp.	Capapé
Raja stellulata	Starry Skate	Jordan
Raja straeleni	Spotted Skate	Poll
Raja texana	Rounded Skate	Chandler
Raja undulata	Undulate Ray	Lacépède
Raja velezi	**Velez Ray**	Chirichigno
Rajella	Skate genus	Stehmann
Rajella annandalei	**Annandale's Skate**	Weber
Rajella barnardi	Bigthorn Skate	Norman
Rajella bathyphila	Deep-water Ray	Holt
Rajella bigelowi	**Bigelow's Ray**	Stehmann
Rajella caudaspinosa	Munchkin Skate	von Bonde
Rajella challengeri	**Challenger Skate**	Last
Rajella dissimilis	Ghost Skate	Hulley
Rajella eisenhardti	Galapagos Gray Skate	Long DJ
Rajella fuliginea	Sooty Ray	Bigelow
Rajella fyllae	Round Ray	Lutken
Rajella kukujevi	Mid-Atlantic Skate	Dolganov
Rajella leopardus	Leopard Skate	von Bonde
Rajella nigerrima	Blackish Skate	de Buen
Raajella paucispinosa	Deepwater Skate sp.	Weigmann
Rajella purpuriventralis	Purplebelly Skate	Bigelow

Scientific name	Common name	Describer
Rajella ravidula	Smoothback Skate	Hulley
Rajella sadowskii	Brazillian Skate	Krefft G
Rajidae	Rays family	Bonaparte
Rajiformes	Skate order	Berg LS
Rhina	Guitarfish genus	Bloch
Rhina ancylostoma	Bowmouth Guitarfish	Bloch
Rhincodon	Whale Shark genus	Smith A
Rhincodon typus	**Braun's Whale Shark**	Smith A
Rhincodontidae	Whale Shark family	Müller
Rhinidae	Guitarfish family	Müller
Rhinobatidae	Guitarfish genus	Müller
Rhinobatos	Guitarfish family	Linck
Rhinobatos albomaculatus	White-spotted Guitarfish	Norman
Rhinobatos annandalei	**Annandale's Guitarfish**	Norman
Rhinobatos annulatus	Lesser Sandshark	Müller
Rhinobatos blochii	Bluntnose Guitarfish	Müller
Rhinobatos cemiculus	Blackchin Guitarfish	Saint-Hilaire
Rhinobatos formosensis	Taiwan Guitarfish	Norman
Rhinobatos glaucostigma	Speckled Guitarfis	Jordan
Rhinobatos holcorhynchus	Slender Guitarfish	Norman
Rhinobatos horkelii	Brazilian Guitarfish	Müller
Rhinobatos hynnicephalus	Ringstreaked Guitarfish	Richardson J
Rhinobatos irvinei	Spineback Guitarfish	Norman
Rhinobatos jimbaranensis	Jimbaran Shovelnose Ray	Last
Rhinobatos lentiginosus	Atlantic Guitarfish	Garman
Rhinobatos leucorhynchus	Whitesnout Guitarfish	Günther
Rhinobatos leucospilus	Grayspotted Guitarfish	Norman
Rhinobatos lionotus	Smoothback Guitarfish	Norman
Rhinobatos microphthalmus	Smalleyed Guitarfish	Teng
Rhinobatos nudidorsalis	Bareback Shovelnose Ray	Last
Rhinobatos obtusus	Widenose Guitarfish	Müller
Rhinobatos ocellatus	Speckled Guitarfish	Norman
Rhinobatos penggali	Indonesian Shovelnose Ray	Last
Rhinobatos percellens	Chala Guitarfish	Walbaum
Rhinobatos petiti	Madagascar Guitarfish	Chabanaud
Rhinobatos planiceps	Pacific Guitarfish	Garman
Rhinobatos prahli	Gorgona Guitarfish	Acero
Rhinobatos productus	Shovelnose Guitarfish	Ayres
Rhinobatos punctifer	Spotted Guitarfish	Compagno
Rhinobatos rhinobatos	Common Guitarfish	Linnaeus
Rhinobatos sainsburyi	Goldeneye Shovelnose Ray	Last
Rhinobatos salalah	Salalah Guitarfish	Randall

Scientific name	Common name	Describer
Rhinobatos schlegelii	Brown Guitarfish	Müller
Rhinobatos spinosus	Spiny Guitarfish	Günther
Rhinobatos thouin	Clubnose Guitarfish	(Lacépède)
Rhinobatos thouiniana	**Shaw's Shovelnose Guitarfish**	Shaw
Rhinobatos variegatus	Stripenose Guitarfish	Nair
Rhinobatos zanzibarensis	Zanzibar Guitarfish	Norman
Rhinochimaera	Chimaera genus	Garman
Rhinochimaera africana	Paddle-nose Chimaera	Compagno
Rhinochimaera atlantica	Straightnose Rabbitfish	Holt
Rhinochimaera pacifica	Pacific Spookfish	Mitsukuri
Rhinoptera	Cownose Ray genus	Van Hasselt
Rhinoptera adspersa	Rough Cownose Ray	Müller
Rhinoptera bonasus	Cownose Ray	Mitchill
Rhinoptera brasiliensis	Ticon Cownose Ray	Müller
Rhinoptera javanica	Flapnose Ray	Müller
Rhinoptera jayakari	Oman Cownose Ray	Boulenger
Rhinoptera marginata	Lusitanian Cownose Ray	Saint-Hilaire
Rhinoptera neglecta	Australian Cownose Ray	Ogilby
Rhinoptera peli	Lusitanian Cownose Ray	Bleeker
Rhinoptera steindachneri	Pacifric Cownose Ray	Evermann
Rhinoraja	Jointnose Skate genus	Ishiyama
Rhinoraja kujiensis	Dapple-bellied Softnose Skate	Tanaka I
Rhinoraja longi	Aleutian Dotted Skate	Raschi
Rhinoraja longicauda	White-bellied Softnose Skate	Ishiyama
Rhinoraja obtuse	Blunt Skate	Gill
Rhinoraja odai	**Oda's Skate**	Ishiyama
Rhizoprionodon	Requiem Shark genus	Whitley
Rhizoprionodon acutus	Milk Shark	Rüppell
Rhizoprionodon lalandii	Brazilian Sharpnose Shark	Müller
Rhizoprionodon longurio	Pacific Sharpnose Shark	Jordan
Rhizoprionodon oligolinx	Gray Sharpnose Shark	Springer VG
Rhizoprionodon porosus	Caribbean Sharpnose Shark	Poey
Rhizoprionodon taylori	Australian Sharpnose Shark	Ogilby
Rhizoprionodon terraenovae	Atlantic Sharpnose Shark	Richardson J
Rhynchobatidae	Wedgefish family	Garman
Rhynchobatus	Guitarfish genus	Müller
Rhynchobatus australiae	White-spotted Wedgefish	Whitley
Rhynchobatus djiddensis	Giant Guitarfish	Forsskål
Rhynchobatus immaculatus	Taiwanese Wedgefish	Last
Rhynchobatus laevis	Smoothnose Wedgefish	Bloch
Rhynchobatus luebberti	**Lubbert's Guitarfish**	Ehrenbaum
Rhynchobatus palpebratus	Eyebrow Wedgefish	Compagno

Scientific name	Common name	Describer
Rhynchobatus springeri	Broadnose Wedgefish	Compagno
Rioraja	Rio Skate genus	Müller
Rioraja agassizii	Rio Skate	Müller
Rostroraja	Bottlenose Skate genus	Hulley
Rostroraja alba	Bottlenose Skate	Lacépède
Schroederichthys	Catshark genus	Springer S
Schroederichthys bivius	Narrowmouthed Catshark	Müller
Schroederichthys chilensis	Redspotted Catshark	Guichenot
Schroederichthys maculatus	Narrowtail Catshark	Springer S
Schroederichthys saurisqualus	Lizard Catshark	Soto
Schroederichthys tenuis	Slender Catshark	Springer S
Scoliodon	Spadenose Shark genus	Müller
Scoliodon laticaudus	Spadenose Shark	Müller
Scyliorhinus	Catshark genus	Blainville
Scyliorhinus besnardi	Polkadot Catshark	Springer S
Scyliorhinus boa	Boa Catshark	Goode
Scyliorhinus canicula	Small-spotted Catshark	Linnaeus
Scyliorhinus capensis	Yellow-spotted Catshark	Müller
Scyliorhinus cervigoni	West African Catshark	Maurin
Scyliorhinus comoroensis	Comoro Catshark	Compagno
Scyliorhinus garmani	Brownspotted Catshark	Fowler HW
Scyliorhinus haeckelii	Freckled Catshark	Miranda-Ribeiro
Scyliorhinus hesperius	White-saddled Catshark	Springer S
Scyliorhinus meadi	Blotched Catshark	Springer S
Scyliorhinus retifer	Chain Catshark	Garman
Scyliorhinus sp. X	**Oakley's Catshark**	Undescribed
Scyliorhinus stellaris	Nursehound	Linnaeus
Scyliorhinus tokubee	Izu Catshark	Shirai
Scyliorhinus torazame	Cloudy Catshark	Tanaka I
Scyliorhinus torrei	Dwarf Catshark	Howell-Rivero
Scylliogaleus	Houndshark genus	Boulenger
Scylliogaleus quecketti	Flapnose Houndshark	Boulenger
Scymnodalatias	Squaliform Shark genus	Garrick
Scymnodalatias albicauda	Whitetail Dogfish	Taniuchi
Scymnodalatias garricki	Azores Dogfish	Compagno
Scymnodalatias oligodon	Sparsetooth Dogfish	Kukuev
Scymnodalatias sherwoodi	**Sherwood Dogfish**	Archey
Scymnodon	Squaliform Shark genus	Bocage
Scymnodon obscurus	Smallmouth Knifetooth Dogfish	Vaillant
Scymnodon ringens	Knifetooth Dogfish	Bocage
Sinobatis bulbicauda	Western Legskate	Last
Sinobatis caerulea	Blue Legskate	Last

Scientific name	Common name	Describer
Sinobatis filicauda	Eastern Leg Skate	Last
Sinobatis melanosoma	Blackbodied Legskate	Chan
Somniosidae	Sleeper Shark family	Jordan
Somniosus antarcticus	Southern Sleeper Shark	Whitley
Somniosus longus	Frog Shark	Tanaka I
Somniosus microcephalus	Greenland Shark	Bloch
Somniosus pacificus	Pacific Sleeper Shark	Bigelow
Somniosus rostratus	Little Sleeper Shark	Risso
Sphyrna	Hammerhead Shark genus	Rafinesque
Sphyrna corona	Scalloped Bonnethead	Springer S
Sphyrna couardi	Whitefin Hammerhead	Cadenat
Sphyrna gilberti	Carolina Hammerhead	Quatro
Sphyrna lewini	Scalloped Hammerhead	Griffith
Sphyrna media	Scoophead Shark	Springer S
Sphyrna mokarran	Great Hammerhead	Rüppell
Sphyrna tiburo	Bonnethead	Linnaeus
Sphyrna tudes	Smalleye Hammerhead	Valenciennes
Sphyrna zygaena	Smooth Hammerhead	Linnaeus
Sphyrnidae	Hammerhead Shark family	Compagno
Springeria	Skate genus	Bigelow
Squalidae	Dogfish Shark family	Blainville
Squaliformes	Dogfish Shark order	Goodrich
Squaliolus	Deepsea Dogfish Shark genus	Smith HM
Squaliolus aliae	Smalleye Pygmy Shark	Teng
Squaliolus laticaudus	Spined Pygmy Shark	Smith HM
Squalus	Dogfish Shark genus	Linnaeus
Squalus acanthias	Spiny Dogfish	Linnaeus
Squalus acanthias ponticus	Spiny Dogfish ssp.	Myagkov
Squalus acutirostris	Dogfish Shark sp.	Chu
Squalus albifrons	Eastern Highfin Spurdog	Last
Squalus altipinnis	Western Highfin Spurdog	Last
Squalus blainville	Longnose Spurdog	Risso
Squalus brevirostris	Japanese Shortnose Spurdog	Tanaka I
Squalus bucephalus	Bighead Spurdog	Last
Squalus chloroculus	Greeneye Spurdog	Last
Squalus crassispinus	Fatspine Spurdog	Last
Squalus cubensis	Cuban Dogfish	Howell-Rivero
Squalus edmundsi	**Edmund's Spurdog**	Last
Squalus formosus	Dogfish Shark sp.	White WT
Squalus grahami	Eastern Longnose Spurdog	Last
Squalus griffini	**Griffin's Spiny Dogfish**	Phillipps
Squalus hemipinnis	Indonesian Shortsnout Spurdog	White WT
Squalus japonicus	Japanese Spurdog	Ishikawa

Scientific name	Common name	Describer
Squalus lalannei	Seychelles Spurdog	Baranes
Squalus megalops	Shortnose Spurdog	Macleay
Squalus melanurus	Blacktailed Spurdog	Fourmanoir
Squalus mitsukurii	Shortspine Spurdog	Jordan
Squalus montalbani	Indonesian Greeneye Spurdog	Whitley
Squalus nasutus	Western Longnose Spurdog	Last
Squalus notocaudatus	Bartail Spurdog	Last
Squalus rancureli	**Cyrano Spurdog**	Fourmanoir
Squalus raoulensis	Kermadec Spiny Dogfish	Duffy
Squalus suckleyi	Spotted Spiny Dogfish	Girard
Squalus uyato	Little Gulper Shark	Rafinesque
Squatina	Angel Shark genus	Duméril AMC
Squatina aculeate	Sawback Angelshark	Cuvier
Squatina africana	African Angelshark	Regan
Squatina albipunctata	Eastern Angelshark	Last
Squatina argentina	Argentine Angelshark	Marini
Squatina armata	Chilean Angelshark	Philippi
Squatina australis	Australian Angelshark	Regan
Squatina caillieti	Angelshark sp.	Walsh
Squatina californica	Pacific Angelshark	Ayres
Squatina dumeril	Sand Devil	Lesueur
Squatina formosa	Taiwan Angelshark	Shen S-C
Squatina guggenheim	Angular Angelshark	Marini
Squatina heteroptera	Gulf Angelshark	Castro-Aguirre
Squatina japonica	Japanese Angelshark	Bleeker
Squatina legnota	Indonesian Angelshark	Last
Squatina mexicana	Mexican Angelshark	Castro-Aguirre
Squatina nebulosa	Clouded Angelshark	Regan
Squatina occulta	Hidden Angelshark	Vooren
Squatina oculata	Smoothback Angelshark	Bonaparte
Squatina pseudocellata	Western Angelshark	Last
Squatina punctate	Angelshark sp.	Marini
Squatina squatina	Angelshark	Linnaeus
Squatina tergocellata	Ornate Angelshark	McCulloch
Squatina tergocellatoides	Ocellated Angelshark	Chen JTF
Squatinidae	Angel Shark family	Bonaparte
Squatiniformes	Angel Shark order	de Buen
Stegostoma	Carpet Shark genus	Müller
Stegostoma fasciatum	Zebra Shark	Hermann
Stegostomatidae	Zebra Shark family	Gill
Sutorectus	Cobbler Wobbegong genus	Whitley
Sutorectus tentaculatus	Cobbler Wobbegong	Peters

Scientific name	Common name	Describer
Sympterygia	Fanskate genus	Müller
Sympterygia acuta	Bignose Fanskate	Garman
Sympterygia bonapartii	Smallnose Fanskate	Müller
Sympterygia brevicaudata	Shorttail Fanskate	Cope
Sympterygia lima	Filetail Fanskate	Poeppig
Taeniura	Stingray genus	Müller
Taeniura grabata	Round Santail Stingray	Saint-Hilaire
Taeniura lymma	Bluespotted Ribbontail Ray	Forsskål
Taeniura meyeni	Blotched Fantail Ray	Müller
Tarsistes	Guitarfish genus	Jordan
Tarsistes philippii	Guitarfish sp.	Jordan
Temera	Sleeper Ray genus	Gray
Temera hardwickii	Finless Sleeper Ray	Gray
Tetrapleurodon geminis	Mexican Brook Lamprey	Alvarez
Tetrapleurodon spadiceus	Mexican Lamprey	Bean TH
Torpedinidae	Torpedo Electric Ray family	Bonaparte
Torpediniformes	Electric Ray order	de Buen
Torpedo	Electric Ray genus	Houttuyn
Torpedo adenensis	Aden Gulf Torpedo	Carvalho
Torpedo alexandrinsis	**Alexandrine Torpedo**	Mazhar
Torpedo andersoni	Florida Torpedo	Bullis
Torpedo bauchotae	Rosette Torpedo	Cadenat
Torpedo californica	Pacific Electric Ray	Ayres
Torpedo fairchildi	New Zealand Torpedo	Hutton
Torpedo formosa	Taiwan Torpedo	Haas
Torpedo fuscomaculata	Black-spotted Torpedo	Peters
Torpedo mackayana	Ringed Torpedo	Metzelaar
Torpedo macneilli	Shorttail Torpedo	Whitley
Torpedo marmorata	Marbled Electric Ray	Risso
Torpedo microdiscus	Smalldisk Torpedo	Parin
Torpedo nobiliana	Atlantic Torpedo	Bonaparte
Torpedo panthera	Panther Electric Ray	Olfers
Torpedo peruana	Peruvian Torpedo	Chirichigno
Torpedo puelcha	Argentine Torpedo	Lahille
Torpedo semipelagica	Semipelagic Torpedo	Parin
Torpedo sinuspersici	Variable Torpedo	Olfers
Torpedo suessii	Torpedo sp.	Steindachner
Torpedo tokionis	Trapezoid Torpedo	Tanaka I
Torpedo torpedo	Common Torpedo	Linnaeus
Torpedo tremens	Chilean Torpedo	de Buen
Torpedo zugmayeri	Baluchistan Torpedo	Engelhardt
Triaenodon	Reef Shark genus	Müller

Scientific name	Common name	Describer
Triaenodon obesus	Whitetip Reef Shark	Rüppell
Triakidae	Houndshark family	Gray
Triakis	Houndshark genus	Müller
Triakis acutipinna	Sharpfin Houndshark	Kato
Triakis maculate	Spotted Houndshark	Kner
Triakis megalopterus	Sharptooth Houndshark	Smith A
Triakis scyllium	Banded Houndshark	Müller
Triakis semifasciata	Leopard Shark	Girard
Trigonognathus	Dogfish genus	Mochizuki
Trigonognathus kabeyai	Viper Dogfish	Mochizuki
Trygonoptera	Stingaree genus	Müller
Trygonoptera galba	Yellow Shovelnose Stingaree	Last
Trygonoptera imitata	Eastern Shovelnose Stingaree	Last
Trygonoptera mucosa	Western Shovelnose Stingaree	Whitley
Trygonoptera ovalis	Striped Stingaree	Last
Trygonoptera personata	Masked Stingaree	Last
Trygonoptera testacea	Common Stingaree	Müller
Trygonorrhina	Fiddler Ray genus	Müller
Trygonorrhina fasciata	Southern Fiddler Ray	Müller
Trygonorrhina melaleuca	Magpie Fiddler Ray	Scott
Typhlonarke	Sleeper Ray genus	Waite
Typhlonarke aysoni	Blind Electric Ray	Hamilton A
Typhlonarke tarakea	Oval Electric Ray	Phillipps
Urobatis	Round Ray genus	Garman
Urobatis concentricus	Spot-on-spot Round Ray	Osburn
Urobatis halleri	**Haller's Round Ray**	Cooper
Urobatis jamaicensis	Yellow Stingray	Cuvier
Urobatis maculatus	Spotted Round Ray	Garman
Urobatis marmoratus	Chilean Round Ray	Philippi
Urobatis tumbesensis	Tumbes Round Stingray	Chirichigno
Urogymnus	Stingray genus	Müller
Urogymnus asperrimus	Porcupine Ray	Bloch
Urolophidae	Ray family	Müller
Urolophus	Stingray genus	Müller
Urolophus armatus	New Ireland Stingaree	Müller
Urolophus aurantiacus	Sepia Stingaree	Müller
Urolophus bucculentus	Sandyback Stingaree	Macleay
Urolophus circularis	Circular Stingaree	McKay
Urolophus deforgesi	Chesterfield Island Stingaree	Séret
Urolophus expansus	Wide Stingaree	McCulloch
Urolophus flavomosaicus	Patchwork Stingaree	Last
Urolophus gigas	Spotted Stingaree	Scott

Scientific name	Common name	Describer
Urolophus javanicus	Java Stingaree	von Martens
Urolophus kaianus	Kai Stingaree	Günther
Urolophus kapalensis	**Kapala Stingaree**	Yearsley
Urolophus lobatus	Lobed Stingaree	McKay
Urolophus mitosis	Mitotic Stingaree	Last
Urolophus neocaledoniensis	New Caledonian Stingaree	Séret
Urolophus orarius	Coastal Stingaree	Last
Urolophus papilio	Butterfly Stingaree	Séret
Urolophus paucimaculatus	**Dixon's Stingaree**	Dixon
Urolophus piperatus	Coral Sea Stingaree	Séret
Urolophus sufflavus	Yellowback Stingaree	Whitley
Urolophus viridis	Greenback Stingaree	McCulloch
Urolophus westraliensis	Brown Stingaree	Last
Urotrygon	Round Stingray genus	Gill
Urotrygon aspidura	Spiney-tail Round Ray	Jordan
Urotrygon caudispinosus	Stingray sp.	Hildebrand
Urotrygon chilensis	Chilean Round Ray	Günther
Urotrygon cimar	Denticled Roundray	López
Urotrygon microphthalmum	Smalleyed Round Stingray	Delsman
Urotrygon munda	Munda Round Ray	Gill
Urotrygon nana	Dwarf Round Ray	Miyake
Urotrygon peruanus	Peruvian Stingray	Hildebrand
Urotrygon reticulate	Reticulate Round Ray	Miyake
Urotrygon rogersi	**Roger's Round Ray**	Jordan
Urotrygon serrula	Stingray sp.	Hildebrand
Urotrygon simulatrix	Fake Round Ray	Miyake
Urotrygon venezuelae	Venezuela Round Stingray	Schultz
Urotrygonidae	Ray family	McEachran
Zameus	Dogfish Shark genus	Jordan
Zameus ichiharai	Japanese Velvet Dogfish	Yano
Zameus squamulosus	Velvet Dogfish	Günther
Zanobatidae	Panray family	Fowler HW
Zanobatus	Panray genus	Garman
Zanobatus schoenleinii	Striped Panray	Müller
Zapteryx	Guitarfish genus	Jordan
Zapteryx brevirostris	Lesser Guitarfish	Müller
Zapteryx exasperata	Banded Guitarfish	Jordan
Zapteryx xyster	Southern Banded Guitarfish	Jordan
Zearaja	Skate genus	Whitley
Zearaja chilensis	Yellownose Skate	Guichenot
Zearaja maugeana	Maugean Skate	Last
Zearaja nasuta	New Zealand Rough Skate	Müller

Lightning Source UK Ltd.
Milton Keynes UK
UKOW03n1942220115

244942UK00001B/9/P